Ecological Census Techniques

Almost all ecological and conservation work involves carrying out a census or survey. This new edition of the best-selling *Ecological Census Techniques* covers everything you need to know in order to plan and carry out a census, and analyse the resulting data. Completely updated, the new edition incorporates two new chapters and additional expert contributions.

An essential reference for anyone involved in ecological research, the book begins by describing planning, sampling and the basic theory necessary for carrying out a census and, in the following chapters, international experts describe the appropriate methods for counting plants, insects, fish, amphibians, reptiles, mammals and birds. In addition, there is a chapter explaining the main methods of measuring environmental variability, and a list of the most common 'censusing sins' provides a light-hearted guide to avoiding errors.

WILLIAM J. SUTHERLAND is a professor in the School of Biological Sciences at the University of East Anglia.

Ecological Census Techniques

a handbook

Second Edition

Edited by

WILLIAM J. SUTHERLAND

University of East Anglia

CAMBRIDGE
UNIVERSITY PRESS

CAMBRIDGE UNIVERSITY PRESS
Cambridge, New York, Melbourne, Madrid, Cape Town, Singapore, São Paulo

Cambridge University Press
The Edinburgh Building, Cambridge CB2 2RU, UK

Published in the United States of America by Cambridge University Press, New York

www.cambridge.org
Information on this title: www.cambridge.org/9780521844628

First published 1996
Second edition 2006

Printed in the United Kingdom at the University Press, Cambridge

A catalogue record for this publication is available from the British Library

ISBN-13 978-0-521-84462-8 hardback
ISBN-10 0-521-84462-2 hardback

ISBN-13 978-0-521-60636-3 paperback
ISBN-10 0-521-60636-5 paperback

Contents

3   **General census methods**   87
JEREMY J. D. GREENWOOD AND ROBERT A. ROBINSON

4   Plants

JAMES M. BULLOCK

List of contributors

MALCOLM AUSDEN
Royal Society for the Protection of Birds, The Lodge, Sandy, Bedfordshire SG19 2DL, United Kingdom

SIMON BLOMBERG
Centre for Mental Health Research and School of Botany and Zoology Australian National University, Canberra, ACT 0200, Australia

JAMES M. BULLOCK
NERC Centre for Ecology and Hydrology, Winfrith Technology Centre, Dorchester, Dorset DT2 8ZD, United Kingdom

ISABELLE M. CÔTÉ
Department of Biological Sciences, Simon Fraser University, Burnaby, B.C., Canada V5A 1S6

MARTIN DRAKE
Consultant Entomologist, Orchid House, Burridge, Axminster, Devon EX13 7DF, United Kingdom

DAVID W. GIBBONS
Royal Society for the Protection of Birds, The Lodge, Sandy, Bedfordshire SG19 2DL, United Kingdom

JEREMY J. D. GREENWOOD
British Trust for Ornithology, The Nunnery, Thetford, Norfolk IP24 2PU, United Kingdom

RICHARD D. GREGORY
Royal Society for the Protection of Birds, The Lodge, Sandy, Bedfordshire SG19 2DL, United Kingdom

TIM R. HALLIDAY
Department of Biological Sciences, The Open University, Milton Keynes MK7 6AA, United Kingdom

JACQUELYN C. JONES
Centre for Ecology, Evolution and Conservation, School of Biological Sciences, University of East Anglia, Norwich NR4 7TJ, United Kingdom

CHARLES J. KREBS
Department of Zoology, University of British Columbia, Vancouver, B.C., Canada V6T 1Z4

MARTIN R. PERROW
ECON Ecological Consultancy, School of Biological Sciences, University of East Anglia, Norwich NR4 7TJ, United Kingdom

DAVE RAFFAELLI
   Environment Department, University of York, York YO10 5DD, United Kingdom
JOHN D. REYNOLDS
   Department of Biological Sciences, Simon Fraser University, Burnaby, B.C., Canada V5A 1S6
ROBERT A. ROBINSON
   British Trust for Ornithology, The Nunnery, Thetford, Norfolk IP24 2PU, United Kingdom
RICHARD SHINE
   Zoology A08, School of Biological Sciences, University of Sydney, Sydney, NSW 2006, Australia
WILLIAM J. SUTHERLAND
   Centre for Ecology, Evolution and Conservation, School of Biological Sciences, University of East Anglia, Norwich NR4 7TJ, United Kingdom

Preface

The first edition of this book was used by a wide range of biologists across the world and we had a lot of kind feedback saying how useful a book it was. In order to bring it back up to date we produced this second edition, in which every chapter has been updated. Some chapters have different authors. I am particularly delighted that Charles Krebs has written the mammal chapter. He obviously has huge experience of mammal studies, unlike me. For the first edition I stepped in to write this chapter at the very end, despite never having carried out a mammal census. Since many projects fail through problems with planning, we have included a completely new chapter on this subject. There is also a new chapter on sampling.

Gratis Book Scheme

The aim of the Gratis Book Scheme is to provide ecological and conservation books to those who would otherwise be unable to obtain them. Almost three and a half thousand books have already been distributed under this scheme to one hundred and sixty different countries. Cambridge University Press will donate two hundred copies of this second edition of *Ecological Census Techniques* for those outside Western Europe, North America, New Zealand, Australia and Japan who would otherwise be unable to obtain a copy. The British Ecological Society has offered to pay for the postage and nhbs.com co-ordinates the distribution. Please see www.nhbs.com/Conservation/gratis-books to request a copy for yourself or someone else. I thank the organisations involved for making this possible. If you are writing a book on conservation or ecology that would be of benefit to others and are interested in donating books then please get in touch with me.

Conservationevidence.com

We know that many of those who use this book are conservation biologists. The website www.conservationevidence.com provides the evidence for the effectiveness of conservation management based on case studies and also provides summaries of the evidence. The site is open to all to use with free access to the information. We greatly encourage readers to add to this website. For each case the information required is what the problem is (e.g. an invasive plant is spreading), what exactly was done (i.e. how it was treated and when) and the outcome (e.g. whether the percentage cover changed or what proportion of treated stumps grew back). Adding such information is invaluable if we are to learn from each other. Without the documented collective experience of conservation biologists we are wasting a lot of money and effort by repeating mistakes and not learning from successes. I hope that you will be able to contribute to this website.

William J. Sutherland
University of East Anglia

1 Planning a research programme

William J. Sutherland

Centre for Ecology, Evolution and Conservation, School of Biological Sciences,
University of East Anglia, Norwich NR4 7TJ, UK

Introduction: reverse planning

This chapter was written in response to my frustration when talking to people who had worked hard collecting data but who had missed opportunities and largely wasted their time as a consequence of poor planning. Planning relates to any research programme, not just those involving carrying out a census.

In thinking about my own research, collaborating with colleagues and advising students I have devised a means of planning research projects, called reverse planning (see Table 1.1). In designing a research programme it is best to think backwards, starting with the question and then considering how to answer it. This method seems clumsy, but many projects either do not have a sensible question or have a sensible question but the planned research will not answer it. Many projects are clearly unrealistic: it is better to identify impossible projects before starting them.

I recommend this process of reverse planning, and especially the key stage of producing graphs, to many students and others each year. Many find this difficult, but this is only because designing projects is difficult. It is better to struggle with the planning than to struggle with poor results.

An advantage of this explicit planning process is that consulting others becomes easier. Most people go into the field with only a hazy impression of what they will do, how they will analyse it and why they are doing it. A pile of potential graphs provides an excellent basis for discussing projects. It is easy for others to make helpful suggestions and point out flaws. Similarly, planning the time allocation and designing data sheets before going into the field also allows others to make useful comments.

I have described this as a straightforward logical process and recommend working through these steps in order. In practice, and especially once experienced, many stages can all be tackled simultaneously. Experienced researchers usually do not carry out this exercise explicitly, but in my experience it is clear that they have subconsciously run through these stages. Even if one is experienced, it is still usually worth running through these stages, starting with the possible figures, then considering the questions, protocol and time to see whether there are any ways of improving the work.

Ecological Census Techniques: A Handbook, ed. William J. Sutherland.
Published by Cambridge University Press. © Cambridge University Press 1996, 2006.

Table 1.1. *How to use reverse planning*

1. What is the specific question?
2. What results are necessary to answer the questions?
3. What data are needed to complete these results?
4. What protocol is required to obtain these data?
5. Can the data be collected in the time available?
6. Modify the planning in response to time available
7. Create data sheets
8. Start and encounter reality

What is the specific question?

Much of the skill in designing a research project is in selecting a suitable question or questions. It should be worthwhile (e.g. of conservation importance or of academic interest), have a clear objective and be practical in the time available. Questions such as 'What is the population ecology of X?' and 'How are development and conservation linked?' are too broad. More suitable questions are those for which there is a specific answer, such as 'What is the population size of X?', 'What are the habitat requirements of X?' and 'Is there more of this species in one habitat than another?'

What results are necessary to answer the questions?

The question or questions will be answered by a series of results. The next stage is to plan the analyses that can answer the stated question. The usual objective of any project is to collect a pile of graphs and tables. To create a PhD thesis you need only a pile of, say, 40–80 interesting graphs or tables, and then just need to write some text to describe the figures, explain the methods used to collect them, describe what the results mean, introduce the issues and then bind it. To write a research paper you need only 3–15 tables or figures (check with the papers in the journal you would like to publish in). Most papers have remarkably few figures, so if a research paper is the objective then it makes sense to concentrate on collecting data for the few critical figures. Thinking about the figures makes planning the data collection much easier. If you are not going to use the data for some such analysis then why are you collecting them?

Good experienced researchers tackle all the stages in Table 1.1 simultaneously and effortlessly explore different combinations of questions and analyses.

I thus suggest concentrating upon the graphs and tables required before starting any project. I strongly suggest sketching the likely presentation. This forces you to think deeply about what you are trying to achieve. The figures can be crudely scribbled with the axes labelled and possible output sketched in. These usually take me about 20 seconds each to draw once I have the idea (the hard bit).

There are five main ways of presenting data.

1. Histograms are used for plotting the frequency of occurrence. A common error is to plot data for individual sites that are better expressed as histograms or tables. For example, suppose that the study measured the fish density along various sections of rivers. Plotting histograms of fish density in each section would be of little use. If the interest is in the density in each section then a table may be more useful, since the exact result can be given. If, however, the aim of the study is to show some general pattern, such as that fish density is related to weed cover, then it is usually better to have a scattergram of fish density against cover.

2. Scattergrams (plotting sample points in relation to two variables, such as the density of invasive plants in quadrats in relation to distance to the forest edge). A common error is to plot as means and errors data that are more usefully shown as histograms. For example, if one is comparing the grass heights of sites that contain a butterfly with those that do not, it is possible to present each as mean grass height ± 1 standard error. It would, however, be more instructive to give the frequency distribution of mean grass height of all sites used as one histogram above a second histogram of the grass heights of unused sites.

3. Tables. These give the exact values for the data. They tend to be used when the reader is interested in the particular cases, whereas graphs are used if the main interest is in the relationship between parameters. For example, use a table if giving the butterfly densities of various sites but a scattergram if showing the relationship between soil pH and butterfly density. Tables are also used if a number of measurements are made, for example when describing the features of a range of sites when there are too many data to be presented as figures.

4. Maps. These are rarely used for presenting data but obviously the best way of showing distribution data. Simply plotting where a species was encountered is of little use. It is necessary to describe the areas that were surveyed. For example the survey may exclude all areas above 500 m because it is known that the species definitely does not occur above this height, so these areas can be shaded in the map. The areas searched should then be shown. It is acceptable to use multiple pie charts on maps, for example to show the proportions of juveniles in various locations or the proportions of injured individuals in various locations.

5. As facts within the text. This is used for giving simple measures, e.g. 'of the twenty individuals found dead, 75% had gunshot wounds'. In practice, the results of less exciting graphs and tables are often stated in the text, e.g. 'there was no relationship between the number of pollinators and sward height $r_s = 0.013$, $n = 23$). However, in planning I suggest that these are still drawn.

Pie charts should be used only for showing data to children or politicians (a table or histogram is usually clearer and the data can be read off precisely). Furthermore, most pie charts have a multicoloured key and it is often a nightmare to work our exactly which shade of pink corresponds to a given slice of the pie. Three-dimensional pie charts (as produced routinely by some computer packages) should not even be shown to children or politicians. The basis of pie charts is that the angle at the centre is in proportion to the frequency represented yet this is not the case in these pointless charts. The one exception regarding the use of pie charts is that they are useful when superimposed on maps to show how proportions vary geographically. For example, the proportion of individuals that are adult or the proportion of trees that are burnt could be counted at various points and each plotted on a map.

My experience of discussing science with students and others is that drawing figures is usually a difficult and painful process. However, it is difficult only because science is difficult. My experience is that, if the person is unable to draw the figures, it is because the person is insufficiently prepared to start the research. Once the science has been resolved, drawing the graphs is simple.

The statistics usually follow from the presentation. Thus scattergrams tend to lead to regression or correlation, histograms tend to lead to t tests, ANOVA or Mann–Whitney tests, and tables often lead to comparisons of frequency data, such as X^2. As Chris du Feu states, 'statistics is drawing pictures then drawing conclusions'.

If the researcher is not a competent statistician then I recommend showing the figure and table sketches to someone who is before collecting the data. It is important to know how the data will be analysed before collecting them. In a few cases the data just cannot be analysed, but more frequently the lack of planning means that the opportunities are missed and the analyses are more restricted than necessary.

What data are needed to complete these analyses?

How much data is likely to be necessary to make these analyses worthwhile? How many individuals have to be marked in order to estimate survival rate? How large an area has to be visited in order to plot the distribution? Of course, this depends upon how strong the relationship or differences are and the variation that occurs. In the absence of any other information I usually suggest that a sample of 20 for the scattergram or each histogram point is a useful, if arbitrary, level to start thinking around. If the differences are likely to be slight, or the variation between samples considerable, then this will be insufficient. If the patterns are likely to be strong and consistent then a smaller sample might well suffice. Chapter 2 shows how large samples need to be to attain the required precision. Understanding how one will analyse that data is important in planning to avoid collecting data that cannot be analysed.

The total data required are thus the data needed to complete all of the planned figures drawn in the previous stage.

What protocol is required to obtain these data?

It is often sensible to devise a protocol, e.g. visit a site within the period 10 a.m. to 4 p.m. and count tortoises along four 100-m transects recording the soil temperature at depth 5 cm at the start of each, analyse the vegetation as percentage cover using a 1-m^2 quadrat at the middle of the transect, estimate the percentage of sand within 1 m^2 every 25 m along each transect. Assess the vegetation and sand cover around each tortoise located. Measure, weigh and sex each tortoise.

It is useful to write down the methods before starting. This is especially true if there is more than one observer, so that the methods can be agreed upon. Providing the full methods in the paper or report is essential in order that others can use the same methods for comparable studies.

Can the data be collected in the time available?

Determining the time necessary to collect the data seems like a boring task. It is. However, many projects are unsuccessful because they run out of time. One example is the expedition that sought to compare two sides of a mountain, but ran out of time and just compared one side of the mountain. If the project is not possible in the time available then it is much better to modify it or do something else than start and have to abandon sections that were never feasible. A common example of such a problem is to collect far more samples than can ever be identified. Apart from the possible ethical and conservation issues, such a project is less efficient than one that collects a sensible number of samples.

I suggest a four-stage process for the basic method for allocating time.

1. Decide upon the time available. This is usually the number of days available. Allow time for illness, bad weather, equipment failure, dealing with others, buying supplies etc. Be realistic.
2. List all the jobs that you would like to do and the number of times, e.g. 'electrofish lake five times along 100-m transect'.
3. Estimate how long each would take in appropriate units for question 1 (e.g. days). Techniques include imagining doing it; estimating the time of each part; considering a similar task or speaking to others who have carried out the task. If necessary use a range of values.
4. Consider whether there is sufficient time.

Using a calendar

This is the same method as just described but the time is divided into periods. This could be hours, days, weeks or months, depending largely upon how long the project is. If using weeks (often a convenient scale), write the weeks along the top and the jobs down the side. Allocate time (e.g. days) to each job, taking into consideration when each should be completed. In reality this will take some juggling (often considerable juggling) to ensure that the work can be completed in time and that the load is distributed sensibly. Again, this juggling is the key part of the exercise.

Example. A project will be going to Sabah to examine moths. Moth sampling takes five days and is done at the start of each month. Mark–release–recapture takes eight days and should be done in as many months as possible. A vegetation survey of seven days can be carried out at any time. A social survey of seven days should be carried out as late as possible, when you are more accepted by the community. The preliminary results will be analysed and presented to the local community. Packing will take 2 days. The project lasts from 1 January to mid March

My answer assuming 5 working days per week and 4 weeks per month:

	Jan.				Feb.				Mar.	
	1	2	3	4	1	2	3	4	1	2
Moth sampling	5				5				5	
Vegetation surveys			2	5						
Social surveys							2	5		
Mark–release–recapture		5	3			5	3			
Present results										3
Pack										2
Totals	5	5	5	5	5	5	5	5	5	5

Calendars with divisions

It is often necessary to subdivide the calendar where there are other constraints, for example if there is a constraint that you can visit the field site for only eight days a month then divide into the calendar 'field site' and 'other'. It is then necessary to juggle the tasks to provide an efficient solution. Others studying frogs or moths may divide the time into days and nights. If various team members have different skills then you could divide for each member and allocate tasks in the most efficient way.

Example. You are going to carry out some mammal surveys and relate the results to vegetation, soil type and land use. You can start on 1 June and end on 1 August. Mammal surveys take place in the morning. You can carry out one transect per morning. Five transects are required per site. The mammal survey is a standard method and must be used. It is important to do this in as many sites as possible. Vegetation surveys take three half days. It is essential to carry out a full vegetation survey for each site with a mammal survey. It would also be interesting but not essential to carry out a bird mist netting survey per site, which takes two mornings per site. The director of the local research institute is away in June. On return he has promised to give you the data on soil type and past management. Meeting him and extracting the data will take two half days. You have been asked to collect lichens in any sites in which this is possible, which would take two half days (morning or afternoon) per site, but this is the lowest priority. The analysis of the project must be completed before leaving and will take about six half days but requires the institute data before it is worth starting. Plan your programme.

My answer. Since the objective is to collect as many mammal surveys as possible and these can be carried out only in the morning, I have cut the bird surveys. The lichen collection can be included without affecting the main objectives so I have included this.

	June								July							
	1		2		3		4		1		2		3		4	
	am	pm	am	pm	am	pm	am	pm	am	pm	am	pm	am	pm	am	pm
Mammals	5		5		5		5		5		5		5		5	
Vegetation		3		3		3		3		3		3		3		3
Institute										2						
Lichens		2		2		2		2								
Analysis												2		2		2
Totals	5	5	5	5	5	5	5	5	5	5	5	5	5	5	5	5

Gantt charts

These are used where a number of individuals are involved as a way of allocating work that has to be carried out in a sequence.

Task	March															Person responsible
	1	2	3	4	5	6	7	8	9	10	11	12	13	14	15	
Obtain aerial photos	×	×														Isabelle
Pitch camp	×	×														Carlos/Andrew
Decide on sites			×	×												All
Cut transects					×	×										Andrew
Cut mist net rides					×	×										Jenny
Count fish							×	×	×	×	×	×	×			Isabelle
Decide on trap location					×											Carlos
Create traps						×	×									Carlos
Check traps							×	×	×	×	×	×	×	×	×	Carlos
Mist net birds							×	×	×	×	×	×	×	×	×	Jenny
Botanical survey													×	×	×	Andrew

This can be combined with allocating time as in the previous methods if each time period is, say, a week and the number of days devoted to each task is then given.

Modifying the planning in response to time available

Allocating time is usually a painful process, since most people have more ambitious plans than are realistic. In practice this is likely to involve rethinking the previous stages. It is this rethinking that is central to the planning of the project. This rethinking usually involves slithering between the various stages until the questions, figures, protocol and time allocation are all compatible. In extreme cases it might be clear that it is not possible to answer a sufficiently worthwhile question in the time available. The options are to start an inevitably doomed project or design another project that might work.

Creating data sheets

Many projects involve data sheets and their creation is often a critical stage in clarifying thinking. The creation of the data sheet and protocol often go together and the protocol then involves filling in the data sheet in a systematic way. Data sheets make it much less likely that any data are missed. This makes it more likely that analysable data will be collected rather than a set of unsystematic observations. It is sensible to create a draft data sheet and test and then alter it before adopting it. It is useful to use a range of forms, e.g. one for the habitat of each site, one for the laboratory soil analysis and a third for recording pollinator behaviour. Completed data sheets do not need to be taken into the field and so are less likely than notebooks to become lost and can even be photocopied and stored in a separate building to reduce the risk of fire or theft. Using abbreviations saves time in the field but can lead to confusion and doubt over the accuracy afterwards. One technique is to confirm any unusual data, for example by underlining to distinguish them from errors.

Quantifying the variation in the field looks simple but actually requires great skill. For example, sites may differ in cattle densities, slope, tree height and the amount of bare ground and the abundance of a rare orchid. The first decision is how each can be measured. One approach is to divide into classes, e.g. high, medium or low cattle density, but this is a poor technique, since it is often not repeatable and is difficult to analyse statistically. It is thus much better to quantify these as continuous variables. The skill is to find simple ways of obtaining sufficient precision. This is a balance between methods that produce data that are too inaccurate to be useful and methods that are more time consuming than the results obtained justify. It is tempting to quantify all variables but this takes time and reduces the sample size or increases the cost, so it is necessary to balance their use against time or cost.

A frequent mistake is to collect data that are too detailed. Experts commonly advise that a huge amount of data is necessary to assess a population or a physical feature with sufficient accuracy and thus a proposal to compare say, twenty sites, is completely impractical. This is often because the expert has a different agenda. The expert is often seeking a precise measure for each site, but for the comparison it may be sufficient that the ranking is reasonably correct. For example, the grazing intensities might vary considerably between sites and you may wish to have some relative measure that will reveal whether light or intense grazing has any consequences but the expert might be thinking about how to provide a precise assessment of grazing pressure for a given site.

Although data sheets avoid many of the problems associated with notebooks, they still have risks. I know of a case in which data that had taken two years to collect were stolen from a car, another in which all data were kept in a black bin bag to keep them dry but thrown out with the rubbish and another in which a departing researcher had all his field data in a briefcase, which was then stolen. In the last case the depressed biologist was later approached by a small boy selling peanuts wrapped in his data sheets. He was able to buy most of the sheets back at considerable cost. One way of reducing this risk is to add the data to the computer as soon as possible but still keep the original data sheets plus separately stored back-up computer files.

Types of data sheet

There are three main types of datasheet, although there can be some overlap.

Single event sheets. These are completed on one occasion, for example for a single survey.

Continuous data sheets. Each new observation is added. For example, for each new individual caught and marked. Date and location are usually added each time.

Updated data sheets. These are usually based around a site, nest or individual. Much of the data is recorded once, but future visits are used to record extra data. Data may be added with each visit.

Example A. Make repeated visits to a series of lakes counting the waterbirds on each and record the diet and nesting success of fish eagles. Lake data sheet to describe the habitat (single event sheet); count data sheet (continuous data sheet); fish eagle diet data sheet (continuous data sheet); nest record card (updated data sheet).

Example B. Carry out a series of point counts recording habitat and collect data on all orchids found and detailed data on flowering time and seed set for the rare species. Data sheet for each point count giving habitat (single event sheet); orchid data sheet (continuous data sheet); rare-orchid data sheet (Updated data sheet).

Box 1.1. **Tips for creating data sheets**

- Place boxes around everything that has to be filled in, especially if others are completing the forms.
- Make the box size appropriate.
- Rearrange the sheet so that space is used sensibly. It may be possible to fit multiple copies (e.g. more than one survey) onto one sheet.
- Order the data on the sheet in a logical order that relates to a sensible sequence in the field.
- Consider how the data will be typed into the database. The data sheet should mirror the database.
- Consider analysis. For example, should data be continuous or categorical?
- Show to others for suggestions of improvements both to the structure of the sheet and to the methods being used.
- Think about what should be done for ambiguous and unusual cases. Write down these criteria.
- If several observers are to use the sheet, consider including an example record at the top of the columns.
- Test and modify. Use all of the following three approaches.

(1) Imagine being in the field and think how the data would be collected.

(2) Test in the field. If necessary, this can be carried out on unrelated systems. There might not be any rainforest nearby, but going into the local wood and considering how the data will be collected will reveal many problems.

(3) Modify after carrying out in reality. It is thus ideal, if it is possible, to modify the data sheets after an initial field visit.

- Leave a column for notes in which unusual events can be recorded. However, it is important that almost everything is quantified and there are not essays written on each.
- Decisions should usually be made in the field (e.g. whether the individual is within the study area). This often requires making definitions (e.g. if the individual is heard but not seen, does it count?).
- Include a box for the observer's name and usually have boxes for date and time started and ended.
- Ask for exceptional cases to be underlined (to distinguish them from errors).

Start and encounter reality

Having planned your project, you have ensured that your project seems feasible and practical and will answer useful questions. Once the data collection starts it is usually clear that the plans are insufficient. It might be impossible to collect certain data. Some data might be far more time consuming to collect than expected. Occasionally some data are much easier to collect than expected. You might notice something really interesting. Under such circumstances rigorously sticking to the plans can be ineffective. However, spontaneous changes of plans rarely work. The sensible solution is to run through the entire reverse planning process again, quickly, and modify it appropriately. Having already worked through a plan in detail and undergone the painful process of planning, it is usually much quicker to change. These decisions can be made almost instantaneously. For example, you notice that some of your study individuals are limping. You then decide to categorise each individual as to whether limping or not and measure snare density in transect in each 1 km^2 of forest. You will correlate the two and plot histograms of percentage limping and snare density at various distances from the forest edge. The transects will take another four days but will replace the vegetation survey that turned out to be more complex than planned because the herbarium specimens proved difficult to use.

Acknowledgements

Thanks are due to Rosie Trevelyan for discussions and suggesting the term reverse planning and to Chris du Feu, Jeremy Greenwood and Rob Robinson for useful suggestions.

2 Principles of sampling

Jeremy J. D. Greenwood and Robert A. Robinson

British Trust for Ornithology, The Nunnery, Thetford, Norfolk IP24 2PU, UK

Table 2.1. *Contents of Chapter 2*

Ecological Census Techniques: A Handbook, ed. William J. Sutherland.
Published by Cambridge University Press. © Cambridge University Press 1996, 2006.

Before one starts

Objectives

The definition of objectives is particularly important in studies of the abundance of animals or plants because it will determine whether one needs to make a full count of the individuals present in an area (or, at least, an estimate of that number) or whether an index of numbers is satisfactory. By an index, we mean a measurement that is related to the actual total number of animals or plants – such as the number of eggs of an insect species found on a sample of leaves of its host plant (as an index of numbers of adult insects) or the number of rabbit droppings in a sample area (as an index of the number of rabbits). Because accurate counts are extremely difficult to make, one may have to make do with an index even when a count would be preferable. Indeed, so long as an index is sufficient for one's purposes, a reliable index is preferable to an unreliable count.

The objectives will also determine the extent to which a population should be divided by sex, age or size in one's study: some work on the structure of communities may require only that total numbers of each species are known, whereas population dynamicists will often need to know the composition of the population by sex and age.

Know your organism

Having defined your objectives, the appropriate methods to apply will be determined by the characteristics of the species you are studying. It is therefore important to know as much as one can about the species and its ecology before planning one's work in detail. Some such knowledge can be obtained by reading the literature and by talking to experts but there is no substitute for carrying out preliminary observations oneself, preferably of the very population that one intends to study. The wisest ecologists tell their students to go out into the field in order to sit and watch their animals before they decide how to study them; wise students accept this advice.

The next step is to conduct a pilot study. This enables one to focus on those aspects of the species' natural history that are particularly relevant to one's proposed study and to test out the field methods; it enables one to estimate the resource requirements for a full-scale study; and it enables one to estimate the likely precision of one's results.

Censuses and samples

It is sometimes possible to count the animals or plants in the whole of the population of interest (all the oaks in a woodland, for example). Such a complete count is a true census. More often, one can study only part of the population, through taking samples. For example; counting the trees in a few one-hectare blocks of forest, to estimate the average number of trees per hectare in the whole forest; estimating the number of fish present in a few pools in order to estimate the total number of fish in all the pools in a catchment; counting the number of whales migrating past an observation post on a few individual days in order to estimate the total number passing through the entire migration season. Provided that the samples are truly representative, one can then extrapolate from them to get an estimate of the size of the whole population. This chapter mostly concerns how to arrange sampling so that the results are as reliable as possible. Cochran (1977) and Thompson (2002) provide further and more technical accounts of this topic.

Even in a small sample area, it may not be possible to observe all of the animals or plants directly. Instead, one may observe (or catch) samples and, using appropriate statistical techniques, estimate the total number in the area from the numbers observed. Such methods are covered in the next chapter.

Know the reliability of your estimates

Precision and accuracy

If one has to be content with an estimate rather than an exact count, it is important to know how reliable it is. Reliability depends on both accuracy and precision. Accuracy is about how close the estimate is to its true value, on average; that is, how small the bias of the estimate is. Precision is about how similar repeated estimates are to each other. One should strive to be both accurate and precise, though this is not easy because some ways of increasing precision may increase bias; and some steps to reduce bias may reduce precision.

Measuring precision

For most of the methods of estimation given below, we also show how to calculate 95% confidence limits. It is usually and approximately true (though this is not a definition of confidence limits) that, if the 95% confidence limits of an estimate of population size are CL1 and CL2, then the chance that the true population size lies between CL1 and CL2 is 95%. It is important to provide confidence limits whenever an estimate is quoted, otherwise the reader has no idea how precise the estimate is. For comparison between different situations, it is often convenient to use the percentage relative precision (PRP) (Box 2.1).

The balance of precision and cost

More precise estimates can always be obtained by extending the study but this will inevitably require additional resources, even if only in terms of the length of time devoted to the work. In planning any project one should consider either the PRP achievable for a given cost or, alternatively, the cost of achieving the level of PRP that one requires. It may turn out that the PRP achievable for the planned level of resources is insufficient to make the study worthwhile. In this case, one can decide either to spend more, in order to achieve the desired PRP, or to abandon the study. In either case, by considering the PRP and costs ahead of the study one will reduce the likelihood of wasting resources on a study that fails to deliver results that are precise enough.

Optimisation of design

In this chapter, we shall consider a variety of different sampling designs. Pay careful attention to the differences between them, because these determine both the bias and the precision, depending

---

Box 2.1. **Percentage Relative Precision (PRP)**

Basics

$$\hat{N} = \text{population estimate}$$
$$CL_1, CL_2 = 95\% \text{ confidence limits of } \hat{N}$$

Definition

The PRP is the difference between the estimated population size and its 95% confidence limits, expressed as a percentage of the estimate. Because confidence limits are sometimes asymmetrically distributed around the estimate, the mean difference between them and the estimate is used. This is calculated simply as

$$PRP = 50 \times (CL_2 - CL_1)/\hat{N}$$

Note

One should use PRPs only as approximate guides. The simple methods presented in this chapter for calculating them are also often approximate, especially when numbers are small.

---

on the circumstances of each particular investigation. Choosing the right method can pay big dividends, in terms of the precision achieved for a given cost. Even within an overall design, variation is possible. For example, if one is sampling insect pests from a crop planted in rows, is it better (in terms of the balance of precision and cost) to take 5 samples from each of 20 rows, 20 samples from each of 5, 1 from each of 100 . . .? We present ways of answering such questions. A facility for automated design, with GIS functionality, is provided in the software package DISTANCE (see Appendix to Chapter 3). It is aimed at surveys using distance sampling (see Chapter 3) but is also relevant to other methods.

Sources of bias

Bias can arise through poor practical techniques. Thus even a direct count of a population is biased if the observer fails to see all the individuals. Other examples of poor practical technique are applying marks (in order to be able to identify an animal subsequently) without considering whether marks are being lost and causing increased mortality through using insensitive trapping techniques. In order that other workers (and you yourself, at a later date) can assess the probability that your working methods were a source of bias, you should always record them as fully and as carefully as possible.

Non-representative sampling can also cause bias. For example, mark–recapture methods may assume that all individuals in the population have an equal chance of being caught, which is equivalent to assuming that the animals caught are representative of the whole population in terms of catchability. But, in reality, they are likely to be the more catchable individuals, so the assumption is wrong. As a result, population size is likely to be underestimated. Since one cannot control the differences in catchability, one has to be careful in one's choice of method. One must consider carefully whether its assumptions are likely to hold in one's study population and, if they are not, to what extent the results are likely to be biased in consequence. Computer-based simulation models are useful here, allowing one to determine the extent of the bias that results from a certain level of departure from the assumptions. Wherever possible, assumptions should be tested and biased estimates avoided. Do not be misled into believing that the more complex methods (which, necessarily, take up more space in these chapters) are generally better than simple ones. Since they rely on more restrictive assumptions, they are often worse. All other things being equal, simple methods are always to be preferred.

What if your estimate is biased?

The best solution is to find another method, or a modification of the existing method, that provides an apparently unbiased estimate. This might not be possible. Alternatively, it might be appropriate to use an index of population size rather than population number itself: the biased population estimate might actually be an appropriate index. At worst, one will be reduced to making the decision as to whether to reject the estimate entirely (and thus conclude that one has no idea as to the size of the population) or to accept it as a rough, but biased, guide. That decision will depend on one's judgement as to the likely magnitude of the bias, the value of having at least some idea of the size of the population (rather than none at all) and the cost of any wrong decisions that could be made as a result of basing conclusions on a false estimate of population size.

Performing the calculations

General

The methods of data analysis presented in this book are all possible using pencil and paper or a pocket calculator. It is better generally to use a spreadsheet, however. This allows one to check the data after one has entered them, it facilitates clear layout of the data and calculations (which is particularly valuable for some of the more complex calculations), it takes much of the tedium out of the analyses, and it allows one to use the same layout repeatedly. Most of the analyses we present can be carried out on spreadsheets such as the widely available Excel®. The ability to drag formulae from one cell to another makes this a powerful tool when one has to perform repetitive calculations; it has numerous functions that perform the simpler calculations, such as calculating an average or a correlation coefficient; other functions provide values of statistics such as Student's *t*, for incorporation into calculations; and the add-in Data Analysis ToolPak provides various other statistical procedures, such as an analysis of variance.

All packages should be used with caution. One should not only read the documentation but, to ensure that they do what one thinks they do, one should also test them out on data sets for which the correct results of the calculation are known. For example, Excel® provides the functions VARA and VARP to calculate the variance based on a sample and on an entire population, respectively.

But to calculate a covariance, it provides only COVARIANCE, which assumes that the data are drawn from the whole population; so, if one has data only from a sample of size m (the usual state of affairs), one must multiply the output from COVARIANCE by $m/(m-1)$ to get a correct estimate.

The more complex analyses are best conducted using specialised software, which we have briefly described in an appendix to Chapter 3.

Precision

In calculations using pencil and paper or pocket calculators it is important to retain sufficient precision in one's figures to avoid rounding-off errors accumulating to such an extent that the final result is imprecise. This is much less of a problem when using computers but if the intermediate stages of one's calculations involve very large or very small numbers one should check that the computer and package one is using are capable of storing these precisely enough.

Simulations

In some situations, the standard statistical methods for estimating numbers and for assigning confidence limits to those estimates might not perform well. It may then be better to use computer-intensive methods such as bootstrapping to obtain the confidence limits and estimate the likely bias of one's estimates, see Manly (1997). Effectively, in doing so, one is simulating the investigation in one's computer to discover how well it performs. Other forms of simulation are often useful: for example, to explore (on the basis of information derived from a pilot study or from background knowledge) how many samples are needed to attain an estimate with the required degree of precision, or which of several possible sampling designs will deliver the most precise results. One can explore the likely consequences of the assumptions of one's methods not being met by simulating in the computer a population in which they are not met. For example, how biased would one's estimate from a mark–recapture study be if the animals differed in capture probability to a specified extent, rather than being equally detectable? Excel® and other spreadsheets can be used for such simulations and a useful Excel® add-in for simulations is available free to download from http://www.cse.csiro.au/client_serv/software/poptools/index.htm. WiSP is a package specifically designed for simulating wildlife surveys (see http://www.ruwpa.st-and.ac.uk/estimating.abundance/. It is written in the free language R (http://www.r-project.org/).

Mathematical terminology

Note that we use the standard terminology of $\log_e$ to refer to natural logarithms but that, to make the layout of equations simpler, we use $\exp(x)$ instead of e^x (= antilog$_e$).

Keep thinking

There is a temptation when making calculations, especially if using a calculator or a computer, to switch off one's critical faculties. Do not succumb! Keep querying the results, at two levels. First, the simple arithmetic: do some checking in your head, like 'Is the sum of 100 numbers between 3 and 13 likely to be my calculated value of 1295?' Second, ecology: ask yourself whether your estimates and their confidence limits seem reasonable. If you are doubtful, rework the calculations or check that you have interpreted the analytical package correctly. This is an effective way of picking up errors.

Sampling – the basics

Defining sample units and the sampling frame

To sample the trees in 1-ha blocks of forest, one overlays a map of the forest with a grid of squares measuring 100 m × 100 m. Each square is a potential sample unit. The 'sampling frame' is the entire list of squares covering the forest. When estimating fish numbers in a catchment, individual pools may be the sample units, all the pools in the area comprising the sampling frame. If the area has been well mapped, one can be sure of being able to list all the pools. But what if the maps are unreliable or non-existent? Finding all the pools in an area may be very difficult, even from the air, especially if the area is a well-wooded one. Properly identifying the entire sampling frame may be the most difficult part of a census study when the sample units are natural objects rather than arbitrary units imposed by the investigator (like the 1-ha blocks).

One method of sampling is to go to a point and count all the organisms within a fixed radius. Another is to go to a point and move off along a line, counting all that are detected within a fixed distance of one's line of travel. The circles round the points and the belts along the routes are obviously the sampling units in these cases. But what is the sampling frame in such cases? In theory, any area, however small, holds an infinite number of points. In practice, of course, the resolution with which one can identify points limits their number: if one can measure one's position with a precision of only 10 m, it is possible to identify only 10 000 different points per square kilometre.

Instead of counting all the organisms within a fixed distance of a point, one may count all those observed, however far away. The sampling unit is then an ill-defined area around the point and the frame is the number of distinguishable points.

The need for replication

An investigator who proceeds to estimate the total number of trees in a forest from the number in one hectare, or who assumed that the total number of fish in a series of 25 similar ponds was just 25 times the number in one of the ponds, or who claimed that he knew the average number of helminth parasites in fish of a particular species because he had caught a single fish and counted its helminths, would be laughed at. Why? Because we know that a single sample unit can provide only an imprecise estimate for the whole study population and, however carefully chosen, it might not be representative. To increase precision and representativeness, one needs more than one sampling unit.

Replication of sampling units has another essential function: it allows the precision of the generalisation to the whole study area to be measured. Thus one can not only estimate the total number of trees in the forest, the total number of fish in all 25 ponds, or the average helminth burden per fish, but also place confidence limits on these estimates.

Methods are described in the next chapter for estimating the precision of population estimates in individual units – for example, for an estimate of the number of fish in a pond based on mark–recapture methods. The precision of the estimate for the whole study area will be greater when the individual sample estimates are more precise. It is not, however, possible to assess overall precision from the precision of estimates for individual sample units. This is because overall precision depends not only on individual precision but also on the amount of variation between individual sample units. The latter can be assessed only if one has sampled more than one unit.

Ensuring that samples are representative

Representativeness is essential

To avoid bias, the samples must be representative of the whole. The potential for bias may sometimes be obvious: if one chooses to study a particularly damp part of a forest, it will have more mosquitoes in it than average, so an extrapolation from its mosquito population to that of the whole forest will overestimate the number for the whole forest. Often it is more subtle: for example, one may choose to work close to roads for convenience, ignoring the multitude of physical and biological impacts that the construction and use of a road may have on its surroundings.

Ensure the sampling frame is complete

Suppose that, in attempting to list all the pools in one's study area, one had missed some. One is more likely to have missed the smaller pools and those better hidden by vegetation. These are likely to have different numbers of fish in them from the numbers in the other pools. As a result, because one can sample only from the list of pools identified, one's estimate of average numbers of fish per pool will be biased. For this reason, it is important to make one's sampling frame as complete as possible. If it is not, one must think hard about possible biases that may result.

Random sampling

Suppose that one has decided to count the trees in ten 1-ha blocks, in order to estimate the number of trees in a whole forest that extends to 50 000 ha. How does one choose which ten of the 50 000 to sample, to ensure that the sample is representative? Surprisingly one does *not* choose ten 1-ha blocks that appear to be 'typical'. Experience shows that it is rare for such judgemental samples to be truly typical. Furthermore, it is not possible to estimate the precision of the overall estimate properly from judgemental samples. Neither does one use 'haphazard' approaches, such as stabbing at a map of the forest with one's eyes closed or, for a smaller-scale investigation, throwing a sample quadrat over one's shoulder; experience again shows that such methods do not provide random samples.

Randomness requires that each potential sample unit has an equal chance of being included in the sample. The best way to ensure this is to give each potential sample unit a number and then choose which numbers to include by using a table of random numbers or a string of random numbers generated by a computer or a calculator. (The Excel$^{®}$ function RANDBETWEEN generates random integers between specified lower and upper bounds. Dragging the function down a column allows one to generate as many random numbers as one wants.) Thus, if one wished to sample five ponds from the numbered list of 25, one would get five random numbers between 1 and 25 and sample the corresponding ponds. (If the same number comes up more than once, ignore the second and subsequent instances and simply extend the list of random numbers to get additional unused ones. In principle, it is possible to include the same sample unit more than once but such "sampling with replacement" is rarely used in population studies.)

Sample units are often naturally arranged in two dimensions. Box 2.2 shows a simple way of choosing random units in this case. A similar technique may be applied if the sample units are based on points rather than areas: the pairs of random numbers are then used to define the distances to the sample points from an arbitrary base point. Only points falling in the study area are utilised.

Particularly when the sample units are natural objects, it may be impossible to enumerate them: for example, all the trees in a forest (on the sampled individuals of which one may be censusing the tree snails). Indirect methods must then be used, such as taking the nearest trees to a set of random points in the forest. Sometimes it requires great ingenuity to get a random sample. (Readers might give some thought as to how to sample leaves at random from a tree.) The key thing is to aim for a method that gives each unit the same probability of being sampled.

Some methods that might appear at first sight to provide random samples do not do so: one needs to be careful. For example, laying down one or more lines at random across an intertidal area and including in the study all those individuals of a bivalve species that happen to be crossed by the line will bias the sampling towards the larger individuals. So will taking into the study those individual trees in a forest whose canopies cover any one of a set of random points. Such size-biased sampling is best avoided; while the resultant data can be used, the methods of interpreting them are beyond the scope of this book. Marginal changes to the protocol may remove the size bias. For example, including individuals whose centres fall within a fixed distance of the line or the point will result in samples that are not biased by size.

Precision in locating samples

Suppose that one was counting 1-m^2 quadrats in a forest of several thousand hectares. One might use random numbers to define the positions of the quadrats to the nearest metre. But it would be practically very costly to locate the positions with that degree of precision on the ground. Given that the definition of the position is random, however, it is permissible to use an imprecise method, such as pacing out or using a GPS accurate to only 20 m. One must, however, take care to consider potential bias. For example, when pacing one may unconsciously take longer or shorter paces as one approaches the sampling point, influenced by what one happens to see. That bias could be overcome by sampling not at the stopping point but at a point that was, say, 30 m away in a random direction.

Box 2.2. **Choosing sample units from a two-dimensional distribution**

Imaginary example: choosing 1-ha sample areas from a woodland. The shaded area in the diagram below represents the woodland and the lines of the grid that is superimposed on it are 100 m apart.

Number the grid rows and columns.

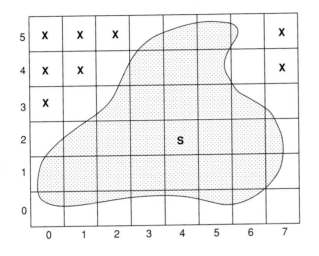

Get a pair of numbers from a random-number table or computer in the usual way. The first defines the column in the grid, the second defines the row. If the numbers were 4 and 2, this would define the sample square as the one marked S in the example.

The process is continued until the required number of units has been defined.

If grid squares that do not contain part of the study area (marked X in the diagram) are chosen by this process, they are ignored.

Note that it is unnecessary to mark out the grid on the ground; as long as one corner is identified on the ground, and the compass orientation of the grid is fixed, it is possible to measure out the locations of the chosen sample units.

Deviations from random

Regular sampling

It is often convenient to take samples at regular intervals – on the same day of the week, at fixed distances along a river, or at all the intersections of a rectangular grid. This results in two problems. First, it will usually produce confidence limits for the overall population estimate that are too narrow. More obviously, though less frequently, it will produce biased results if the regularity of the sampling coincides with a natural regularity in the distribution of the organisms. For example, waterbird numbers using a lake may be less than average on Sundays because hunting activity is most intensive on Saturdays; if counts are always made on Sundays, they will underestimate the average number present over the whole week. (Such data can be used to generalise only about Sundays, not all days of the week.) There may be similar, and often more subtle, regularities in spatial distribution, such as the patterns in soil produced by geomorphological processes or the regularities of dispersion resulting from animals' territorial behaviour or the allelopathic interactions of plants. Regular sampling should generally be avoided if possible.

People often believe that regular sampling is advantageous because it distributes the samples all over the study area, thus ensuring that, in aggregate, they are truly representative. Since such distribution can be achieved by stratified sampling (see below), regular sampling is not required.

Regular sampling is justified in one situation: when one is trying to map variation in abundance of an organism across a study area as well as to estimate its total abundance. The advantages for mapping of a regular distribution of the sample sites may then outweigh its disadvantages for population estimation.

More samples in some places

For reasons beyond the control of the investigator, some parts of the study area might not be sampled as intensively as others: for example, sample units that are far from base or require helicopters to get to them. This sort of non-randomness can be circumvented by stratification (see below), especially if it is identified and planned for in advance. It is also not uncommon that part of the study area is more difficult or more costly to survey than the rest. Again, stratified random sampling is the solution (see below).

The problem of inaccessible areas

Parts of a study area may be totally inaccessible. If so, they must effectively be excluded from the study. Having made a proper estimate of numbers in the accessible part of the area, one may then make a less formal estimate of the numbers in the inaccessible area, based on whatever knowledge is available. An estimate of the total population is then possible but the fact that only part of the population has been studied properly must be reported clearly. Thus a (stratified) random sample of Britain outside large urban areas, where the survey methods could not be applied, gave an estimate of the total number of social groups of badgers *Meles meles* in non-urban Britain as $41\,900 \pm 4400$ (95% confidence limits); from other knowledge of badger populations in large urban areas (thought to total only several hundred), the total number of social groups in Britain may be estimated as around 43 000.

Points on grids

Suppose that one is counting the organisms within 10 m of points that are the intersections of a grid with the lines 100 m apart (or, to put it differently, points located to a precision of only 100 m). Clearly, much of the study area will not fall into the relatively small circles around the relatively distantly placed points and will stand no chance of being sampled. The solution is to use a finer grid.

Gaps between adjacent samples

If the sample units are squares, adjacent units can fit together exactly. If they are circles around points, there are bound to be gaps between them (unless they overlap). However, these gaps are sufficiently small that the fact that they can never fall in the sample is unimportant in practice.

Other gaps may be larger. Suppose that one is sampling invertebrate animals on a mudflat. In taking a sample from a point, one may considerably disturb the immediately surrounding area. Thus the area within a certain distance of each sample is one within which one should not take another sample. Unless this results in a substantial proportion (more than 10%) of the study area being excluded from the sampling, this causes no problems. Otherwise, an adjustment has to be made in the calculation of confidence limits of the resultant population estimates.

Unavoidable bias

Sometimes it is impossible to avoid some bias during sampling. Large-scale projects, such as national surveys that depend on volunteers being prepared to go out to survey sample localities, rarely result in all of the chosen localities being surveyed. Volunteers may, for example, not be interested in surveying certain habitats or places where they are likely to find few interesting animals or plants. This could cause serious bias. Even if the reason for failing to cover some areas is that landowners refuse access, bias can result, for access may be more likely to be refused for certain types of land. All surveys where inclusion of individual randomly chosen samples is optional are liable to bias.

The example of a parasitologist surveying the helminths in fish presents another class of problem. The fish that are caught for study are not, in the formal sense, a random sample of the fish population. Indeed, if helminth burdens alter catchability, they may be a biased sample. They can be regarded as a representative sample of the more easily catchable individuals in the population. Since it is impossible to take formally random samples from most natural populations, one has to make do with such catchable samples and use one's knowledge of natural history to judge how well they represent the populations from which they are drawn. If one catches animals using two different methods and the results differ, this is evidence that at least one of the methods is biased, though it does not indicate which; similarity of the results does not indicate that bias is absent, since there may be similar bias in the two methods.

The shape and size of sampling units

Shape

In principle compact sample units are best since they have short edges relative to their areas, thus minimizing the number of decisions one has to make about whether to include or exclude an organism, herd or territory that overlaps the edge. But strips may be easier to work, for example

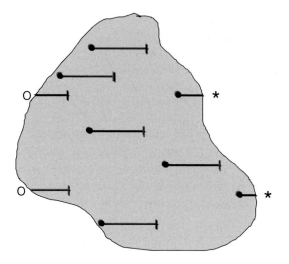

Figure 2.1   Random strips across an area. Each starts at a filled circle and ends at a bar. Note how some have to wrap around from points marked with an asterisk to points marked with an open circle.

in aerial or ship-borne surveys, on steep ground (when it may be difficult to walk other than along the contours), or on cliffs (when straight down a rope may be the only option).

If strips are to be used, it is generally best to have all of them in parallel, to avoid overlap (Figure 2.1). Strips of fixed length from random starting points are a good option. If a strip extends beyond the boundary of the study area, stop at the boundary and insert the missing length at the opposite side of the study area (Figure 2.1). Do not extend the missing length backwards from the starting point: this would result in undersampling of the parts of the study area close to the boundary. If, by chance, the random starting point of one strip falls on a previously chosen strip, leave the second strip out and choose a new one. Unless the samples cover a large proportion of the study area, this departure from strict randomness will have little effect.

In aerial surveys, it is often more effective to fly the entire width of the study area for each strip, rather than breaking off when a fixed length has been flown.

In any study, it is important to minimise the distance travelled between sample units relative to the time spent surveying the units, so the pattern of travel between sample sites should be thought about. One method of arranging strip samples if one has to return to one's starting point to pick up a vehicle, etc., is return from the end of the strip on a parallel route; one surveys that as well and adds the data for the outward and return legs together, so getting bigger samples per hour in the field. Square or triangular routes are an alternative. One may need to space the legs to make the observations on them independent.

Size: constant or variable?

The 'size' of a sample unit may be determined by its area or volume but there are other measures of size, such as the length of a line transect, the amount of time spent making counts, the number of traps in a grid, or the number of points in a set of point transects or point counts. Where the investigator can control the size of the sample units, it is best to choose units of uniform size. This generally maximises the precision obtained for a given effort. One has to accept variation in size if sampling units are naturally defined (such as the fish in ponds or the ponds themselves) or if it is convenient to use existing features to define sampling units rather than an artificial grid (such as using fields rather than superimposed grid squares in an agricultural area). Strips spanning the width of the study area will differ in size as the width varies. It may be tempting to subdivide such strips into lengths short enough to be constant, but the data from the subdivisions of a strip must be combined because they are not independent, so the unequal size of the sample units persists. (If habitat varies along a strip, subdivision of the strip combined with habitat recording may be useful for relating numbers to habitat; but that goes beyond what we have space for here.)

Sometimes, one may be interested in the mean numbers of organisms per sample unit (such as numbers of fish per pond or numbers of helminths per fish), so one can ignore differences in size of the units. But if one is interested in densities (numbers per unit area), one must allow for the differences in size.

In the rest of this chapter we assume that sample sizes are equal, unless explicitly stated otherwise.

What size is best?

If sample units are too small, the number of organisms in them may be systematically biased by an 'edge effect'. For example, observers often tend to include individuals that lie on (or even just outside) the sample boundary and trap-based methods may be biased by animals immigrating during the trapping programme. However, if sample units are so large that each one requires a huge effort, it will not be possible to take many samples during the whole study, thereby reducing the precision of the overall estimate of average numbers. The balance to be struck between a few large units and many small ones will vary from case to case, especially in relation to the size and mobility of one's study organism. It should be considered at the planning stage of any study. The discussion on cluster sampling (see below) is relevant to such considerations.

Estimation of means and total population sizes

Means and confidence intervals

The calculation of the mean and variance of numbers per sampling unit (or of an index or of density) is straightforward (Box 2.3). To calculate the standard error and confidence limits of the mean one needs to consider the relationship between the samples and the whole study area from which they have been drawn. It is generally easy to consider the area as comprising M potential sampling units, of which m are actually sampled: the imaginary example in Box 2.2 is a clear-cut case. In other situations there is a potentially infinite number of units to be sampled (the number of points in an area is theoretically infinite if their positions are located with infinite precision) or, more commonly, the number is very large and unknown (such as the number of wheat shoots in a field). When there is a finite number (M)

Box 2.3. **Means and confidence limits from a set of samples**

Basics

> $m =$ number of sample units studied
> $M =$ total number of potential sampling units in the study area (may be effectively infinite – see the text)
> $\hat{N}_i =$ estimated number of organisms in the *i*th sample

Note: the same method can be applied to indices or densities.

Examples

(1) Number of grain aphids *Sitobion avenae* on 50 random shoots of winter wheat *Triticum aestivum* in an English field in May 1988. (Data from N. Carter, personal communication.)

Data ($f =$ number of shoots with $\hat{N}$ aphids):

$\hat{N}$	0	1	2	3	4	5	6	7	8	9	10	11	12	13	14	15	16	21	26	33
f	14	4	12	3	3	2	1	0	3	0	1	0	2	0	1	0	1	1	1	1

(2) Number of newly emergent shoots of bramble *Rubus fruticosus* in ten random 1.28-m × 1.28-m quadrats within a 9-m × 9-m study plot in a British woodland (Hutchings 1978; Diggle 1983):

$\hat{N}$	0	1	4	4	7	7	7	8	11	13

Calculation of mean and variance

This can be done 'by hand' (following the methods given in standard statistics textbooks) or using statistical packages on a computer.

For the above examples, the estimates are

(1) $N = 4.66,$ $s^2 = 48.27$
(2) $N = 6.20,$ $s^2 = 16.62$

Calculation of standard error when *M* is effectively infinite (i.e. the sampling fraction, *m/M*, is small)

As in example 1, where there was a huge (and uncounted) number of shoots:

$$s_{\hat{N}} = \sqrt{s^2/m}$$
$$= \sqrt{48.27/50} = 0.98 \text{ (example 1)}$$

Calculation of standard error when *M* is finite (i.e. the sampling fraction, m/M, is large)

As in example 2, where there were only 49 potential sample units in the study area:

$$s_{\bar{N}} = \sqrt{s^2(1 - m/M)/m}$$
$$= \sqrt{16.62(1 - 10/49)/10} = 1.15 \text{ (example 2)}$$

Calculation of confidence limits for the estimated mean

$$95\% \text{ confidence limits} = \bar{N} \pm t \times s_{\bar{N}}$$

where t is Student's t for $P = 0.05$ and $m - 1$ degrees of freedom.

Example 1: limits $= 4.66 \pm (2.01 \times 0.98) = 2.69$ and 6.63
Example 2: limits $= 6.20 \pm (2.26 \times 1.15) = 3.60$ and 8.80

---

of potential sampling units, then the sampling fraction (m/M) needs to be taken into account in calculating the standard error and the confidence limits of the mean. When the sampling fraction is small (that is, M is large compared with m), this correction can be ignored.

Total population size

Provided that the data come from a random sample, the means and confidence limits calculated from the sample units can be generalised to the whole study area. Hence, if $\bar{N}$ is the mean number of animals or plants per sample unit, the best estimate of the total size of the population in the whole study area is

$$\hat{N}_T = M\bar{N}$$

Similarly, multiplying the standard error (or confidence limits) of $\bar{N}$ by M provides the standard error (or confidence limits) of $\hat{N}_T$.

How many units to sample

Generally speaking, the precision of the overall estimate depends on the square root of the number of replicate samples. Thus, to halve the width of the confidence interval, one needs to quadruple the number of replicates. Notice that, unless the sampling fraction is large, it is the number of samples that determines the precision of one's estimate, not the sampling fraction. One does *not* need to take more samples from larger study areas.

Box 2.4 shows how to calculate both the sample size that one needs to take in order to obtain a percentage relative precision (PRP) (see Box 2.4) of the required magnitude (or less) and also the PRP attainable for a predetermined sample size.

Box 2.4. **Estimating the sample size needed for random sampling to attain a fixed PRP**

Basics

Q = the required percentage relative precision (Box 2.1)
$\bar{N}$ = mean number of organisms per sample unit
s = standard deviation of number of organisms per sample unit
m' = sample size required for there to be a 95% chance of obtaining a PRP of Q or less

Method 1: $s/\bar{N}$ estimated or guessed in advance

Calculate a first approximation to m':

$$m_0 = (200/Q)^2 (s/\bar{N})^2$$

If $m_0 < 25$, $m' = m_0 + 2$;   if $50 > m_0 > 25$, $m' = m_0 + 1$;   if $m_0 > 50$, $m' = m_0$.
Imaginary example: estimated $s/\bar{N} = 0.8$; $Q = 20\%$:

$$m_0 = (200/20)^2 (0.8)^2 = 64; \qquad m' = 64$$

To have a 95% chance of achieving a PRP of 20%, one should plan to take a sample of 64.

Method 2: requiring a preliminary survey

m_1 = number of sample units in the preliminary survey
$\hat{N}_1$ = mean estimated from this preliminary sample
s_1 = standard deviation estimated from this preliminary sample
$m_0 = (200/Q)^2 (s_1/\hat{N}_1)^2 (1 + 2/m_1)$

If m_0 is less than 25 or 50, it should be adjusted as in Method 1 (above).

Imaginary example:

$m_1 = 10$
$\hat{N}_1 = 3.2$
$s_1 = 4.0$
$Q = 25\%$
$m_0 = (200/25)^2 (4.0/3.2)^2 (1 + 2/10) = 120$

Since $m_0 > 50$, $m' = m_0 = 120$. Thus, following a preliminary sample of 10, a further 120 sample units are needed, if the required precision is to be obtained with a probability of 95%.

Adjustment for large sampling fraction

M = total number of sample units available in the study area

When the required sampling fraction (m'/M) is large, the required sample size can be reduced to

$$m' = m'M/(M + m')$$

This adjustment is worth making if m'/M is greater than 0.1 – i.e. if the number of units needed is more than 10% of the total number available in the whole study area.

Precision attained for a fixed cost

C_T = the fixed total cost
c = cost of taking one sample
Q' = precision (PRP) attained (with 95% probability) for a fixed total cost

Method 1 ($s/\bar{N}$ estimated or guessed in advance):

$$Q' = (200s/\bar{N})\sqrt{c/C_T}$$

Imaginary example: $C_T = £10\,000$, $c = £5$, prior estimate of $s/\bar{N} = 2.0$:

$$Q' = (200 \times 2.0)\sqrt{5/10\,000} = 9\%$$

Method 2 (requiring a preliminary survey):

$$Q' = (200s_1/\hat{N}_1)\sqrt{c(m_1 + 2)/(m_1 C_T)}$$

Imaginary example: $C_T = £10\,000$, $c = £5$, $m_1 = 10$, $s_1 = 4.0$, $\hat{N}_1 = 3.2$:

$$Q' = (200 \times 4.0/3.2)\sqrt{5(10 + 2)/(10\,000 \times 10)} = 6\%$$

Note that making the prediction requires some knowledge of the standard deviation and mean of the number of animals or plants per sample unit, or at least of their ratio. (The latter is often fairly stable across populations, so one can use values obtained in other studies.) Two methods of estimating the required sample size are shown in Box 2.4. The first requires some estimate of the $s/\bar{N}$ ratio (or a guess at its value); the second requires a preliminary survey to estimate s and $\bar{N}$, followed by a top-up survey to take the total sample size to the required number (the final estimate of $\bar{N}$ and its confidence limits being based on both parts of the survey combined).

Unequal sample units

When sample units are unequal in size, the mean number per sample unit can be calculated in the usual way. Indeed, so can the total population size. However, the confidence limits of the estimate of total population size must be calculated differently.

Begin by plotting a scatter diagram of the numbers counted in each sample unit against the area of the units. If the points fall on a line that goes roughly through the origin and if the scatter of the points about the line increases as the area increases (as in Figure 2.2), then proceed as in

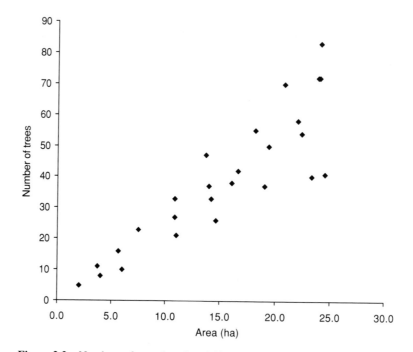

Figure 2.2   Numbers of trees found on 26 islands of various sizes.

Box 2.5. If the distribution of the points is markedly different from this, other methods may be appropriate (Cochran 1977, Thompson 2002).

When distributions are not normal

The normal distribution is an abstraction on which many statistical procedures are based. Unfortunately, the frequency distributions of counts, population indices and densities are rarely normal. Figure 2.3 shows the data for the first example in Box 2.3, which are clearly not normal. Such departure from normality has no effect on the estimation of the means and variances but it does affect the standard error and confidence limits, which may therefore be misleading if one applies standard methods to non-normal data. Standard statistical methods, such as *t*-tests and analysis of variance, are also based on the assumption of normality. However, they are fairly robust with respect to departures from that assumption and can usually be used safely unless the distribution is grossly non-normal.

It may be possible to fit a statistical distribution other than the normal to the data; the Poisson and negative-binomial distributions are often applied to count data. The former, for which the variance of numbers per sample is equal to the mean, applies when organisms are randomly distributed. More often they are to some degree clustered, with the variance greater than the mean; the negative binomial may then fit the data. If such alternative distributions can be fitted, then confidence limits appropriate to them may also be fitted.

A common solution to the problem of non-normality is to transform the data, though this is advisable only if the data are grossly non-normal. This allows one validly to carry out statistical

Box 2.5. **Estimating total numbers when sample areas are unequal**

M = total number of sites comprising the study area

A = total area of these sites

m = number of sites sampled

a_i = area of the ith site

n_i = numbers on the ith site

$\hat{D}$ = estimated density of organisms per unit area

$= \sum_i n_i / \sum_i a_i$

$\hat{N}$ = estimated total number in the N sites

$= A\hat{D}$

Example

Trees counted on islands in a lake (Figure 2.2):

Total number of islands (M) = 142

Number of islands surveyed (m) = 26

Total area of all islands (A) = 2130 ha

Total area of islands surveyed ($\sum_i a_i$) = 392.1 ha

Total number of trees counted ($\sum_i n_i$) = 1009

Density of trees

$\hat{D} = 1009/392.1 = 2.5733/\text{ha}$

Estimated number of trees on all 142 islands

$\hat{N} = 2130 \times 2.5733 = 5481$

Variance of $\hat{N}$

Calculate, either 'by hand' (using the methods described in standard textbooks) or using a computer package

$\text{var}(n)$ = variance of n (454.402 for the trees)

$\text{var}(a)$ = variance of a (51.465 for the islands)

$\text{cov}(an)$ = covariance of a and n (135.352 for the trees and islands).

The variance of $\hat{N}$ is then

$$\text{var}(\hat{N}) = [\text{var}(n) + \hat{D}^2\text{var}(a) - 2\hat{D}\,\text{cov}(an)]M^2(1 - m/M)/m$$

For the trees:

$$\text{var}(\hat{N}) = (454.402 + 2.5733^2 \times 51.465$$
$$- 2 \times 2.5733 \times 135.352)142^2(1 - 26/142)/26$$
$$= 62\,463$$

The standard error of $\hat{N}$ is the square root of this (in this case 250).

Confidence limits for $\hat{N}$

Very approximate confidence limits of $\hat{N}$ are provided by adding and subtracting twice the standard error to and from $\hat{N}$. Bootstrapping will provide much better confidence limits.

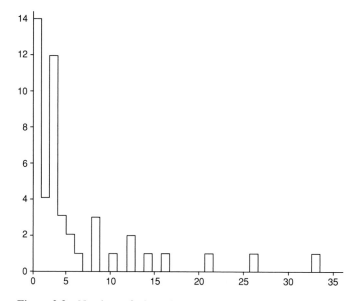

Figure 2.3   Numbers of wheat shoots (in a sample of 50) with various numbers of aphids *Sitobion avenae* in an English field in May. (Data from N. Carter, personal communication.)

analyses that depend on the observations being normally distributed. For data for which the variance is approximately equal to the mean, taking the square roots of the data is appropriate; for those for which the variance is greater than the mean, take logarithms instead (adding 1 to each value before doing so if any of the values is zero). It is commonly assumed that one can calculate means and confidence limits by back-transforming the estimates calculated for the transformed data: for example, if one has used the square-root transformation, one squares the estimates of the transformed data. Unfortunately, such back-transformations may be unreliable and should never be regarded as other than approximate.

Another problem with using transformations of counts is that count data often include a par-ticularly large number of zero values (e.g. Figure 2.3). No transformation can normalise such a distribution. There are three ways of dealing with such problems. One is to increase the size of the sample units, if possible. This will often reduce the number of zero counts. Another solution, which has the same effect, is to combine related sampling units; for example, if the several sample counts from each of the farms included in the survey that produced the data in Figure 2.4(a) are combined, the mean counts per farm are substantially closer to a normal distribution, with far fewer zero values (Figure 2.4(b)).

The excess of zero counts often arises because parts of the study area are unsuitable for the species, so one's samples comprise a mix from suitable sites (some with zero counts, just by chance) and from unsuitable sites (all with zero counts). Zero-inflated Poisson or zero-inflated negative-binomial distributions may be fitted to such data, though this is likely to require the skills of a professional statistician.

Finally, as mentioned in the section 'Simulation' (above), one can obtain confidence limits by using computer-intensive 'randomisation' methods, such as the jack-knife or bootstrapping (Manly 1997).

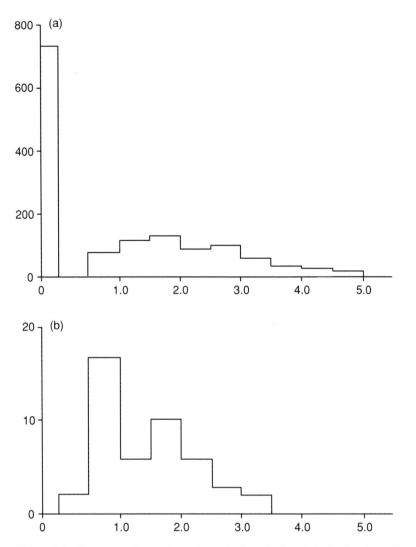

Figure 2.4 Frequency distributions of the number of pairs of chaffinches *Fringilla coelebs* per kilometre of hedgerow on 46 British farms: (a) individual values for 1331 100-m sample lengths of hedgerow; (b) mean values for each farm. (Data from the British Trust for Ornithology, personal communication.)

The layout of samples

Cluster sampling

What is cluster sampling?

Suppose that one is counting ferns within 10-m × 10-m sample quadrats in a forest measuring approximately 30 km × 50 km. Because of the size of the forest, it takes a long time to reach each randomly chosen sample location. Once there, however, counting the ferns takes just a short time. Would it not be more efficient to take a number of samples at each point – perhaps counting all the 10-m × 10-m quadrats in a 40-m × 40-m grid (Figure 2.5(a))? This is cluster sampling – taking a set of samples at each of a number of random positions. Other examples would be taking mud samples at the corners of a square quadrat laid down in random positions on a shore (Figure 2.5(b)) or placing standard sets (e.g. a 2 × 5 array at 1-m spacing) of pitfall traps in random locations. Clustering may be temporal rather than spatial: for example, obtaining an average index of frog numbers for a month by counting the number of frog calls in four 15-min periods on each of four evenings (the evenings being randomly positioned during a month) rather than by counting the calls in one period on each of 16 random evenings.

The problem with cluster sampling is that the sample units within each cluster are not independent of each other: the numbers of ferns in adjacent 10-m × 10-m quadrats are likely to be more similar than those in quadrats that are further apart; the numbers of frog calls in 15-min periods on the same evening are likely to be more similar than those in 15-min periods on different evenings. As a result, if one were to use the standard methods of assessing the precision of means, which assume that the samples are all independent, one could be seriously misled. The only solution to this problem is to combine the data for all the samples in the cluster and then treat the cluster totals as single sample measurements, using the usual statistical methods. Thus one's sample size is the number of clusters, not the total number of individual sample units.

If this is the best one can do with cluster sample data, why not just take one sample unit in each cluster? The answer is that the relative precision for a cluster is greater than the precision of a single unit, since it is based on a larger area or a greater amount of time. We have already seen that combining sample units may overcome the problem of having large numbers of zeros in the data set.

Note that, throughout this book, we assume that all clusters in a study are equal in size (i.e. they contain the same number of basic sample units). Clusters that differ in size can be used but data analysis is less easy.

Considerations of cost and precision

The overall total or mean for the study population will be assessed more precisely if one samples more clusters and takes more samples in each of them, but there will be a limit imposed by the resources available. Suppose that one has 100 hours in which to complete the fieldwork in a study, that it takes $4\frac{1}{2}$ hours to locate each cluster, and that each individual sample takes another half hour. It would then be possible to study

1 cluster, with 191 samples; or
4 clusters, with 41 samples each; or
10 clusters, with 11 samples each; or
20 clusters, with one unit each (which would be a simple random sample).

Which would be best? That is, which would give the most precise estimate of the overall mean? (Or, which amounts to the same thing, which would attain a given level of precision for the lowest

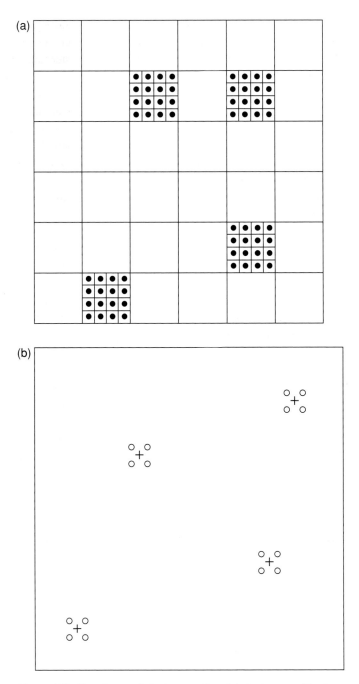

Figure 2.5   Two forms of cluster sampling: (a) cluster areas (the large squares) are chosen at random and all sample units (small squares) in each are sampled; (b) points are chosen at random (+) and samples (○) taken in a fixed pattern relative to each.

cost?) The decision depends on having carried out preliminary survey work to determine variation in population estimates (or densities or indices) between clusters and between the sample units in the same cluster. Box 2.6 shows how, by carrying out a pre-survey based on clusters of the chosen size, one can determine whether cluster sampling at that scale is better or worse than simple random sampling. For the example shown, the latter would consist of counting skylarks on just a single 200-m transect in each 1-km × 1-km study area. Note that the basis of this comparison is that, for a given total cost, one could sample more 1-km × 1-km areas if one were walking only a single 200-m transect in each. In the example in Box 2.6, it is arbitrarily assumed that the cost of visiting a 1-km × 1-km square to do any work at all is about 50 times greater than the cost of walking one 200-m transect section once one has located the square – since it takes so much longer to travel about England from square to square than it does actually to walk the transects. Note that, while the example deals with an index, estimates of density or of numbers in each sample unit could be substituted without changing the methods.

Box 2.6. **Choosing between cluster sampling and simple random sampling**

Basics

Treat each cluster as a 'group' or 'treatment' in a one-way (single-factor) analysis of variance, with the individual sample units providing the data in the form of estimates of population sizes for each unit. Carry out the analysis of variance in the usual way to estimate the ratio of the between-cluster mean square to the within-cluster mean square (F). This is usually used for significance testing, but we need it for a different purpose.

Obtain an estimate of the costs of the fieldwork:

c_M = basic cost of sampling a cluster, ignoring the cost of taking each individual sample

c_U = additional cost per sample unit

Thus, if there are U sample units per cluster, the total cost of sampling each cluster is

$$c_M + Uc_U$$

Decision criterion

Cluster sampling is superior if

$$c_M/c_U > F - 1$$

Example

Counts of skylarks *Alauda arvensis* seen on 200-m sections of transects on 16 1-km × 1-km areas of English farmland in spring 1994, there being 10 sections systematically placed on

---

each area: the 1-km × 1-km areas are the clusters and the 200-m sections are the sampling units. (Data from British Trust for Ornithology, personal communication.)

The data are too voluminous to quote. An analysis of variance provided a ratio of the estimates of the between-cluster and within-cluster mean squares:

$$F = 4.63$$

In this example, an approximate value of c_M/c_U is 50.

Thus cluster sampling, following the decision criterion above, is superior to simple random sampling in this case, for a fixed total cost.

Note

The decision is independent of the number of sample units per cluster (and thus, for a fixed total cost, the number of clusters). However, changing the number of samples per cluster is likely to change their spatial layout, which will change F. So this sort of comparison is specific to a particular cluster size and layout. In general, larger clusters will result in smaller F values, so, while a design with many small clusters may be superior in a particular case to simple random sampling, a design with just a few large clusters might not be. See Box 2.8, for choosing between different cluster designs.

Caution

Are variances of clusters homogeneous? See Box 2.7.

Box 2.7. **Homogeneity of variances**

The methods that we present in Boxes 2.6, 2.14 and 2.15 depend on the assumption that the variances of the counts are the same in different groups. (In Box 2.6, 'groups' are clusters). It is a good idea to calculate the variance of each group independently, to check that they are not too disparate.

What does 'too disparate' mean in this context? For a rough guide, we would say that if the largest variance is more than three times greater than the smallest, one should be concerned, unless the total sample size (over all groups combined) is less than 30. Substantially larger differences may occur when sample sizes are smaller, just by chance.

A more formal test of homogeneity of variances, to be found in standard statistical texts, may then be applied. If it shows the differences between groups in terms of the magnitude of their variances to be clearly significant, then either one should flag up that the calculated confidence limits are likely to be biased or one should estimate the limits by bootstrapping rather than using the methods in the boxes.

Box 2.8 shows how to compare two different cluster designs; more than two can be compared in the same way. The clusters may differ merely in size (as in Box 2.8) but they may also differ in layout: for example, each cluster could be 25 plants growing in a long section of a single row of a crop or 30 plants growing in short sections of three adjacent rows.

Further analysis of the data on which the examples in Boxes 2.6 and 2.8 are based shows that the costs (for a given precision) of sampling based on clusters of various sizes, as percentages of that for cluster sizes 6 and 7 (which are the best sizes), are

Cluster size	1	2	3	4	5	6	7	8	9	10
Relative cost	330	214	160	135	107	100	100	107	108	111

Thus, in the range 5–10, it makes little difference exactly which cluster size is used. This is a typical result. The exact magnitude of the difference between the basic cost of a cluster and the additional cost of each sample unit is also rather unimportant if the difference is large. This means that it is fairly safe to apply conclusions about optimal cluster size from work in one area or at one time to similar places or times, since it is unlikely that the differences between places or times will have much effect on the optimum.

How many clusters to sample

One works out how many clusters one needs to sample by applying the methods of Box 2.3, treating clusters as individual samples (since this is how one treats them to work out overall means and confidence limits for the whole study area).

Multi-level sampling

What is multi-level sampling?

Because the data on individual sample units within a cluster are combined, information on the variance between sample units is lost in cluster sampling; the precision of one's estimates is thereby reduced. One can overcome this by sampling only some of the units within each cluster – and by doing so at random. Thus, one might randomly locate 40-m × 40-m grids in a forest and take a random sample of some of the sixteen 10-m × 10-m quadrats within each of them (Figure 2.6). This achieves the same sort of cost savings as cluster sampling but allows one to make more precise estimates. Both the costs and the precision of multi-level sampling tend to be intermediate between those of simple random sampling and those of cluster sampling. It is often superior to both.

A slightly different reason for multi-level sampling is when it is costly to determine the entire range of possible low-level samples (the sampling frame). For example, suppose that one wished to take 5-m stretches of streams as the samples from an entire catchment of several thousand square kilometres. To map all the tens or hundreds of thousands of kilometres of streams in the whole area would be hugely expensive. Instead, one might take a random sample of 4-km × 4-km squares, map the streams in these, and take a random sample of 5-m stretches of steams within each square.

Box 2.8. **Choosing between different cluster designs**

Basic

> U_1, U_2 = the numbers of sampling units per cluster, in each of the two designs
> s_1^2, s_2^2 = the variances of the cluster totals (or means), in each of the two designs
> c_1, c_2 = the costs of sampling the whole cluster, in each of the two designs

Decision criterion

The design that gives the greatest precision for a fixed total cost, or the least total cost for a fixed precision, is the one with the lowest value of

$$c_i s^2 / U_i$$

For a comparison of just two designs, the ratio of the values of this expression is the ratio of costs (for the same standard error of the overall mean) and also the ratio of standard errors (for the same cost).

Example

For the skylarks in Box 2.6, the results for the ten-segment clusters were compared with what would obtain if only five segments were surveyed per cluster (i.e. five 200-m segments per 1-km × 1-km square).

The variances for the two sets of clusters were

$$s_1^2 = 43.90 \quad \text{(10-segment)}$$
$$s_2^2 = 11.46 \quad \text{(5-segment)}$$

If c_M is the basic cost of visiting a 1-km × 1-km square and c_U the additional cost of each transect segment, then

$$c_1 = (c_M + U_1 c_U) = c_U(U_1 + c_M/c_U)$$
$$c_2 = (c_M + U_2 c_U) = c_U(U_2 + c_M/c_U)$$

In this example, we take c_M/c_U as 50, so

$$c_1 = 60 c_U \qquad c_2 = 55 c_U$$

Thus

$$c_1 s_1^2 / U_1^2 = (60 c_U \times 43.90)/100 = 26.3 c_U$$
$$c_1 s_2^2 / U_2^2 = (55 c_U \times 11.46)/25 = 25.2 c_U$$

Sampling five 200-m segments in each 1-km × 1-km square is thus slightly superior to sampling ten, though the costs (for a fixed precision) are only 4% less.

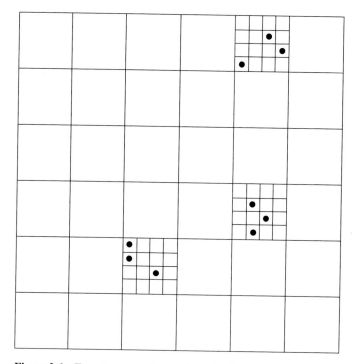

Figure 2.6   Two-stage sampling: major units (large squares) are randomly chosen and minor units (small squares) sampled at random within each.

Subsampling is another form of multi-level sampling. Thus it would rarely be convenient to take the whole of the material from a 10-cm soil core for cultivation of microflora. In the lab, one would take random subsamples from the material from each core for this purpose.

We shall refer here to the larger sample units (e.g. the 40-m × 40-m grids in Figure 2.6) as major units and to the units within them (e.g. the 10-m × 10-m quadrats) as minor units. Our treatment will be restricted to two levels of sampling, though the principles can be extended to as many levels as one wishes: for example, 40-m × 40-m grids may be located within random 1-km × 1-km squares in a forest to measure the vegetation and one might then take 10-cm-diameter soil cores randomly within 1-m ×1-m quadrats within the 40-m × 40-m grids to study the soil microflora, with subsampling from the cores in the lab – an example of five-level sampling. The number of levels depends on practical convenience.

Optimal distribution of sampling

The precision of the overall mean depends both on the variance between the mean values for major units and the variance between minor units within major units (the second variance being assumed to be constant over all major units); it depends also on the number of units sampled at each level. Given a fixed limit to the cost of the work (or a certain level of precision required), there is clearly an optimal distribution of sampling effort between major and minor units, which depends both on the variances and on the costs of sampling at the two levels. To determine the optimal number of minor units to sample within each major unit, as shown in Box 2.9, one requires some knowledge at least of the ratio of the two variances and of the ratio of the costs at the two levels. These may be obtained through a preliminary survey or through general knowledge of the situation one is studying. As with cluster sampling, the difference in realised precision between the absolute optimal sampling pattern and a pattern close to it is often slight, so precise knowledge of the variances and costs is not required.

Once the optimal number of minor units per major unit has been determined, one can work out the corresponding number of major units, depending on whether one is working to a fixed cost or to a required level of precision (Box 2.9). Note that, for the latter, one requires estimates of the two variances individually, not just their ratio.

Box 2.9. **The optimum allocation of sampling in two-level sampling**

This is an illustrative example. The precise form of the calculations may be different in different cases (such as the stream-sampling case described in the text). The principles are, however, constant.

Basics

M = number of major units available for sampling (may be effectively infinite)
m = number of major units actually sampled
U = number of minor units available within each major unit (assumed to be the same for all major units)
u = number of minor units actually sampled in each major unit (assumed to be the same for all major units)

We need to calculate

u_{opt} = optimum value of u

From this, we can estimate the value of m needed to carry out the task for a fixed cost (m_c) or to achieve a required precision (PRP, Box 2.1) (m_p).

Example

An ecologist wished to estimate the density of worms of a particular species on a shore extending over 150 000 m^2. In a preliminary survey, he established five randomly placed 2-m × 2-m quadrats and took four random 20-cm × 20-cm cores from each. Thus the sampling structure was

$M = 150\ 000/(2 \times 2) = 37\ 500$
$m = 5$
$U = (2 \times 2)/(0.2 \times 0.2) = 100$
$u = 4$

The means and variances of the numbers of worms per core within the quadrats were

Quadrat	Mean	Variance
1	5.00	8.667
2	5.50	3.000
3	7.75	6.917
4	2.50	5.667
5	8.50	19.000

These can be used to calculate (in the usual way) three further quantities:

$$\bar{\bar{N}} = \text{mean of the means} \quad = 5.850 \text{ worms/core}$$
$$s_M^2 = \text{variance of the means} = 5.675$$
$$s_u^2 = \text{mean of the variances} = 8.650$$

The investigator also found that, on average, it took half an hour to locate each quadrat and six minutes to take each core and count the worms. These can be used as measures of costs:

$$c_M = 0.5 \text{ hours} = \text{basic cost per major unit sampled}$$
$$c_u = 0.1 \text{ hours} = \text{additional cost for each minor unit sampled}$$

Estimation of u_{opt}

$$u_{\text{opt}} = \sqrt{\frac{c_m}{c_u} \bigg/ \left(\frac{s_M^2}{s_u^2} - \frac{1}{U}\right)} = \sqrt{\frac{0.5}{0.1} \bigg/ \left(\frac{5.675}{8.650} - \frac{1}{100}\right)} = 2.78 \approx 3$$

Thus three cores should be taken in each quadrat.

Note that if either

$$u_{\text{opt}} > U$$

or

$$s_M^2 s_u^2 < \frac{1}{U}$$

then one should sample all of the minor units within each major unit (thus simplifying the design to simple random sampling of major units).

Sample size for a fixed cost

Suppose that the limit of expenditure on the survey is to be C. Then

$$m_c = C/(c_m + c_u u_{\text{opt}})$$

If $C = 8$ hours in our example,

$$m_c = 8/(0.5 + 0.1 \times 3) = 10$$

Thus, the optimum allocation of sampling effort (to achieve the most precise estimate of the mean population density) in this case, if eight hours are available, is to use ten quadrats and take three cores in each.

Sample size to achieve a required percentage relative precision (Q)

$$m_p = \left[s_M^2 + s_u^2 \left(\frac{1}{u} - \frac{1}{U} \right) \right] \bigg/ \left[\frac{s_M^2}{M} + \left(\frac{\bar{N}}{200} Q \right)^2 \right]$$

If the required PRP (see Box 2.1) in our example is 20% and assuming that we take the optimum number of cores in each quadrat (three, in this case), then

$$m_p = \left[5.675 + 8.650 \left(\frac{1}{3} - \frac{1}{100} \right) \right] \bigg/ \left[\frac{5.675}{37500} + \left(\frac{5.850}{200} 20 \right)^2 \right]$$

$$= 24.7 \approx 25$$

Thus the optimal allocation of sampling effort in this case, to achieve a PRP of 20%, is to use 25 quadrats and take three cores in each.

Estimates from two-level sampling

Box 2.10 shows how to estimate the mean numbers of organisms per sample, with confidence limits, for two-level sampling. It assumes that all of the minor units are of the same size, that each of the major units contains the same number of minor units, and that the same number of minor units is sampled from each major unit. All these assumptions can be relaxed but the calculations become less straightforward and it is generally best to have such even distribution of sampling effort. Note that the number of minor units within major units can be effectively infinite, if the samples are point counts, for example, since there is an infinite number of points within an area.

Box 2.10 is presented in terms of numbers, including estimation of total numbers in the whole study area. As usual, indices and densities can be treated in the same way.

Stratified sampling

What is stratified sampling?

It is often obvious even before a study takes place that there are systematic variations in population density across the study area, or one may suspect that densities are likely to differ in different habitats. If so, then it is usually valuable to divide the study into subareas that differ in density or habitat and to sample randomly within each. Such subareas are strata. They are not the same as the major units of two-level sampling, which are random subdivisions, not subdivisions systematically chosen to reflect the variation within the study population.

Box 2.10. **Calculating overall means, totals and confidence limits from two-level sampling**

Example

As in Box 2.9. Suppose that the ecologist carried out a main survey using ten quadrats with three cores in each (to achieve the most precise estimates within a total cost of eight hours – see Box 2.9). Thus

$$M = 37\,500$$
$$m = 10$$
$$U = 100$$
$$u = 3$$

Suppose that the various means and variances (see Box 2.9) were

$$\bar{\bar{N}} = 5.240 \text{ worms/core}$$
$$s_M^2 = 5.392$$
$$s_u^2 = 10.017$$

Estimation of overall mean and total numbers

The best estimate of the overall mean number of animals per minor unit is simply $\bar{\bar{N}}$. The best estimate of the total number in the entire study area is $\bar{\bar{N}}MU$.
In the example,

$$\bar{\bar{N}}MU = 5.240 \times 37\,500 \times 100 = 19.65 \text{ million worms}$$

Confidence limits

The standard error of $\bar{\bar{N}}$ is

$$s_{\bar{\bar{N}}} = \sqrt{(1 - m/M)s_M^2/m + (1 - u/U)s_u^2/(uM)}$$

In this example,

$$s_{\bar{\bar{N}}} = \sqrt{(1 - 10/37\,500)5.392/10 + (1 - 3/100)10.017/112\,500}$$
$$= 0.7343$$

Approximate 95% confidence limits are

$$\bar{\bar{N}} \pm (t \times s_{\bar{\bar{N}}})$$

using Student's t for 5% significance and for $m(u - 1)$ degrees of freedom.

In this example, the limits are

$$5.240 \pm (2.086 \times 0.9289) = 3.71 \text{ and } 6.77 \text{ worms/core}$$

The standard error and confidence limits for the total number are obtained by multiplying the values of $\bar{\bar{N}}$ by MU

In this example, the confidence limits for the total are thus 13.9 million and 25.4 million worms on the whole beach.

The strata do not need to be individually continuous. One might, for example, divide a study area into high- and low-density areas that happened to be intermingled (Figure 2.7).

Stratification allows separate estimates of the means and variances to be made for each stratum. Its main value, however, is that it allows the overall mean to be estimated with much greater precision. For example, the confidence limits for the example used in Box 2.11 are half as wide as they would have been if the same number of samples had been taken without stratification. Using

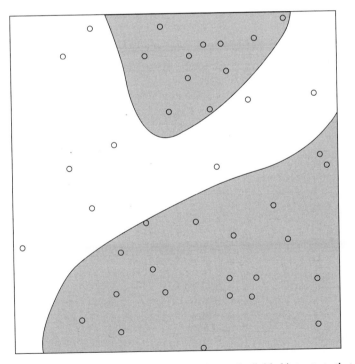

Figure 2.7   Stratified sampling: the study area is divided into strata that differ in mean density of the organism being investigated and samples are taken randomly within each. In this example there are two strata, one of high density (shaded), which in this case happens to encompass two separate areas, and one of low density; the investigator has chosen to sample more intensively in the high-density stratum.

simple random sampling, such an improvement in sample size would have required a fourfold increase in sample size. Stratification is also valuable if the costs of sampling are different in different parts of the study area; places further from the base of operations may, for example, entail greater travel time and costs. Investigations in which different organisations are responsible for various parts of the study area are also well treated as stratified studies – each organisation's area being a stratum.

Box 2.11. Estimation of overall mean and confidence limits from a stratified random sample

First steps

Calculate the mean and variance of numbers per sample units for each stratum independently.

The table below shows results from a survey of the number of main setts of badgers *Meles meles* in Great Britain (Reason *et al.* 1993). Using environmental characteristics, the country was divided into eight land-class groups. Urban areas were omitted. The sample units were 1-km × 1-km squares.

The tabulated values are

M_h = total number of units in stratum h

m_h = number of units sampled from stratum h

W_h = 'weight' of stratum h

   = $M_h / \sum_h M_h$

$\bar{N}_h$ = mean number counted per unit in stratum h

s_h = standard deviation of counts in stratum h

$g_h = M_h(M_h - m_h)/m_h$

$A_h = g_h s_h^2$

$B_h = (g_h s_h^2)^2/(m_h - 1)$

Land-class group	M_h	m_h	W_h	$\bar{N}_h$	s_h	$W_h \bar{N}_h$	g_h	A_h	B_h
I	44 762	627	0.197 91	0.472	0.9014	0.093 41	3 150 831	2 560 119	10 469 984 234
II	28 513	352	0.126 07	0.174	0.5066	0.021 94	2 281 121	585 435	976 450 589
III	39 004	425	0.172 45	0.172	0.4329	0.029 66	3 540 554	663 508	1 038 309 564
IV	27 180	308	0.120 17	0.205	0.5089	0.024 64	2 371 367	614 135	1 228 538 813
V	25 775	224	0.113 96	0.031	0.1947	0.003 53	2 940 076	111 453	55 702 676
VI	23 814	143	0.105 29	0.014	0.1673	0.001 47	3 941 966	110 333	85 727 716
VII	31 550	313	0.139 50	0.073	0.3008	0.010 18	3 148 650	284 892	260 138 973
VIII	5 573	63	0.024 64	0.016	0.1269	0.000 39	487 416	7 849	993 699
Sums	226 171	2455	1.000 00			0.185 23		4 937 724	14 115 846 264

Estimate of mean number

The estimate of the mean count per sample unit (i.e. number of setts per km²) is

$$\bar{N} = \sum_h W_h \bar{N}_h$$

$$= 0.1852 \text{ in the example}$$

The variance of this estimate is

$$V_N = \sum_h A_h / (\sum_h M_h)^2$$

$$= 4\,937\,724/226\,171^2 = 9.6528 \times 10^{-5} \text{ in the example}$$

Approximate 95% confidence limits for the estimate are $\bar{N} \pm t\sqrt{V_N}$, where t is Student's t for $P = 0.05$ and d degrees of freedom. The value of d is calculated as

$$d = (\sum_h A_h)^2 / \sum_h B_h$$

$$= 4\,937\,724^2 / 14\,115\,846\,264 = 1727$$

In the example, the limits are

$$0.1852 \pm 1.9613 \times \sqrt{9.6528 \times 10^{-5}} = 0.1660 \text{ and } 0.2045 \text{ setts per km}^2$$

(Given that the confidence limits are only approximate, one can usually assume a value of 2 for t unless sample sizes are less than 60. The calculation for d is given for the badger example for the sake of completeness.)

Estimate of total number

To obtain an estimate of the total number in the whole study area (and its confidence limits), multiply the mean (and confidence limits) by $\sum_h M_h$.

Thus, rounding to an appropriate level, the estimate of the number of badger setts in non-urban areas of Great Britain is $41\,900 \pm 4400$.

How to stratify

If administrative organisation imposes the stratification, one accepts the stratification that it presents. Otherwise one should choose to subdivide the study area in the way that minimises the within-stratum variance in population density (and, correspondingly, maximises the differences between strata). How can one do this ahead of the fieldwork? One way is to base it on a rough preliminary survey. For example, one may be able to estimate numbers roughly by simple observation of what animals or plants can be seen, without precise counts; or one may get a rough idea of densities using simple point counts, in order to stratify an area prior to using more precise methods such as mark–recapture; or one might even simply divide the area into places apparently with and without the species. Alternatively, one can stratify according to habitat or ecological characteristics with which the organism's density is likely to be correlated. These methods will not provide such effective stratification as if one were able to use precise knowledge of the variation in population density, but they will provide much better results than simple random sampling. So long as there is some difference between the stratum means, however small, stratification is advantageous.

Stratification may lose its advantages if the number of strata is large relative to the total number of sampling units (so that each stratum contains just a few units). It is usually sufficient to use only 3–6 strata.

If the costs of sampling are different in various parts of the study area, such cost differences can be used to define strata, especially if the cost differences are clear. For example, if one is studying an area including both mainland and islands, the latter may be much the more costly to sample because of the need to use boats to reach them; stratifying into mainland and island areas is likely to be useful. If there is variation in both density and cost, then both need to be considered, but variation in density is generally the more important.

Estimation of overall mean and totals

To estimate the overall mean, one needs to estimate the mean and variance (in the usual way) for each stratum separately. One also needs to calculate the stratum weight (the proportion of the total study area that the stratum comprises) and the stratum sampling intensity (the proportion of the potential sample units in each stratum that were actually sampled). Box 2.11 shows how these values can be used to obtain estimates of mean density and of the total population of the study area, with confidence limits.

Optimal allocation of sampling effort

How many samples should one take from each stratum? If the costs of sampling and within-stratum variance in density are the same in each stratum, then sampling with the same intensity in each stratum is best – that is, it gives the greatest precision for the least cost. If the variances are unknown (and unguessable) in advance, one should assume that they are all the same. If the variances are known to differ, however, one should sample more intensively in the strata that are the more variable. If costs differ, one should sample more where the costs per sample are lower. Box 2.12 shows exactly how the effort should be distributed.

In practice, the exact distribution of sampling effort makes little difference to the precision attained. This means that one can use a rough idea of the variances in each stratum to guide the allocation of sampling effort.

Post-stratification

Sometimes one does not have the prior knowledge to stratify in advance but one gathers information during the survey that allows the simple random sample to be divided into strata afterwards. For example, damp and dry soils may be too intermingled to map them in advance, yet the dampness of the soil at each sample site can be readily assessed at the same time as the plants are surveyed. If plant numbers are influenced by soil moisture, it is worth classifying the sites after the survey and analysing the data as a stratified sample. Another reason for post-stratification is when part of the study area is less intensively sampled than originally planned, perhaps because

Box 2.12. **Optimum allocation of sampling effort in stratified sampling**

Estimate of overall mean number per sample unit

To work out the optimum allocation of sampling effort over strata, one needs estimates of the standard deviation of numbers in each stratum and the costs of taking samples in each stratum (c_h), either from a pilot survey or by informed guesswork.

To illustrate the method, we use the example in Box 2.11, with purely arbitrary values for the c_h. Tabulate these values, with the stratum sizes and weights, extending the table with three further columns of calculations. The first two of these are self-explanatory. For the third,

$$m'_h(\text{opt}) = (M_h s_h / \sqrt{c_h}) / D$$

where D is the sum of the penultimate column.

The final column is the proportion of samples that should be taken in stratum h. Thus, to optimise the distribution of sampling effort in the badger survey, 39% of the samples should have been taken in land-class I, 14% in land-class II, etc.

Land-class group	M_h	W_h	s_h	c_h	$M_h s_h$	$M_h s_h / \sqrt{c_h}$	$m'_h(\text{opt})$
I	44 762	0.1979	0.9014	4	40 348.5	20 174.2	0.3912
II	28 513	0.1261	0.5066	4	14 444.7	7 222.3	0.1400
III	39 004	0.1725	0.4329	4	16 884.8	8 442.4	0.1637
IV	27 180	0.1202	0.5089	4	13 831.9	6 916.0	0.1341
V	25 775	0.1140	0.1947	4	5 018.4	2 509.2	0.0487
VI	23 814	0.1053	0.1673	9	3 984.1	1 328.0	0.0258
VII	31 550	0.1395	0.3008	4	9 490.2	4 745.1	0.0920
VIII	5 573	0.0246	0.1269	9	707.2	235.7	0.0046
Sums	226 171	1.0000				51 573.0	1.0000
						$= D$	

Note, for checking, that the sums of the W_h and $m'_h(\text{opt})$ columns must both equal 1.0.

Total number of samples to be taken when the budget is fixed

Calculate one further column (the first column in the table below, with sum E).

The total number of samples to be taken is then

$$m = (C - c_0) D / E$$

where C is the total budget available and c_0 is the cost of setting up the survey over and above the cost of taking the individual samples.

Suppose that it had been estimated for the badger survey that 20 000 (C) man-hours were available, of which 1000 (c_0) would be used in setting up the survey, analysing the results and publishing them.

Then

$$m = 19\,000 \times 51\,573/21\,411 = 2576 \text{ samples}$$

assuming that they were taken from the strata in the proportions indicated by m'_h(opt).

Land-class group	$M_h s_h \sqrt{c_h}$	$W_h s_h$	$W_h s_h / \sqrt{c_h}$	$W_h s_h \sqrt{c_h}$	$(W_h s_h)^2$
I	80 696.9	0.178 40	0.089 20	0.356 80	0.031 83
II	28 889.4	0.063 87	0.031 93	0.127 73	0.004 08
III	33 769.7	0.074 66	0.037 33	0.149 31	0.005 57
IV	27 663.8	0.061 16	0.030 58	0.122 31	0.003 74
V	10 036.8	0.022 19	0.011 09	0.044 38	0.000 49
VI	11 952.2	0.017 62	0.005 87	0.052 85	0.000 31
VII	18 980.5	0.041 96	0.020 98	0.083 92	0.001 76
VIII	2 121.6	0.003 13	0.001 04	0.009 38	0.000 01
	214 110.9		0.228 03	0.946 68	0.047 79
	= E		= F	= G	= H

Total number of samples to be taken to attain a required precision

Further columns of calculation are required, as in the table above.

If V is the required variance of the estimated mean, then the total sample size required is

$$m = FG/(H/M + V)$$

If working in terms of percentage relative precision, substitute V (as usual) by $(\bar{N}Q/200)^2$, where Q is the required PRP (see Box 2.1) and $\bar{N}$ is the overall mean.

Thus, for a PRP of 5 in the badger survey:

$$m = (0.228 \times 0.947)/[(0.047\,79/226\,171) + (0.185 \times 5/200)^2] = 9966 \text{ samples}$$

would be required, assuming that they were taken from the strata in the proportions indicated by m'_h(opt).

When the calculated optimum is impossible

Sometimes the above calculations suggest that the optimum sample size from a stratum is greater than the number of units in the stratum. In that case, one should survey the whole stratum and devote one's further sampling effort to the remaining strata in the proportions indicated by m'_h(opt) above. This may happen if the variance for a stratum is particularly large, so it is even better if one can split the stratum in question, to produce strata with smaller variances.

of illness, transport problems, weather, etc. Especially if one suspects that the abundance of the study organism may be different in the undersampled area, it may be worth dividing the data into two strata (or more, if appropriate), 'fully sampled' and 'undersampled'.

Post-stratification differs from pre-stratification in that the allocation of sampling effort over the strata is random, rather than planned. However, current advice seems to be that confidence limits calculated as in Box 2.11 will be satisfactory for most purposes.

Box 2.11 assumes that stratum sizes are known. This might not always be the case. For the soils example, the overall proportions of damp and dry soils in the study area would probably not be known but would have to be estimated from the proportions in the sites actually sampled, perhaps supplemented with further random sampling aimed just at determining soil characteristics. If the size of the strata has to be estimated, rather than being known, confidence limits for the total population are probably best estimated by bootstrapping. They will be wider than those obtaining when stratum sizes are known. As a result, the variance of the estimated mean is rather greater than with true stratified sampling.

Strip transects within or across strata

As noted above, it may be more convenient to have long, narrow quadrats than square ones. Such line transects may run across areas within which there are considerable variations in numbers of the organism being studied. That variation may be clinal, with a gradient in numbers from one edge of the study area to the other. How should the line transect be placed in relation to the cline?

Figure 2.8(a) shows one answer to this question: to divide the cline up into strata, the divisions lying parallel to the contours of population density, and to take randomly positioned line transects within each stratum, their long axes also lying parallel to the contours of density. Figure 2.8(b) presents an alternative: do not stratify but simply take randomly positioned line transects along the direction of the cline (cutting across its strata).

Which arrangement is superior? If one wishes to obtain a description of the cline, then the second method is better; but if one wishes to get an overall estimate of average density or numbers, the answer is less clear-cut. If one has sufficient prior knowledge to be able to stratify effectively, then the two methods are about equally efficient statistically; the choice should thus be determined largely by practical convenience. If, however, the strata do not fit the variations in density closely then the first method is not as good as the second.

Adaptive sampling

Why adaptive sampling is needed

The numbers of some species vary greatly from place to place, with high densities in just a few places. The majority of random samples will then contain few individuals and thus provide limited information, so it is more efficient to concentrate effort on the high-density areas than to take fully random samples. If the areas of high and low density are identifiable in advance, even if only roughly, then a stratified sampling regime can be applied, with a greater sampling intensity

CLINE

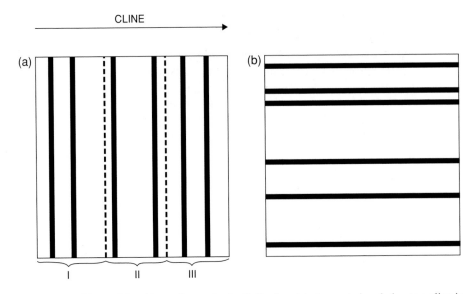

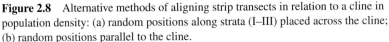

Figure 2.8   Alternative methods of aligning strip transects in relation to a cline in population density: (a) random positions along strata (I–III) placed across the cline; (b) random positions parallel to the cline.

in the high-density stratum. But that might not be possible. Adaptive sampling is an alternative approach.

What is adaptive sampling?

Adaptive sampling is a way of adjusting the pattern of sampling in the light of early results, so as to concentrate effort in areas of high density and thus make the work more efficient. It relies on the fact that a sample unit containing many individuals is likely to have neighbours that also contain many. The marked non-randomness of the final sample is taken into account in the analysis.

   There are various forms of adaptive sampling. The one presented here, adaptive cluster sampling, is the one most likely to be useful to ecologists. However, the methods have been little used in ecology so far and their efficacy relative to more traditional sampling methods has not fully been tested by practical experience. What is clear is that adaptive cluster sampling is more efficient and, in particular, provides more realistic confidence limits than does simple random sampling whenever organisms are very patchily distributed.

How to take an adaptive cluster sample

One starts with a simple random sample, as in Figure 2.9(a), in which a sample of 10 plots out of a possible 100 has been taken, yielding counts of 0, 0, 0, 0, 0, 0, 0, 0, 0 and 20. One must then define a criterion to distinguish plots with high counts (which are likely to have neighbours with higher than average numbers) from the rest. In this example, it is obvious: a non-zero count. In other

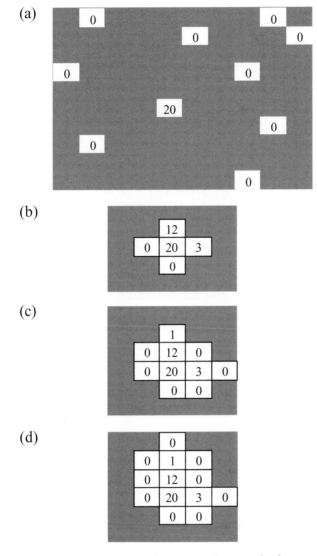

Figure 2.9 The results of taking ten random samples from a study area (a), and the results of sequentially extending the network based on the sample unit that yielded a count of 20, based on the criterion of extending from any plot with a non-zero count (b)–(d).

cases it might not be so obvious: if the counts were 0, 0, 0, 1, 1, 1, 1, 7, 20 and 47, for example, one might set the criterion as 'above 6' or 'above 7'; it would probably make little difference in practice.

Having defined the criterion, one then surveys all the plots that share a border with those plots that satisfy the criterion. In our example, only one plot does so; Figure 2.9(b) shows the outcome of surveying its neighbours. Two of those neighbours also satisfy the criterion, so one surveys their neighbours (Figure 2.9(c)). One continues until none of the new plots entering the sample satisfies the criterion (Figure 2.9(d)).

Having carried through the extended sampling based on each of the original sample plots that satisfied the criterion, one ends up with a set of clusters. Some (like that in Figure 2.9(d)) have several sample plots within them, a number of 'edge units' (which do not satisfy the criterion) surrounding a 'network' of units that do satisfy the criterion. Thus, the cluster in Figure 2.9(d) comprises a network of four units, with nine edge units. The rest of the clusters are those units in the original sample that did not satisfy the criterion; they are regarded as networks of size 1. Thus Figure 2.9 produces nine networks of size 1 and one of size 4 (not of size 13 – remember that the edge units are ignored).

Sometimes two or more of the original random sample units may enter the same eventual cluster. The cluster is counted once only, despite that.

Analysis

An estimate of the total population, with confidence limits, can be obtained using the methods in Box 2.13. The confidence limits are approximate and better limits may be obtained by bootstrapping. Note how wide the confidence limits in the example in Box 2.13 are, because the distribution of that population was very patchy. They would have been even wider if simple random sampling had been used.

Box 2.13. **Estimating total population size from an adaptive cluster sample**

Example

In a study of an orchid, a meadow was divided into 5000 potential sample units, of which a random 50 were surveyed. No flowering spikes were found in 46 of these, so the criterion for extending networks was taken to be having found any flowering spikes at all in the sample unit.

In the four networks based on the four units in the original sample that held any spikes, the numbers of spikes observed were 5, 11, 17 and 24, the networks in which these occurred having sizes 2, 3, 2 and 4 cells, respectively.

Sample-size information:

M = number of units into which the study area is divided (5000 in the example)

m = number of units in the original random sample (50 in the example)

$C1$ = number of possible ways of sampling m units from M (see notes on 'Combinations', below; 2.284×10^{120} in the example).

Note that, in the calculations, all the clusters (46 in the example) that yielded counts of zero can be omitted because they contribute to the results only through their being included in m.

Estimation of means

To keep track of the calculations, number the networks that had non-zero counts. The numbering is arbitrary but it is convenient to arrange the networks in size order.
   Tabulate the following:

k = arbitrary network number
M_k = number of sample units in the kth network (the network size), remembering not to include the edge units
n_k = count from the kth network

Calculate

$C2_k$ = number of possible ways of sampling m units from $M - M_k$
$p_k = 1 - C2_k/C1$ (this is the probability of the kth network being included in a random sample of size m)
n_k/p_k
$A_k = (1/p_k^2 - 1/p_k) n_k^2$

For the example, the data and the results are

k	M_k	n_k	$C2_k$	p_k	n_k/p_k	A_k
1	4	24	2.194×10^{120}	0.039 42	608.90	356 140
2	3	11	2.216×10^{120}	0.029 71	370.28	133 038
3	2	17	2.238×10^{120}	0.019 90	854.19	715 113
4	2	5	2.238×10^{120}	0.019 90	251.23	61 861
sum					2094.60	1 266 152

Estimate of the mean number per sample unit:

$$\bar{n} = \sum_k (n_k/p_k)/M$$
$$= 2094.60/5000 = 0.4169 \text{ in the example}$$

Variance and confidence limits of the mean

Calculate all values of

$C3_{kh}$ = number of possible ways of sampling m units from $M - M_k - M_h$, where h takes the same set of values as k, with the calculation being for every possible combination of k and h. Values for the example are

h	1	2	3	4
M_h 4		3	2	2
k M_k				
1 4	2.107×10^{120}	2.129×10^{120}	2.150×10^{120}	2.150×10^{120}
2 3	2.129×10^{120}	2.150×10^{120}	2.172×10^{120}	2.172×10^{120}
3 2	2.150×10^{120}	2.172×10^{120}	2.194×10^{120}	2.194×10^{120}
4 2	2.150×10^{120}	2.172×10^{120}	2.194×10^{120}	2.194×10^{120}

Note that, especially when using a spreadsheet, it is convenient for the purposes of the calculation to list the M_k and M_h values along the margins of the table, as is done above, because these are the raw material for calculating the $C3_{kh}$ values.

Note also that the table is symmetrical about the diagonal and that all rows with the same M_k are identical, as are all columns with the same M_h. This is a useful check on the calculations.

Calculate all values of

$$p_{kh} = 1 - (C2_k + C2_h - C3_{kh})/C1$$

Again, the marginal $C2_k$ and $C2_h$ values feed into the calculations. The p_{kh} table has the same diagonal symmetry and row and column identities as the $C3_{kh}$ table:

h	1	2	3	4
$C2_h$	2.19×10^{120}	2.216×10^{120}	2.238×10^{120}	2.238×10^{120}
k $C2_k$				
1 2.194×10^{120}	0.001 524	0.001 148	0.000 769	0.000 769
2 2.216×10^{120}	0.001 148	0.000 865	0.000 580	0.000 580
3 2.238×10^{120}	0.000 769	0.000 580	0.000 388	0.000 388
4 2.238×10^{120}	0.000 769	0.000 580	0.000 388	0.000 388

Calculate the values of

$$B_{kh} = [1/(p_k p_h) - 1/p_{kh}]n_k n_h$$

– except those for which $k = h$.

It is convenient to place p_k, p_h, n_k and n_h in the margins of the B_{kh} table. Note how the diagonal is blank because values for which $k = h$ are left out:

k	p_k	h \ n_k	1	2	3	4
		p_h	0.039 42	0.029 71	0.019 90	0.019 90
		n_h				
			24	11	17	5
1	0.039 42	24		−4 442	−10 298	−3 029
2	0.029 71	11	−4 442		−6 294	−1 851
3	0.019 90	17	−10 298	−6 294		−4 292
4	0.019 90	5	−3 029	−1 851	−4 292	
						Sum = −60 414

The variance of the mean of n is

$$\text{var}(\bar{n}) = \left(\sum_k A_k + \sum_k \sum_h B_{kh} \right) \Big/ M^2$$
$$= (1\,266\,152 - 60\,414)/5000^2 = 0.048\,23 \text{ in the example}$$

Approximate 95% confidence limits for the mean are given by

$$\bar{n} \pm t \sqrt{\text{var}(\bar{n})}$$

where t is Student's t for $P = 0.05$ and for $m - 1$ degrees of freedom.
  For the example, the limits are

$$0.4169 \pm 0.4413 = -0.02 \text{ and } 0.86$$

Since a negative mean is impossible, we would quote zero as the lower limit.

Combinations

The number of ways in which one can draw a random sample of x items from a total population of X is referred to as the number of combinations of x from X. For example, if one has three items (A, B and C) and takes samples two at a time, the number of combinations is three (AB, AC and BC): the order in which the items occur is irrelevant, that is AB is regarded as being the same as BA.
  Mathematical terminology for combinations is

$$C(X, x) \quad \text{or} \quad xC_x \quad \text{or} \quad \binom{X}{x}$$

The easiest way of getting $C(X, x)$ is to use a computer-based function, such as Excel® function COMBIN(X, x). If one needs to calculate a value for oneself, then use

$$C(X, x) = X! / [x!(X - x)!]$$

where $x!$ ('x factorial') is the product of x and all positive integers less than x, that is,

$$x(x - 1)(x - 2) \ldots 1$$

Repeated counts at the same site

Why take repeated counts?

Repeating the counts at one's sample sites is a way of improving the precision of one's estimates of the average total population. It may also be used to study regular changes in numbers, as may happen over daily, monthly, or annual periods. Also, it is used in long-term surveillance (see pp. 67–85). For any of these purposes, the most cost-effective design is to repeat the counts simultaneously (or effectively so) at every sample site. The presentation below assumes that has been done.

Improving precision

If one makes observations at each random sample site on a number of random occasions, then one is effectively increasing one's sample size and thus the precision of one's estimate of average numbers present in the whole study area. However, if it costs no more to take a sample at a new site than to take a further sample at the same site, it is better to increase the sample size by covering more sites once only than by repeating the surveys at each site. If there is a cost in setting up a site (such as getting maps, seeking permission to survey, making a reconnaissance visit), then there is a balance to be struck between the number of sites and the number of visits to each. Box 2.14 shows how to calculate this.

Box 2.14. **Multiple visits to sample sites**

Each of m randomly chosen sites is visited at a number (t) of randomly chosen times, the times being the same for all sites.

Example data

Location	Time						
	1	2	3	4	5	6	7
A	19	16	29	15	33	28	8
B	20	20	39	33	33	15	23
C	44	33	26	54	59	29	21
D	18	42	51	35	28	49	21

In the table above $m = 4$ and $t = 7$.

Estimate of mean count

Calculate the arithmetic mean of the mt counts in the usual way. In this case, the mean of the 28 counts is

$$\bar{N} = 30.04$$

Confidence limits

Conduct a two-way analysis of variance (without replication) on the counts, with location and time as the two factors. For current purposes, we are not interested in knowing whether there are significant differences between locations or times, which is the usual purpose of conducting analyses of variance. Rather, we need the three mean squares from the analysis, to calculate the error variance of $\bar{N}$, as follows:

$$\mathrm{var}(\bar{N}) = V^2/(mt)$$

where $V^2 = [MS(\text{sites}) + MS(\text{times}) - MS(\text{error})]$

For the example data,

$$\mathrm{var}(\bar{N}) = (422.131 + 193.952 - 116.825)/28$$
$$= 17.8306$$

Approximate confidence limits for $\bar{N}$ are then

$$\bar{N} \pm t \times \sqrt{\mathrm{var}(\bar{N})}$$

where t is Student's t for $P = 0.05$ and d degrees of freedom.

The value of d may be obtained from the following expression, though this will generally underestimate d and thus lead to estimates of confidence limits that are somewhat too wide (G. E. Thomas, personal communication):

$$\hat{d} = V^4 \left/ \left(\frac{MS^2(\text{sites})}{m-1} + \frac{MS^2(\text{time})}{t-1} + \frac{MS^2(\text{error})}{(m-1)(t-1)} \right) \right.$$

$$= 499.258^2 \left/ \left(\frac{422.131^2}{3} + \frac{193.952^2}{6} + \frac{116.825^2}{18} \right) \right.$$

$$= 4 \quad \text{(rounded up from 3.8)}$$

Hence, the confidence limits for the example are

$$30.04 \pm 2.776\sqrt{17.8306}$$
$$= 18.3 \text{ and } 41.8$$

Optimum sampling strategy

The optimum balance between making counts at different sites and at different times depends on the variation between sites relative to the variation between times and on the costs of establishing new sites and making an individual count. If

$C = $ cost of setting up a sampling programme at a site
$c_t = $ cost of making each individual count

then the optimum number of times to count each site is

$$t_{opt} = \sqrt{[MS(\text{times})/MS(\text{sites})]C/c_t}$$

Suppose that the data in our example were from a pilot survey, allowing the mean squares to be estimated from the analysis of variance.

Suppose also that, in this example, it costs 500 units of available resource to set up each site and 50 to make each count. Then

$$t_{opt} = \sqrt{(193.952/422.131)500/50} = 2.1$$

That is, the optimum use of resources in this case would be to sample just twice at each site.

If the total budget for the work is B, then the number of sites that can be covered if t_{opt} counts are made at each is

$$m_{opt} = B/(C + t_{opt}c_t)$$

Suppose that $B = 10\,000$ in the example, then

$$m_{opt} = 10\,000/(500 + 2 \times 50) = 17$$

Caution

Are the variances homogeneous across all sites and across all times? See Box 2.7.

Studying changes in numbers

If one is interested in how numbers change over, say, the course of a year, one takes samples at fixed times. In general, these should be equally spaced over the whole study period but, if one knows in advance when numbers are most likely to change, then one should concentrate surveys around those times. Box 2.15 shows how to estimate the population at each time.

The optimum balance between how many sites to survey and how many times to survey each depends not only on the costs of establishing new sites and of repeated visits, and on relative magnitudes of the temporal and spatial variations in numbers, but also on the extent to which the temporal pattern is similar over all sites and the pattern of the change can be paralleled in the pattern of sampling. Simulations based on pilot studies are the best way of determining the optimum design.

Mean or peak population

The peak population present in an area is estimated less precisely than the mean, being based on fewer data. However, it may be more important for some purposes to know the peak rather than

Box 2.15. **Variation in numbers over time**

Each of m randomly chosen sites is visited at a number (t) of fixed times, to study the changes in numbers between times.

Example data

Location	Time					
	1	2	3	4	5	6
A	36	21	29	15	28	13
B	15	25	39	38	29	0
C	29	38	26	54	55	14
D	13	47	56	40	23	24

In the table above,

$$m = 4 \text{ and } t = 6.$$

Estimate of mean counts at each time

These are calculated individually. For the example data, the means for the six times are

$$23.25, 32.75, 37.50, 36.75, 33.75, 12.75$$

Confidence limits

The variances for the estimated means are all the same and are estimated by

$$\text{var}(\bar{N}_k) = Q^2/(mt)$$

where

$$Q^2 = [MS(\text{sites}) + (t-1) MS(\text{error})]$$
$$= 1014.251$$

where the mean squares are obtained from a two-way analysis of variance, as in Box 2.14.
   For the example data,

$$\text{var}(\bar{N}_k) = (243.486 + 5 \times 154.153)/(4 \times 6)$$
$$= 1014.251/24 = 42.26$$

Approximate confidence limits for any $\bar{N}_k$ are then

$$\bar{N}_k \pm t \times \sqrt{\text{var}(\bar{N}_k)}$$

where t is Student's t for $P = 0.05$ and d degrees of freedom.

The value of d may be obtained from the following expression, though this will generally underestimate d and thus lead to estimates of confidence limits that are somewhat too wide (G. E. Thomas, personal communication):

$$\hat{d} = (m-1)Q^4/[MS^2(\text{site}) + (t-1)MS^2(\text{error})]$$

$$= 3 \times 1014.251^2/(243.486^2 + 5 \times 154.153^2)$$

$$= 17$$

Hence, the confidence limits for the first time in the example are

$$23.25 \pm 2.11 \sqrt{42.26} = 9.5 \text{ and } 37.0$$

Caution

Are the variances homogeneous across all sites and across all times? See Box 2.7.

the mean. For example, an area that holds the whole population of a species for a short time each year, even if it holds no individuals at all for the rest of the year, may be more important than a site that holds half the population for almost the whole year. The judgement as to whether to concentrate on peak or mean is an ecological one, to which statistics can contribute little.

Comparing two or more study areas

Basics

Suppose that one wishes to compare the populations in two or more areas. The minimum sample size for a valid comparison is two sample sites per area. If one takes only one sample per area, then the differences between those samples may reflect average differences between the study areas but equally they may simply be a reflection of local variation between sites within areas. Using analysis of variance, these two sources of variation can be separated, allowing one to produce estimates (with confidence limits) of the mean numbers per sample site in each area (Box 2.16).

Note that these area-specific estimates of density are valid even if the analysis of variance indicates that the differences between areas are not statistically significant. As ecologists, we know that the null hypothesis for the significance test (that the average population densities in the different areas are absolutely identical) is unlikely. Rather than adopt that unlikely conclusion, it is better to estimate the area-specific densities even when we cannot prove from the data that they are different.

Note also that, while the analysis of variance in Box 2.16 provides an easy test of the significance of the overall differences between areas, formal comparisons of particular pairs of means are less straightforward. They depend crucially on whether or not they were planned in advance. Standard statistics textbooks discuss such pairwise comparisons.

Box 2.16. **Differences between areas**

Example data

Between six and eight random samples were taken from each of four study areas (A, B, C and D), giving these counts:

A	B	C	D
5	4	58	25
18	23	46	22
43	21	38	12
43	2	43	3
22	10	26	0
11	0	10	18
0	9		1
	27		33

Area means

Individual area means, $\bar{N}_i$, are estimated in the usual way. For the example data, they are

20.29, 12.00, 36.83, 14.25

Testing the differences between areas

Carry out a one-way analysis of variance in the usual way.
  For the example data, this results in

	d.f.	Mean square	F
Between areas	3	825.183	4.14
Within areas	25	199.191	

The value of F (with 3 and 25 degrees of freedom in this example) provides a test of the null hypothesis that there are no differences between sites. In this case it is significant ($P = 0.016$).

Confidence limits of area means

The variances of each $\bar{N}_i$ are estimated by

$$\mathrm{var}(\bar{N}_i) = MS(\text{within area})/m_i$$

where m_i is the sample size for that area.

For the example data they are

$$28.4558, 24.8988, 33.1984, 24.8988$$

Approximate confidence limits are then

$$\bar{N}_i \pm t \times \sqrt{\text{var}(\bar{N}_i)}$$

where t is Student's t for $P = 0.05$ and for the 'within-areas' degrees of freedom (25 in the example, giving $t = 2.0484$).

For the example, the results are

	A	B	C	D
LCL	9.36	1.78	25.03	4.03
Mean	20.29	12.00	36.83	14.25
UCL	31.21	22.22	48.64	27.47

where LCL is the lower confidence limit and UCL the upper one.

Differences between pairs of means

Unless one has a-priori reasons for comparing particular pairs of means, the conduct of such comparisons is not straightforward. Should you need to make such comparisons, consult a statistician or a statistics textbook.

What if the *F*-test result is not significant?

If the *F*-test result from the analysis of variance is significant, there is no problem. The normal rules of statistical inference allow you to conclude that the $\bar{N}_i$ represent the best estimates of the mean counts in the study areas.

If you set out to estimate the means for each of the areas, then the same inferences can be drawn, even if the *F*-test result is not significant: $\bar{N}_i$ is the best estimate of the mean for the ith area. The difference is that one now has no good evidence from the data that the means are actually different.

If, however, your prime purpose was to test whether the means were significantly different, then a non-significant *F*-test result means that you accept the null hypothesis of 'no difference'. It is then not logical to make the inference that the $\bar{N}_i$ represent the best estimates of the area means. One must infer that all the means are the same, with the best estimate being the simple arithmetic mean of all the samples, ignoring the areas from which they came.

Matters of design

It is generally most efficient, in terms of precision achieved, for a given cost, to take equal numbers of samples in each area.

It is essential that the same field techniques are used to estimate numbers in the various study areas. If they are not, then apparent differences between areas in mean numbers may simply reflect differences in effectiveness of the various techniques. If more than one observer is involved and the technique being used is sensitive to the field skills of the observer, then this needs to be taken into account. This can be done either by calibrating every observer against a standard or, better, by having each observer carry out the same number of surveys in each area. If this is not possible, a statistician should be consulted about the best way to distribute observer effort over areas.

It is also important not to confound spatial with temporal differences: if the samples from one area are taken in one week and those from another in the next week, differences could arise from changes in numbers between the two weeks rather than from genuine differences between the areas. The simplest way of avoiding this is to take the samples in random order, irrespective of study area. However, it is statistically more efficient to proceed as follows: one surveys one (random) plot in each study area, visiting the areas in random order; one then surveys another set of random plots, one from each area and again in random order; this is repeated as much as one needs to attain the desired sample size. Thus, if one had three study areas and planned to take four samples from each, the order might be (using letters for areas and numbers of replicates): B1, A1, C1, B2, C2, A2, B3, A3, C3, C4, B4, A4. This is an example of the design known as 'randomised complete blocks', with the 'blocks' in this case being the first, second, third, etc. sets of samples, each containing one sample from each area. One advantage of this over a completely randomised design is that the latter can inadvertently result in an uneven distribution of samples from study areas (such as A1, A2, A3, A4, B1, B2, B3, B4, C1, C2, C3, C4, to give an extreme example). More importantly, it removes some of the variation resulting from temporal changes, so making the estimates more precise.

Analysis

Box 2.17 shows how data from a randomised block design may be analysed.

When areas differ in size

If one is interested in the total population of each area rather than in the mean numbers per sample plot in each area, then one needs to take differences in size of the areas into account. As usual, the total population of an area is estimated as M times the mean for the area, where M is the size of the area relative to the size of the sample plots. (Similarly, multiply confidence limits and standard error of the mean by M to get those for the total.)

Modelling spatial variation in numbers

Stratification as modelling

Stratification (pp. 43–51) is an advance on purely random sampling because it assumes that the density of the population at a location depends on the stratum to which the location belongs, so that information about the relative size of the strata can be used to improve the overall population estimate. Allowing each stratum to have a characteristic density is a form of modelling the

---

Box 2.17. **Differences between areas, with a temporal sequence of samples arranged in randomised blocks**

Design

For cases such as that in Box 2.16, the order in which the individual samples are taken is disregarded. If time may have an important influence, it is better to organise the sampling in randomised blocks (see the main text). Thus the data can be laid out in a two-dimensional table, like those in Boxes 2.14 and 2.15, with area and block being the two dimensions.

Analysis

The analysis is parallel to that presented in Box 2.15, in the following sense. Box 2.15 involved estimating time-specific means, with replication over random sites. This box involves estimating area-specific means, with replication over random blocks. Hence the analysis proceeds exactly as in Box 2.15, substituting 'time' with 'area' and 'site' with 'block' to obtain estimates of area-specific means, with their confidence limits.

---

relationship between population density and the environment. Various other sorts of modelling allow better estimates of total population size than simple random sampling. They are described in outline here.

Correlations with mapped variables

Suppose that the study area has been mapped for some variable relevant to one's organism, such as altitude above sea-level, percentage of woodland in each 1-km square, or the presence or absence of open water. Using population estimates from random sites, one can model the relationship between numbers and the mapped variable. One can then use the values of the latter over the whole study area to predict local values of population density across the whole area and thus estimate the total population. See Thompson (2002) for details.

The method is similar to 'double sampling', which is introduced in the next chapter, though in the latter the auxiliary variable is often available from only a sample of locations, not the whole area.

Using covariates

Various methods presented in the next chapter involve estimating the population size at individual sites from samples of the animals or plants at those sites. Traditionally, the estimates for the various sites have been calculated independently. One could then model the relationship of numbers with some auxiliary mapped variable, as in the last section, but it is more efficient to analyse the data from all sites as a coherent set, introducing the auxiliary variable as a covariate, rather than using such a two-stage approach. Royle *et al.* (2004) demonstrate this in the context of distance sampling, but the principles apply to other indirect methods of population estimation.

Spatial patterns: kriging

The population size of a species at one location is generally correlated with its numbers at nearby locations. The pattern of variation in numbers across an area can therefore be modelled from a series of point estimates distributed across the area, using various methods of spatial modelling, such as 'kriging', which take the pattern of spatial correlations into account and allow one to predict the mean (or total) for the whole study area (Thompson 2002). The most efficient distribution of points for obtaining a precise estimate of the total population is a triangular array; a square array is rather less efficient; a square array with the points in alternate rows shifted by half the side of the square is intermediate.

Surveillance and monitoring

The difference between surveillance and monitoring

Surveillance means repeatedly surveying something (population size, in our case), in order to measure how it changes. Surveillance programmes are often described as 'monitoring', but this term should be restricted to something that goes well beyond mere surveillance.

True monitoring entails setting targets, in our case in terms of the size or change in the population that is regarded as desirable. The surveillance programme is then carried out with the primary objective of determining whether the targets are being attained. It is preferably carried out in such a way as to provide an understanding of the reasons for failure, should the targets not be achieved.

Monitoring and adaptive management

What is adaptive management?

If scientists have some understanding of why a population is declining more than is considered desirable, they can advise on changes in the management of the population that are likely to slow or reverse the decline. Changing the management regime in the light of experience and deeper understanding, adaptive management, is clearly a rational approach to both the exploitation and the conservation of wildlife.

Some adaptive management regimes can be quite clear-cut. For example, the management of a rare species on a reserve set up to benefit that species and which is managed by a conservation body; or when a hunted population is managed by a co-operative body involving all the hunting and conservation interests, backed by a firm legislative framework. Other cases are less formal. For example, conservation scientists may monitor wildlife populations generally in a country, against a target that there should be no overall declines and no drastic declines in individual species. Should there be such declines, there may be no formal mechanism in place for taking action but the information can be made widely available and, as a result, changes to broad policy areas may eventually happen. Reforms to the Common Agricultural Policy of the European Union have been driven partly by the results of such general wildlife monitoring in a number of member states.

Science in the community

These examples show that, while scientists have an important role in wildlife monitoring, they are not the only players. The science is useful only if it is combined with the technical expertise of others, such as land managers and economists, and if it is used as part of the evidence on the basis of which decisions are made by the wider community. Decisions made without considering the scientific evidence are unsound; science that is conducted in isolation from the economic, social and political context is of limited value and is likely to be ignored.

Target setting, the first stage in the monitoring process, well illustrates the need for interaction between those with different types of expertise and interests. Conservationists, farmers, foresters, hunters and the tourist industry may have very different views as to how many wolves, for example, they would like to see in their country. Some of their views may be based on misunderstandings of the ecology, such as the number of sheep that wolves may kill; science can help to resolve such disputes. Some differences may be based on misunderstanding of other technical issues, such as the relative socio-economics of farming and tourism; economists (or other relevant experts) can help to resolve such disputes. Finally, some differences may result from differences between people in their economic interests or in their aesthetic and cultural values; resolving those disputes must be a political process for the whole community.

Surveillance and research

Monitoring involves a number of processes and a series of questions (Figure 2.10). When a target has been agreed, surveillance must be undertaken to determine whether it is attained. This means that, periodically, question 1 (Figure 2.10) must be asked. If the target has been reached, all we have to do is maintain the surveillance (so that we can check whether it continues to be reached in the future). If it has not, then we need to ask whether the departure from target is within the normal range, for there is no point in becoming concerned about short-term population fluctuations that are entirely normal. If it is, then we would normally simply continue the surveillance, though we may intensify the surveillance to increase our ability to pick up quickly changes that take the population beyond the normal range. However, as with many of the questions in Figure 2.10, one's answer may be that one is not sure, rather than a clear yes or no, rendering the decision as to what to do a matter of judgement. For question 2, the judgement will become easier as surveillance progresses because this will allow the normal range to be better established.

The third question, 'Do we know why not?', arises if the departure from target is judged to lie outside the normal range. We may know, or have a good idea of, why it does because the ecology of the species is well understood as a result of previous research. Furthermore, a well-designed monitoring programme involves surveillance not only of the focal organism but of other species and environmental conditions that are likely to affect it, potentially providing pointers as to the cause of failure to reach the target. Deeper understanding of the causes of population change

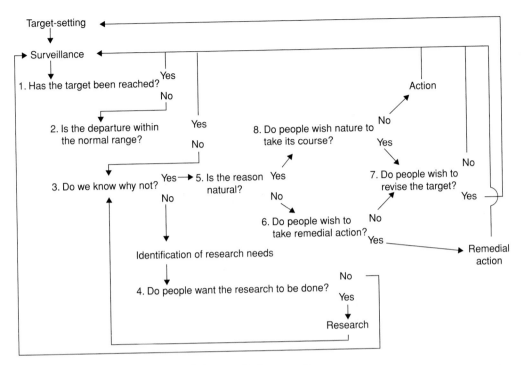

Figure 2.10   The population-monitoring process.

will come if one measures reproductive and survival rates as well as population size, as in the Integrated Population Monitoring programme of the British Trust for Ornithology (Figure 2.11). If many species are monitored at once, contrasting the ecological characteristics of species that are increasing, stable and declining in population can also provide pointers, such as the identification of agricultural intensification as important through the observation that the average rate of decline in population of farmland bird species in Britain was greater than the average for woodland species (Fuller *et al*. 1995). Surveillance will be most useful if carried out on a timescale relevant to the demography of the species. Thus, if one censuses a species at intervals of five years, one cannot tell whether a decline in numbers has been caused by the pesticide spillage in year 1, the drought in year 2, the outbreak of disease in year 3, or the flooding that destroyed nests in year 4. Even for an annually breeding species, it may be valuable to conduct censuses at intervals of less than a year, to measure reproductive success or mortality at critical times.

If we do not know why the target has not been achieved, we must identify the research needs and ask whether people want the research to be done – 'people' in this case meaning those who will fund it or those to whom they are responsible. If finding out the cause of the problem is not their priority, at least the surveillance must continue, in case the problem gets worse. If the research is undertaken successfully, it allows us to return to question 4.

Once the reason for the failure to achieve the target has been established well enough for practical action, people typically ask whether the reason is natural or the result of human activity,

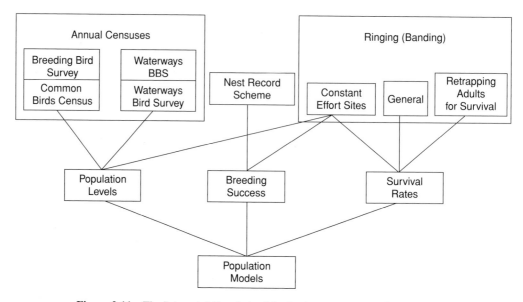

Figure 2.11 The Integrated Population Monitoring programme of the British Trust for Ornithology. Various schemes in the top row of boxes provide data that allow demographic rates to be estimated and thus population models to be built.

on the grounds that Nature may be allowed to take its course. This is becoming less relevant as people realise that few parts of the Earth's ecology can be regarded as entirely natural, given the ubiquity and scale of human impacts, and that some people may not wish to see Nature take its course in some cases. On this basis, questions 6 and 8 are effectively mirror images.

From research to action

Decisions on questions 6 and 8 (and, indeed, on question 7, which often follows them) are matters to which science can make only a limited contribution. They are largely to be decided on the basis of the values, both material and non-material, of the wider community: do people wish action to be taken to conserve a 'desirable' population or to eliminate an 'undesirable' one; and, if so, are they prepared to pay the price? If they are, then action can be taken to reverse the departure from target, guided by the scientific understanding of the reasons for the departure. Surveillance then continues, to establish the outcome of the action that has been taken.

If people do not wish action taken, the logical thing to do may seem to be to revise the target – a positive answer to question 7. But that would be an admission of defeat that people might not be prepared to concede, so it may be reasonable to retain the target even if the community is not prepared at present to take the action needed to attain it. That keeps the original target in view, which may lead to a change of heart about taking action in future (especially if the situation gets worse).

Sampling design for surveillance

Use the same sample locations

In general, if one wishes to measure changes precisely, it is much better to sample repeatedly from the same sites than to take a different random sample on each sampling occasion. The reason is that the difference between independent random samples taken at two times comprises not only the differences related to time but also the differences between locations. In contrast, if the same locations are used on both occasions, the between-location differences are cut out. The estimates of change based on independent random samples may therefore have much wider confidence limits than those of estimates based on sampling the same locations.

Maintaining the same sample plots throughout the surveillance may have further benefits: for example, obviating the need to re-map the area or seek permission for access on each occasion. Equally, there may be costs of doing so, such as mapping or marking the plots so that they can be located again. However, unless the spatial variation in population density is on a very small scale, going back to approximately the same location will still be better than taking completely independent samples, so there is no need to put great effort into precisely locating the study plots.

Of course, if one's method of study is likely to affect the population, then one should not keep going back to the same place. This does not necessarily mean adopting independent sampling on each occasion. For example, if one's samples are soil cores (a case where the animals *and* their habitat are destroyed by the sampling process), one might take each subsequent sample 2 m away (in a random or fixed direction) from the previous one, sufficiently far for it not to be affected by the previous sampling but close enough for it to be effectively the same locality in terms of population variation.

Unplanned loss of sample sites

In any large-scale, long-term monitoring programme, it is inevitable that some sites will be lost to the study. Permission for access may be withdrawn, land may be swept away by flood, or volunteer observers may resign from the programme. It is easy enough to replace lost sites with an equal number of new random sites and it is fairly easy to use statistical methods that allow for some turnover of sites. It is more difficult to prevent such losses leading to bias: landowners may withdraw permission for access or volunteers give up the work if populations are falling; land destroyed by flood is likely to be close to rivers and so hold atypical populations of some species. It is therefore important to measure differences between sites that drop out of the sample and the other sites in terms of the sizes or rates of change of populations, during the period just before they drop out, so that the magnitude of such biases can be assessed. Of course, if sites are pulled out of the programme because their owners intend carrying out management that will be bad for the animals or plants being monitored, the resultant bias cannot be directly assessed.

If sites drop out it is generally better to replace them with new randomly selected sites rather than with sites chosen to match the lost sites, since the matching process can introduce bias. If the sites that drop out tend to share certain characteristics, such as wetland sites being more likely to drop out than those in dry country, then those characteristics should be used to stratify the sites and replacement should be with sites from the same stratum.

Rotational sampling

There may be regional variations in the changes that take place in a population occurring over a large area. These may be efficiently picked up by adopting a programme of stratified random sampling, the regions being the strata.

Another approach is useful when the population occupies a number of sites that is sufficiently limited that one wishes to keep all sites under surveillance, though too numerous for each site to be surveyed on each sampling occasion. This is 'rotational sampling', also known as sampling with partial replacement or panel sampling. Various designs have been devised, all involving the planned dropping of some of the sample sites on each occasion and their replacement by new sites (Urquhart & Kincaid 1999). For example, each site may be surveyed for a fixed number of occasions before dropping out, and then re-enter the sample after a fixed time has elapsed. The analysis of the data depends on the exact design used. The optimal design depends on the relative importance of measuring the overall change in the population and of detecting changes at individual sites.

Monitoring the effects of changes in management

Changes in the management of individual sites or of the whole study area might not be known in advance to the ecologist monitoring the population, but an understanding of their impacts may be gained if the management is kept under surveillance together with the population. If a management change to the whole study area is known about in advance, surveillance may be intensified immediately before and after the change, to increase the chance of understanding its effects. The best situation is when the change will be implemented at only some sites: one can then keep a random sample (or all) of those sites under surveillance, together with a random sample of the sites at which management has not been changed, to provide quasi-experimental data on the effects of the change.

Describing long-term changes

Short-term variation and long-term trends

Natural populations may vary considerably in size in the short term, during the course of an annual cycle or from year to year. In terms of monitoring for the purpose of managing populations, such short-term variation is generally of less immediate interest than longer-term trends. This section is concerned with how such longer-term trends can be described and assessed.

If one is studying just a single population, it may be possible to census it precisely. More commonly, the field methods merely yield an estimate of the population at a site; and, even if one can census a single site precisely, one is often making an estimate of the numbers in a large area on the basis of data from a sample of sites. So, in addition to the variation in actual numbers from time to time, there is sampling variation to contend with.

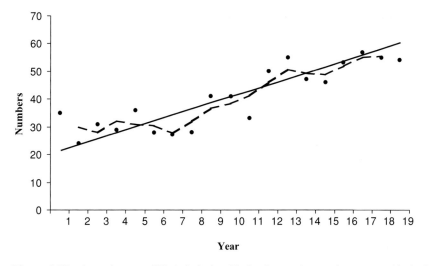

Figure 2.12   Annual counts (filled circles), with the three-point moving average (dashed line) and a straight line fitted by regression (continuous line).

Unified analysis across sites and times

Estimating population density using distance sampling is considered fully in the next chapter. Briefly, it involves estimating density from the sample of animals or plants actually observed, taking into account the fact that not all those present will be observed. To do this, a 'detectability function' has to be estimated from the data. It is possible to treat the data for each site and each time independently. However, precision is improved if one assumes that the detectability function is the same for all times and places in a study area, analysing all the data as a coherent package (Buckland *et al.* 2001, 2004). Similarly, the 'robust model' for mark–recapture studies allows one to integrate data across time at each study site.

Of course, such coherent analyses are valid only if the relevant conditions are constant. One cannot assume that detectability functions are the same across sites that differ markedly in habitat characteristics, for example.

Although such coherent analyses are useful if they can be carried out, one usually has independent counts or estimates from one's different sites and different times. The methods below assume such independence.

Data from a single site

If one has data from a single site, the simplest way of describing the overall trend is to fit a straight line to the data using the usual technique of least-squares regression, the counts (or estimates) being regressed on time (Figure 2.12). It is perhaps biologically more realistic to assume that the rate of change of the population is constant not in absolute terms but in relative terms, implying an exponential relationship between numbers and time. Regressing the logarithm of counts on time is then appropriate.

In both these cases, the slope of the regression line is a simple measure of the average rate of change in numbers over the span of time covered by the data. However, there may be medium-term trends that are obscured by fitting such simple models. In this case, it may be better to simplify the picture in a rather less extreme way, by using a smoothing technique, such as replacing each data point by the 'three-point moving average' (that is the average value for that occasion and the occasions immediately before and after; see Figure 2.12). This is a useful compromise between the complexity of the raw data and the oversimplification of assuming a constant rate of change. However, there is an infinity of ways to smooth data, with no general reason for choosing one rather than another.

Whenever a straight line or a smoothed curve is fitted, it is important also to plot the original data points, so that the goodness of fit of the line can be judged.

Simple but inefficient methods for multiple sites

Most monitoring is conducted at a sample of several or many sites rather than just one. The simplest way of analysing multiple sites is to regress (linearly or exponentially) counts on date, or to plot smoothed averages. The only difference from the corresponding single-site analyses is that there are now multiple counts on each date. This is inefficient, however, for it ignores the fact that (at least in well-designed programmes) most of the sites counted on any one occasion were also counted on the previous occasion.

The continuity across time is partly taken into account in the 'chaining method'. Here, one takes all the sites counted both on occasion 1 and on occasion 2, to calculate a simple index of change between those occasions. One does the same for occasions 2 and 3, 3 and 4, etc., and then works out the overall picture by chaining these successive indices together. Even this is inefficient, however, because it does not use the continuity over more than two occasions. Furthermore, if there is turnover in the composition of the sample (sites dropping out and being replaced), statistical chance can cause apparent long-term trends in the resultant index even when there are no real trends.

Generalised linear models

Generalised linear models (GLMs) (Crawley 1993) are extensions of regression methods that allow for not all sites being counted on every occasion, which is the typical pattern in large-scale, long-term surveillance programmes. The underlying statistical model for GLMs is that the count at site x on occasion y is given by

$$C_{xy} = s_x t_y$$

where s_x is a constant 'site effect' (the same on all occasions) and t_y is a time effect (the same at all sites). This multiplicative model is made additive by taking logs:

$$\log(C_{xy}) = \log(s_x) + \log(t_y)$$

Various statistical packages allow GLMs to be calculated. Box 2.18 provides some general guidance on using them. One may generate t_y values for each year independently, reflecting the full

Box 2.18. **General guidance for GLMs for count data**

Estimates for each occasion

To obtain population estimates for each sampling occasion you should generally have three columns of data, one containing the counts (the dependent variable), one containing a column identifying (as a numerical variable) which site each count comes from and a third (a sequential numerical variable) identifying the occasion of the count; site and occasion are the independent variables. They should be specified as categorical variables. Choose the linear model option (or similar) from the statistics menu and specify the columns for the dependent and independent variables. Choose a log link function and a Poisson error distribution.

   The log link applies the log transformation needed to make the model additive (see the main text). The Poisson error distribution is appropriate for the sampling error of count data, though if the counts are more variable than Poisson (with variance greater than the mean) this should be accommodated by including an overdispersion factor.

   This procedure should generate output with an estimate of the average count for each site (taking into account the time-intervals it was surveyed) and an estimate of abundance (averaged over all sites counted) for each time-interval. These estimates will probably be on a log scale, because of the link function, unless one has chosen an appropriate option to transform them to the linear scale. The linear estimates provide an index of abundance, usually relative to the first or last year.

Estimating a trend line

To produce a trend over time, all that is required is to specify the time-interval variable as an ordinary numerical variable (i.e. not categorical). This will yield a single estimate for time, which is the slope of the trend in numbers over time. Note that this slope is linear on the log scale on which the model is fitted (because of the link function).

TRIM

The package TRIM (Trends and Indices for Monitoring Data) (Pannekoek & van Strien 2001) has been designed specifically for fitting these Poisson-type models to generate both annual indices and linear trends, and is downloadable from www.ebcc.info.

Sites fixed or random?

If the sites counted represent all or most of the sites at which the study species occurs, so that one may be interested in numbers at each site, then 'sites' enters GLM calculations as a fixed factor. Where only a subsample of potentially occupied sites is counted, it is more

76 Principles of sampling

appropriate to treat the 'site' as a random factor because one is not interested in the sites chosen *per se* but as being a representative selection of all potential sites. In GLM packages it is usually assumed that the factors are fixed, so one needs to specify that 'site' is a random factor if that is the case.

variation from year to year, or make the assumption that there is a linear trend and estimate the slope of this trend.

Generalised additive models

Whereas GLMs produce either estimates for individual occasions or estimates of a linear trend, generalised additive models (GAMs) allow one to produce intermediate models, like the smoothed curves referred to earlier (Fewster *et al.* 2000). The extent of smoothing can be varied to suit the data by altering the number of degrees of freedom (d.f.) assigned during the analysis. Figure 2.13 provides an example, showing how an intermediate level of smoothing (d.f. = 11, in this case) shows the major trends by removing the minor fluctuations and how heavy smoothing (d.f. = 3, in this case) rather obscures anything except the very broad picture that numbers were much the same in 2000 as in 1995, whatever the fluctuations in between. One may adopt a formal criterion, such as Akaike's Information Criterion, to choose the best-fitting model. Or one can decrease the degree of smoothing systematically until all the main patterns in the data appear to have emerged, such that any further decrease serves only to roughen the output. In any fitted

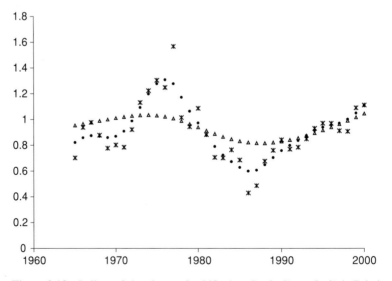

Figure 2.13 Indices of abundance of goldfinches *Carduelis carduelis* in Britain, using generalised additive model fits to Common Bird Census data:
✱ indicates the full model (with 35 degrees of freedom (d.f.), one fewer than the number of years of data); • indicates a model with 11 d.f.; and Δ a model with 3 d.f.
(Data from the British Trust for Ornithology; model-fitting by S. N. Freeman, personal communication.)

GAM curve, significant turning points (when upward trends turn downward and vice versa) can be identified objectively (by calculating the second derivative of the curve).

Route regression

In route regression a linear (or log-linear) trend is fitted independently to the data for each site. The estimate of the overall trend at each time is then obtained from the average of the trends for all sites being counted at that time (Link & Sauer 1997, 1998). The investigator can choose to use an unweighted average or an average weighted by the abundance of the focal species at the site, by the precision of the trend estimate at the site, by the area of the site, etc. (There is no general agreement as to which is best.) Many investigators prefer GLMs and GAMs over route-regression models because they allow more complex trends to be fitted.

Process models

The models described above are empirical models: they are essentially descriptive, with little biology going into the structure of the model. Process models, in contrast, are ones that incorporate the ecological processes that cause numbers to vary. They are thus more realistic (though this might not be helpful, if the complexity of reality obscures the broad picture). They can also provide more precise estimates, because they incorporate additional data. For example, data on recruitment to or losses from the population (or both) can be incorporated with census data to provide a demographic model that yields population estimates that are more precise than those derived from census data alone (Besbeas *et al.* 2002, 2005).

Alerts and indicators

Why alert systems are needed

Many monitoring schemes involve the surveillance of several species simultaneously. It is then useful to be able to pick out those species for which one detects departures from their targets that are most significant in terms of the purposes of the monitoring. Effective means of doing so need to take into account the uncertainty associated with most estimates of population size and change, in particular providing an optimum balance between too many false alarms and failing to react to real problems. If the system is too sensitive, resources are wasted in reacting to apparent problems that are either of short-term nature or just the result of imprecise measurement. If it is not sufficiently sensitive, real problems might not be dealt with early enough. The results of surveillance need to be considered both in the short term, to pick up problems early (with safeguards against false alarms), and in the long term, to pick up slow but continued departure from targets.

Alert systems are designed to provide warning of potential problems, taking all these considerations into account. In terms of Figure 2.10, they provide means of systematically addressing question 1 and sometimes question 2. Questions 3 and onwards are then applied to those species for which alerts have been raised.

Baillie *et al.* (2005) present a well-established alert system, that for birds in the United Kingdom.

More on targets

When several or many species are being monitored and an alert system is being applied generally to them all, it is useful to have similar targets for them all, rather than species-specific targets. For example, the overall target may be 'Restore numbers to 1970 levels, for all species'. Such general targets are both easier to work with and easier for the wider community to understand than are numerous species-specific targets.

Completely general targets may be inappropriate in some cases. Alongside a general target of restoring numbers to 1970 levels, one may wish to increase the numbers of some rare species that had already declined substantially by 1970 to well above their 1970s levels. If these species are included in the general monitoring programme, then one may need to have a range of targets within the system. Commonly, however, such species are too scarce to be included within the general monitoring. Rather, they require species-specific surveillance programmes, to which species-specific targets can be attached, with the alert system adapted as appropriate.

In 2000, the British government set itself the target of reversing the decline of farmland birds by 2020. Some regarded this as an ambitious target, given the apparently inexorable decline in farmland bird species in Britain since the mid 1980s and the difficulties of reforming the European Union's Common Agricultural Policy. Others noted that, if the rate of decline then prevailing were to continue until 2020, by which time many species would be almost extinct, a bottoming-out would be almost inevitable; they therefore felt the target to be easily achievable. There is always a balance to be struck between ambition and achievability: targets that are too ambitious may soon be perceived as unachievable, so that people may give up even trying to approach them; on the other hand, targets that are easy to achieve may, when they are achieved, leave people feeling that more could have been done had they been more ambitious.

Because of the huge impacts that mankind is having on the Earth's ecology and on other species, most targets for wildlife are aimed at maintaining or even increasing numbers. But we should not forget that there may be some species whose numbers we wish to reduce.

Timescales

Alert programmes may assess each species on several timescales. British birds, for example, are assessed on scales of the last 5 years, the last 10, the last 25 and the entire data span. Tables of trends over these spans allow a more refined approach than just a single timescale would do, without the broad picture being obscured by the detail that is present in the full set of annual population indices. They allow one to concentrate on long-term trends without overlooking recent changes that might be the start of a change in the long-term pattern.

Built-in adjustments: filters

In the monitoring programme for waterbirds in Britain, population changes are assessed at many individual sites, as well as nationally. As a result, large numbers of alerts could potentially be raised every year. To remove the less important of these, a system of filters has been proposed, incorporating both conservation concern and biology (Atkinson *et al.* 2006, Austin *et al.* 2004). A strong filter is applied to populations that are of no particular conservation concern, a weaker one to those of local importance, an even weaker one to those of national importance and none to those of international concern; departures from target have to be very marked to pass through the stronger filters. Strong filters are also applied to species that show great variation in local population size from year to year (because apparent short-term declines may be temporary), to species that are short-lived (they are potentially more capable of rapid recovery than are long-lived species) and to highly mobile species (in which local declines may result from temporary population movements).

Similar filters could be developed for other programmes covering many species and populations.

Statistical significance

A typical system may raise an alert if there is a decline in a population of more than a certain percentage over a set time period. Because rates of decline are usually only estimates, one needs to assess the strength of the evidence for any decline that appears to exceed the alerting criterion. The traditional approach is to carry out a test of statistical significance, against the null hypothesis that there has been no change. If a significant result is achieved, the estimated population change is taken to be the actual change.

As in most ecological research, the critical probability for significance is taken to be 95%. Some workers suggest that, because alerts will usually be raised only if numbers fall, with increases being ignored, a one-tailed test is appropriate. Correspondingly, they advocate plotting 90% confidence intervals on the graphs of population size. As an alternative, one may plot 84% confidence intervals, for it is then approximately true that, if the confidence limits of two occasions do not overlap, then they are significantly different at the 95% level (one-tailed test) or 90% level (two-tailed test) (Payton *et al.* 2003).

Some workers argue that testing against the null hypothesis that there has been no change is both perverse and overly conservative. It is perverse because the alert criterion is that change is no greater than a certain percentage; surely, they say, *this* should be the null hypothesis that one subjects to a (one-tailed) test. The proponents of this view argue that the appropriate null hypothesis is that the decline in the population is greater than the level set by the alerting criterion, so that an alert is raised if the result of a test of this hypothesis is non-significant. Under this approach, alerts are more likely to be raised for scarce (and hence less precisely measured) species than for common ones. The more traditional test is unduly conservative because one may fail to reject the null hypothesis of no change not only when there has been little change but also if the confidence limits of one's estimate of change are too wide. Because scarce species are likely to produce only small samples (and thus wide confidence limits), this leads to the paradoxical result that the same estimated decline may raise an alert for a common species but not for a scarce one (Figure 2.14).

Yet other workers would recommend that the significance-testing approach is wholly inappropriate for alerting systems on the grounds that it constrains our conclusions to fall into clear-cut categories: simply 'significant alert' and 'no significant alert'. They prefer a Bayesian approach, in which the conclusions can be expressed in terms of an estimated probability of the departure from target falling within particular ranges. For example, under the Red, Amber and Green alerts system used in the United Kingdom, a 25%–50% decline over 25 years leads to an Amber

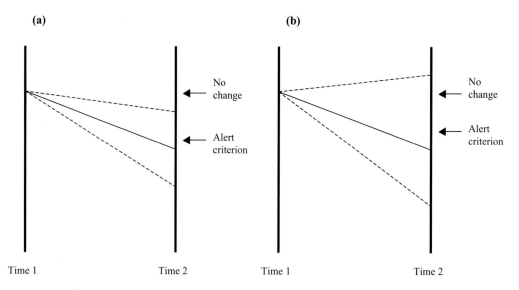

(a)   **(b)**

Time 1   Time 2   Time 1   Time 2

Figure 2.14 Alternative hypothesis tests for alerts. Vertical lines represent two occasions; continuous lines represent estimated population declines, with dashed lines representing confidence limits. The two cases illustrate declines of the same magnitude (exceeding the alert criterion) but the confidence limits are wider in (b) than in (a). If the null hypothesis used is that the population has not declined, the result is significant for (a) but not for (b), so no alert is triggered for (b). If the null hypothesis used is that the decline exceeds the alert criterion, the result is significant for neither (a) nor (b), so an alert is triggered for (b) as well as for (a).

alert, a decline of more than 50% to a Red alert. Under this system, the Lapwing *Vanellus vanellus* would have qualified for an Amber alert in both 1997 and 1999; but a Bayesian analysis assigned probabilities to Amber and to Red of 0.96 and 0.04, respectively, in 1997 but 0.55 and 0.45 in 1999, flagging up a substantial change in the evidence of the decline of this species between those two years (King *et al.* 2006). However, what the proponents of Bayesian methods would claim as a less constrained form of result than obtains under the traditional system is considered by others to be too complex to be placed in front of non-technical members of the community, who may prefer to be told simply whether there is or is not cause for concern over a species. On the other hand, non-scientists have proved perfectly capable of understanding the uncertainties of climate-change scenarios when these have been explained to them; there is no reason to suppose that they could not understand the uncertainties associated with ecological monitoring.

Indicators

In the present context, indicators are summary measures based on the populations of various species that are considered to show how things that people may regard as important are faring. Thus the 'Wildbird Indicator' is used by the British government as an indicator of quality of life, partly on the basis that people like to see birds themselves and partly because the numbers of birds

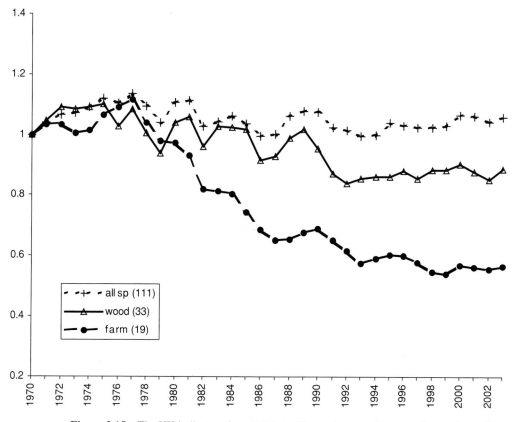

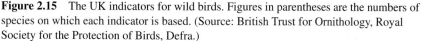

Figure 2.15 The UK indicators for wild birds. Figures in parentheses are the numbers of species on which each indicator is based. (Source: British Trust for Ornithology, Royal Society for the Protection of Birds, Defra.)

in the country may be an indicator of wildlife more generally (Figure 2.15). A similar indicator is used by the European Union (www.ebcc.info).

An indicator based just on farmland birds is used by the British government to assess its success in achieving its target of reversing the decline in farmland birds by 2020. Such habitat-specific indicators can also throw up general problems in the management of particular habitats. Thus the steeper decline of farmland birds than of woodland species (Figure 2.15) has signalled general problems of the management of farmland (Chamberlain & Vickery 2002) as opposed to the more varied problems facing some woodland species (Fuller *et al.* 2005).

An indicator will be useful only if the species included within it are chosen for their relevance to what the indicator is meant to indicate. For example, a farmland indicator should comprise only those species for which farmland is a major habitat. (Because species' habitat preferences may vary geographically, the list of relevant species may vary from country to country.) It is important that the criteria for inclusion are seen to be objective because the information provided by indicators is often politically sensitive.

For general monitoring purposes, rare species should be excluded. They tend to have particular habitat requirements, which make them poor general indicators. They are often the subjects of special management, which makes them poor indicators of the more general management of an environment. Also, their rarity may mean that their populations may be imprecisely monitored (unless they are so rare that their whole population is readily counted). Care needs to be taken in respect of species that were once common enough to include in the index but which have declined in population so much that they can no longer be properly censused. Simply removing them from the index introduces serious bias. The alternative is to assign some arbitrary low value to their populations if they have become too rare to be censused.

Data from various species can be combined in various ways. A method of giving each species equal weight is to divide its population in each year by the population in the first year (for which the index is thus 1.0 in all cases), and then take the average of these indices in each year as the indicator for that year.

The arithmetic mean of such species indices is not a useful indicator. Consider an indicator based on two species, one of which doubles in number between the first and second years, while the other halves. The two indices in the second year are thus 2.0 and 0.5, the mean of which is 1.25. Yet a doubling and a halving should cancel out. The geometric mean (Box 2.19), which has a value of 1.0 in the second year in this case, is a better measure.

Box 2.19. **Calculating a geometric mean**

If your statistics package does not provide this function (e.g. GEOMEAN in Excel®), proceed as follows.

Take the logarithm (natural or base 10) of each of the numbers.
Calculate the arithmetic mean of the logs.

The geometric mean is then the exponent (antilog) of the arithmetic mean of the logs.

	Number	$\log_e$(number)
	1.05	0.049
	0.82	−0.198
	3.17	1.154
	1.34	0.293
	1.25	0.223
	2.03	0.708
	0.99	−0.010
	0.38	−0.968
Arithmetic mean	1.38	0.156
Geometric mean	1.17	

Multi-species population indicators tend to give all species equal weight and to treat all increases as desirable and all decreases as undesirable. But there are some species whose presence may be regarded as indicating deterioration in the 'quality' of a habitat, such as those species characteristic of other communities or invasive aliens. In principle, each such species can be incorporated in the indicator by taking the inverse of its index each year. Thus, if it is twice as abundant in year 10 as in year 1, one uses 0.5 as its contribution to the index in year 10, instead of 2.0.

Planning and managing a monitoring programme

Planning

The first step, which must not be rushed, is to decide one's objectives. Because monitoring, once one goes beyond mere surveillance, is a process that draws in an array of interested parties, it is essential that all those parties participate in discussions about objectives.

Some broad issues must then be addressed, such as what particular failures to achieve targets are likely to be of greatest concern and what actions are likely to be taken should those targets not be hit. The answers to those questions will influence the design of the monitoring programme, particularly the surveillance.

Statistical design issues must be addressed. One must identify the design that is most efficient, providing the most precise estimates given the resources available, while avoiding undue bias. It is important not to be too precious about design, for a perfect design might not be practically possible; so long as one thinks carefully about the likely scale of bias (both when planning and when interpreting the results), it is almost always better to have biased surveillance than no surveillance at all. Various aspects of the design may conflict. For example, one may have a choice between using highly standardised methods implemented by a small number of professional fieldworkers or less standardised methods implemented by a large number of volunteers. Results from the latter approach will be more precise (because of the greater sample size) but potentially more biased (because of the looser standardisation). Careful thought is required as to what is needed for one's particular purpose.

The principles outlined in this chapter are not the only ideas to be used in the planning process: the practical experience of those running similar programmes will be of particular value, as will pilot studies to test out the methods and to provide information that is useful in deciding on final details of the design.

The most important principle of planning is to do it. The second most important principle is not to spend so much time planning that the start of the programme is significantly delayed. With rare exceptions, you will be unable to gather information about the present population in future, so each year's delay in starting is a permanent loss to the long-term programme.

Reviewing

Any long-term programme should be kept under review, to discover whether the methods can be improved in the light of practical experience (which may expose deficiencies), changing circumstances (such as public attitudes or resources), changing objectives and improved theoretical

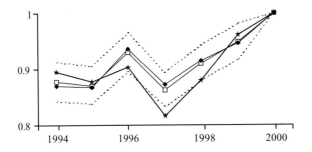

Figure 2.16   Comparison, during a seven-year overlap period, between two indices of great tit populations in southern Britain. Filled diamonds show the Breeding Bird Survey index, stars the Common Bird Census index and open squares an index based on both data sets (with 95% confidence intervals as dashed lines). The indices are all based on an arbitrary value of 1.0 in the year 2000.

insights. It is generally useful to have a standing committee of experts, who know the programme well, to keep the work under constant review. This should be backed up by periodic reviews by outside experts, who may be able to see deficiencies more clearly because they are not so close to the routine, as well as being able to bring in fresh insights. These 'experts' should not just be ecologists and statisticians but should include representatives of all parties interested in the programme.

Publication of results in peer-reviewed journals is another valuable form of review, as well as providing validation of the programme.

Changing methods

However careful the planning has been, results of the first few years of a monitoring programme may show that the methods can be improved. A judgement must then be made as to whether the potential improvement is sufficient to outweigh the loss of those first few years' data (they will effectively be lost because, if the methods change, the old and new data are likely not to be comparable).

After such early adjustments, it is best to stick with the same methods for a substantial period, to reap the benefits of long-term continuity. It is certainly important to prohibit frequent minor adjustments, which destroy that continuity.

At some stage, however, it may become clear that substantial improvements are possible. The changes should be planned and piloted even more carefully than was the original methodology, so that the benefits of making the change are properly assessed. Most importantly, there should be overlap between the old and new surveillance programmes, so that the old and new data sets can be linked to maintain the long-term continuity (Figure 2.16).

Maintaining consistency

Changes in personnel (and even unreliable memories) are always liable to result in changes in the detail of practical procedures. For this reason, it is important to use methods that are likely to be robust against individual interpretation and to record fully the exact methods used. When methods depend on subjective judgement (such as the interpretation of the maps in the territory-mapping method described in Chapter 9), new personnel need to be carefully trained and experienced personnel should regularly check that their interpretations remain consistent over time.

Acknowledgements

We thank many colleagues, especially at the British Trust for Ornithology (BTO), for answering our queries and supplying copies of papers. We especially thank S. N. Freeman (BTO) and G. E. Thomas (University of Dundee) for statistical advice.

References

Atkinson, P. W., Austin, G. E., Rehfisch, M. M. *et al.* (in press). Identifying declines in waterbirds: the effects of missing data, population variability and count period on the interpretation of long-term survey data in the UK. *Biological Conservation*.

Austin, G. E., Jackson, S. F. & Mellan, H. J. (2004). *WeBS Alerts 2000/2001: Changes in Numbers of Wintering Waterbirds in the United Kingdom, its Constituent Countries, Special Protection Areas (SPAs) and Sites of Special Scientific Interest (SSSIs)*. BTO Research Report No. 349. Thetford, British Trust for Ornithology.

Baillie, S. R., Marchant, J. H., Crick, H. Q. P. *et al.* (2005). *Breeding Birds in the Wider Countryside: Their Conservation Status 2004*. BTO Research Report No. 385. Thetford, British Trust for Ornithology (http://www.bto.org/birdtrends).

Besbeas, P., Freeman, S. N., Morgan, B. J. T. & Catchpole, E. A. (2002). Integrating mark–recapture–recovery and census data to estimate animal abundance and demographic parameters. *Biometrics* **58**, 540–547.

Besbeas, P., Freeman, S. N. & Morgan, B. J. T. (2005). The potential of integrated population modelling. *Australian and New Zealand Journal of Statistics* **47**, 35–48.

Buckland, S. T., Anderson, D. R., Burnham, K. P., *et al.* (2001). *Introduction to Distance Sampling: Estimating Abundance of Biological Populations*. Oxford, Oxford University Press.

Buckland, S. T., Anderson, D. R., Burnham, K. P. *et al.* (2004). *Advanced Distance Sampling – Estimating Abundance of Biological Populations*. Oxford, Oxford University Press.

Chamberlain, D. & Vickery, J. (2002). Declining farmland birds: evidence from large-scale monitoring studies in the UK. *British Birds* **95**, 300–310.

Cochran, W. G. (1977). *Sampling Techniques*, 3rd edn. New York, John Wiley & Sons.

Crawley, M. J. (1993). *GLIM for Ecologists*. Oxford, Blackwell Scientific Publications.

Diggle, P. J. (1983). *Statistical Analysis of Spatial Point Patterns*. London, Academic Press.

Fewster, R. M., Buckland, S. T., Siriwardena, G. M., Baillie, S. R. & Wilson, J. D. (2000). Analysis of populations trends for farmland birds using generalised additive models. *Ecology* **81**, 1970–1984.

Fuller, R. J., Gregory, R. D., Gibbons, D. W. *et al.* (1995). Population declines and range contractions among lowland farmland birds in Britain. *Conservation Biology* **9**, 1425–1441.

Fuller, R. J., Noble, D. G., Smith, K. W. & Vanhinsbergh, D. (2005). Recent declines in populations of woodland birds in Britain: a review of possible causes. *British Birds* **98**, 116–143.

Hutchings, M. J. (1978). Standing crop and pattern in pure stands of *Mercurialis perennis* and *Rubus fruticosus* in mixed deciduous woodland. *Oikos* **31**, 351–357.

Link, W. A. & Sauer, J. R. (1997). Estimation of the population trajectories from count data. *Biometrics* **53**, 488–97.

  (1998). Estimating population change from count data: application to the North American Breeding Bird Survey. *Ecological Applications* **8**, 258–268.

Manly, B. F. J. (1997.) *Randomization, Bootstrap and Monte Carlo Methods in Biology*. 2nd edn. London, Chapman and Hall.

Pannekoek, J. & van Strien, A. (2001). *TRIM 3.0 for Windows (Trends and Indices for Monitoring Data)*. Voorburg, Statistics Netherlands.

Payton, M. E., Greenstone, M. H. & Schenker statistical N. (2003). Overlopping confidence intervals or standard error intervals: what do they mean in terms of statistical significance? *Journal of Insect Science* **3**, 34–39.

Reason, P., Harris, S. & Cresswell, P. (1993). Estimating the impact of past persecution and habitat changes on the numbers of badgers *Meles meles* in Britain. *Mammal Review* **23**, 1–15.

Royle, J. A., Dawson, D. K. & Bates, S. (2004). Modeling abundance effects in distance sampling. *Ecology* **85**, 1591–1597.

Thompson, S. K. (2002). *Sampling*, 2nd edn. New York, John Wiley & Sons.

Urquhart, N. S. & Kincaid T. M. (1999). Trend detection in repeated surveys of ecological responses. *Journal of Agricultural, Biological and Environmental Statistics* **4**, 404–414.

3 General census methods

Jeremy J. D. Greenwood and Robert A. Robinson

British Trust for Ornithology, The Nunnery, Thetford, Norfolk IP24 2PU, UK

Table 3.1. *Contents of Chapter 3*

Ecological Census Techniques: A Handbook, ed. William J. Sutherland.
Published by Cambridge University Press. © Cambridge University Press 1996, 2006.

Introduction

In the previous chapter we saw how to derive conclusions about a whole population from information from sample areas. In this chapter, we are concerned with how to get the information from the individual sample areas (or from the whole population if it can be completely covered). To be clear, we shall refer to those areas as 'sample areas' or 'study areas'; if we need to refer to the whole population of interest, we shall use that explicit phrase.

There are three approaches to measuring population within a study area. Most obviously, one may be able to carry out a complete count. Unfortunately, this is often impossible because one cannot be sure that one has detected all the individuals in the population. This has led to the development of various methods that involve counting just some of the individuals that are present and estimating (either explicitly or implicitly) the detectability of individuals, so that the total population of the sample area can be estimated from the sample of animals or plants actually observed by allowing for the detectability being less than perfect. However, it is often not easy, indeed it may even be impossible, to make such estimates. In such cases, one may settle for the third approach, namely obtaining an index of the population. That is, a measure that is related to population size but is not an estimate of the actual population size.

The diversity of methods can be confusing. First, it is important not to confuse methods that are different but have similar names and, conversely, to recognise the same method masquerading under different names in various accounts. Second, how does one choose which method is the best for a particular purpose? Do not make the mistake of assuming that more complex methods will provide better measures; they often do not. It is tempting to choose a method that has been used widely by others but this is sensible only if it has been used for cases similar to one's own. Even then, one should not blindly follow fashion but should assess the method for oneself. Are the assumptions underlying a method reasonable for one's study population? How precise are the various methods likely to be? How biased are they? What are their costs? Of course, if you wish to compare your results with those of other studies, there is always an advantage in using the same methods – unless those methods were clearly inadequate.

It is important to be aware of the extent to which your methods may affect the population or the individuals within it. Methods that are overly intrusive (by, for example, increasing mortality rates or causing animals to flee the area) will make your conclusions worthless and may be ethically unacceptable.

Some methods may provide information additional to a measure of population size. For example, the Jolly–Seber mark–recapture method also provides estimates of rates of recruitment to and loss from the population (which may allow the building of demographic models that provide insights into the causes of population change). The likely value of such information should also be part of one's assessment of the best method to use for one's purposes.

The appendix to this chapter describes software for the analysis of data for many of the methods covered here.

Complete counts (1): general

Not as easy as it seems

It is sometimes possible to count directly the number of individuals in a study population, such as the number of acacia trees in a savannah area, the number of limpets on a section of rocky shore, or the number of elephants in a herd. But beware! Individual organisms (even trees and elephants) can turn out to be remarkably easy to overlook and many such direct counts turn out, when more careful studies are made, to be grossly incomplete. If one is familiar with a study area and frequently encounters the animals in it (especially if they become recognisable as individuals), one has a better chance of finding them all in a census; but such intimate knowledge should not lead one to be complacent about the possibility of missing animals that always take avoiding action as one enters their domain or that live in rarely visited parts of the study area. Plants of the particular species that one is trying to count may be overlooked easily amongst dense vegetation, especially if they are not flowering, and some perennials do not produce above-ground parts every year.

It is important to avoid counting when weather conditions reduce the detectability of animals. Counting may be easier if animals are counted when concentrated together – such as migratory fish as they pass through a restricted part of a river, flocking birds on their roosts, or seals on their breeding beaches. The very crowding of the animals may, however, give rise to difficulties of counting under such conditions. One tip is to divide the area to be counted into sections, using either natural features or a grid that has been laid out specially for the purpose of counting. Photographs in which the animals can be counted at leisure are also useful. Plants and relatively immobile animals may be marked to indicate that they have already been counted.

Even if a count is incomplete, it may be better than no knowledge at all. For example, if managers of harvested populations set a limit on the size of the harvest that is based on the assumption that the population size is the same as the count, they can be confident (within the limits of their population model and their capacity to regulate the harvest) that the population will not be over-harvested. The 'minimum number known to be alive' (the usual terminology) is obtained from counts that are as complete as possible or from intensive trapping.

Sampling the habitat

Small animals may be counted by removing a sample of the habitat (such as a volume of water, a core of soil, or some leaves of a plant) and sorting it later in the laboratory, in comfortable conditions, with good light and perhaps using optical aids. It is important not to let animals escape when one collects such samples and to store the samples so that, even if the animals die, they remain identifiable. Extracting the animals from such samples may be time-consuming, so various mechanical aids such as desiccation funnels are often used (see Chapter 5). It is important always to check, for the particular animals and substrate involved in one's study, that all animals are being extracted – by checking what remains in some of the samples after they have passed through the extractor. (An extractor that is less than 100% efficient may, of course, be used to provide an index of the animal population so long as it extracts a constant proportion of the animals from the sample.)

When animals are unexpectedly numerous, a mechanical extractor may produce such large numbers that it is not cost-effective to count them all. Subsamples may then be taken, to reduce the labour. It is neither necessary nor desirable that these should be random samples: what is important is that they should be representative. Because animals floating in a dish, for example, tend to cluster together (often in the corners or towards the centre), it needs care to achieve representative subsampling. It is usually necessary to mark out a grid and to take subsamples from all parts of it. Southwood (1978, p. 145) shows a useful grid for subsampling one-sixth of the total from a circular dish.

Attempted complete enumeration

There are some species for which it is possible to recognise individuals. This means that, even if one never sees all the members of a population at once, one can count the number in the population simply by keeping a record of each of the individuals seen. Eventually, unless there is a constant stream of immigrants or of new-born animals, one will reach a stage when no new animals are being seen; the list of the individuals seen is then a complete list of the population. (If one cannot distinguish individuals with certainty, one's count is a minimum number known to be alive.) More commonly, one can mark animals when they are seen for the first time; by counting the number of animals marked and carrying on until one is encountering no unmarked animals, the total number present can again be determined. (Marking methods for various sorts of animals are described in subsequent chapters.)

An alternative way of counting the population over time is to catch the animals and not release them – as might be done in a pest-control programme. Again, if the trapping is continued until no further animals are being caught, the total catch tells one what the population size was at the start of the study. Immigration is a particular problem with this method, since new animals are likely to move into the living space vacated by the trapped animals.

The problem with enumeration is the effort that is required if one is to be confident that one has observed every animal in the population. For example, if one has encountered only ten individuals, one has to carry on the observations until there have been a total of 50 separate encounters (an average of five per individual) before one can be 90% confident that there are, indeed, only ten animals present in the population. Methods that allow one to estimate the number of unobserved animals are more efficient.

Complete counts (2): plotless sampling

These methods, which have been discussed in detail by Diggle (1983), are suitable for sessile organisms that are so sparsely distributed that laying down sampling areas and counting all the individuals in each is inordinately time-consuming. Censusing the individuals of one of the less common tree species in a forest would be an example.

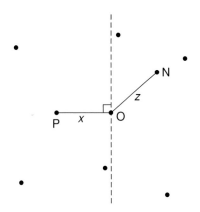

Figure 3.1   The T-square method. The black circles represent individuals of the study species. O is the nearest individual to a random point P; N is O's nearest neighbour on the opposite side of the dashed line (which is the line perpendicular to the line OP). For further explanation, see the text.

The general idea is either to pick m individuals at random and to measure the distance from each to its nearest neighbour or to go to m random points and measure the distance from each to the nearest individual of the species in question. If the individuals are randomly distributed, the density of the population can be estimated from such measurements. Unfortunately, animals and plants are rarely randomly distributed and these methods are seriously biased if the distribution tends to be either aggregated or more regular than random. The biases are, however, in opposite directions (for aggregated distributions, the nearest-neighbour method overestimates density, whereas the point-to-object method underestimates it; for regular distributions, the converse is true). This makes T-square sampling, which combines both types of measurement, useful.

Figure 3.1 illustrates T-square sampling. Choose a random point (P); measure the distance (x) from the point to the nearest individual of the species in question (O); lay down a line at right angles to the line joining O and P (the dashed line in Figure 3.1); measure the distance (z) from O to the nearest individual (N) on the opposite side of the line to P. Make a number (m) of such measurements – ten is the recommended minimum – each based on a new random point. It is important that the core area in which one places the m random points should be somewhat smaller than the total study area, to allow for the possibility that the nearest individuals to some of the points may lie outside the core area.

The estimation of density from these measurements is straightforward (Box 3.1). Although, as explained above, the method is reasonably robust in the face of non-random distributions, Box 3.1 illustrates a test of randomness, which it is prudent to apply routinely, so that potential biases are not forgotten.

The effective area sampled at each point in plotless sampling varies according to local population density. (That is why estimates are biased when individuals are non-randomly distributed.) Thus it is best to regard each of the estimates obtained as being the density in an undefined area centred on the m random individuals or points; their average is the mean density in the whole study area.

Box 3.1. **T-square sampling**

Example

Japanese black pine *Pinus thunbergii* in a core 4-m × 4-m quadrat, within a 5.7-m × 5.7-m quadrat (Numata 1961). Ten (=*m*) pairs of measurements were made (x_i = point-to-tree distances, z_i = nearest-neighbour distances, all in metres). Calculated values in the third column are used to test for randomness of the distribution.

	x_i	z_i	$x_i^2/(x_i^2 + z_i^2/2)$
	0.1	0.3	0.1818
	0.3	0.2	0.8182
	0.1	0.4	0.1111
	0.8	0.5	0.8366
	0.3	0.1	0.9474
	0.7	0.7	0.6667
	0.6	0.7	0.5950
	0.9	0.7	0.7678
	0.7	0.3	0.9159
	0.5	0.3	0.8475
Sum	5.0	4.2	6.6880

Estimation of density

$$\hat{D} = m^2/(2.828 \sum_i x_i \sum_i z_i)$$
$$= 10^2/[2.828(5.0)(4.2)] = 1.68 \text{ trees/m}^2$$

(This compares well with the value 1.625 trees/m^2 obtained from a complete census.)

Test for random distribution

Calculate

$$t' = \left\{ \sum_i \left[x_i^2 / (x_i^2 + z_i^2/2) \right] - m/2 \right\} \sqrt{12/m}$$
$$= (6.6880 - 10/2)\sqrt{12/10} = 1.85$$

If t' is greater than +2, the distribution is significantly more regular than a random distribution; if it is less than −2, it is significantly clumped. In either case, $\hat{D}$ may be biased.

This is an approximate test. If you are really interested in the possible non-randomness of the distribution of organisms, consult Diggle (1983).

Sample counts (1): mark–recapture methods

Fundamentals of mark–recapture

The basic idea

Suppose that one catches a sample of animals from a population, marks them and releases them. Suppose further that, after allowing the marked animals time to become thoroughly mixed into the rest of the population, one takes another sample. It is reasonable to assume that the proportion of marked animals in the second sample is the same as that in the population at large. This idea can be expressed symbolically as follows. If n_1 is the number of animals first marked and released, if n_2 is the size of the second sample and m_2 the number of marked animals in that sample, and if N is the total population size, then we expect that $m_2/n_2 = n_1/N$. It is obvious that, since n_1, n_2, and m_2 are known, N can be estimated. All mark–recapture methods rest on this basic idea, though most entail animals being caught and marked on several occasions.

Marking

Methods of marking various sorts of animals are described in subsequent chapters. Some individual marking methods are obvious: an individually numbered ring (band) can be applied to a bird's leg or an individual number can be written on a snail's shell. Others are less obvious. For example, one can apply a spot of colour to one of various positions on an animal's body. If there are C colours available and L different positions, this provides $(C + 1)^L - 1$ combinations ('+1' because the absence of a coloured mark in a position can be used; '−1' because an individual 'marked' with no colour in any position cannot be distinguished from an animal that has not been caught before). The values of C and L need only be modest for this number to be large: five colours and eight locations provide approximately 1.7 million combinations, enough for most purposes. However, smaller values (especially of C) may provide too few combinations for individual markings: if one has only one colour then, even with eight possible locations for placing the mark, there are only 321 combinations. If the marking possibilities are limited in this way, one may be reduced to batch-specific marking, which requires far fewer combinations – only one for each capture (excluding the last one).

Simpler marking systems are sometimes sufficient. It may be possible to apply only a single mark, so that all one can say about an animal in a catch is whether it has never been caught before or whether it has been caught at least once. If multiple marks can be applied, but always the same mark, it is possible to say how many times the animal has been captured but not when those occasions were. The value of marking, in terms of information provided, thus increases from single marks, through multiple and then batch-specific marks, to individual marks.

All methods assume that marks are not lost. This can be checked by marking animals in two different ways. If any of them have only one mark when they are subsequently recaptured, this shows that the other one has been lost. If one assumes that the chances of losing the two types of mark are independent, then the proportion of type-A marks lost between marking and recapture can be estimated from the numbers of animals recaptured with both marks (R_{AB}) and with only type B (R_B). It is $R_B/(R_B + R_{AB})$. Such estimates of the rates of loss of marks can be used to correct population estimates made by some of the methods described below. If marks are lost and no correction is applied, population size will be overestimated. The same bias will arise if marks are overlooked when animals are recaptured. This is a problem in fisheries biology, for example, where the scientist may mark a sample of fish but then rely on commercial fishing for the recapture information.

Marking is also assumed not to affect mortality or behaviour. This assumption must be checked carefully because population estimates are seriously biased if marks do affect the animals. The mortality rates of animals marked in different ways can be compared to determine whether any of the methods are particularly harmful, while direct observations of the behaviour of marked and unmarked animals can often be made. Statistical tests should also be applied to the recapture data themselves, since changes in mortality or behaviour that cannot directly be observed may show up in the pattern of recaptures.

Marking and resighting

If the marks can be observed without recapturing the animals, then a second capture is unnecessary. All one needs to do is to observe a sample of animals and count how many are marked and how many unmarked. For this method to work, the marks must be conspicuous enough that they are not overlooked but not so conspicuous that they increase the probability of the animal being taken by a predator. The resightings can be made over a period rather than on just one occasion but they should not be repeated in the same part of the study area since this will make it likely that the same animals are counted twice. If one makes counts in different parts of the study area, one can test the assumption (which is fundamental to all mark–recapture methods) that the marked animals have mixed completely with the rest of the population by testing whether the proportions resighted are the same in the different areas (Box 3.2). Observations should not extend over so long a period that animals may have entered or left the population. If the resightings made for different parts of the population do extend over a period, one may check the assumption of closure by testing for a declining trend in the proportion of marked individuals in the population (Box 3.3).

Box 3.2. **Testing whether marked animals have mixed fully with the rest of the population prior to resighting surveys**

If the marked animals have mixed fully with the others, then the proportions of marked animals will be equal in all subdivisions of the area. This can be tested using the usual contingency-table approach.

Lay out the observations of numbers of marked and unmarked animals seen during resighting surveys as in this example:

		Subdivision			
	1	2	3	4	Total
Marked	25	16	12	31	84
Unmarked	80	40	42	85	247
Total	105	56	54	116	331

Then, for every value in the table, including the marginal totals, calculate $x \log_e x$ (where x is the original number):

Transformations to $x \log_e x$				
Marked	80.472	44.361	29.819	106.454	372.189
Unmarked	350.562	147.555	156.982	377.625	1360.819
Total	488.666	225.420	215.405	551.416	1920.501

Calculate

A = sum of all the transforms in the body of the table (excluding the marginal totals)

$= 80.472 + 44.361 + \cdots + 377.625 = 1293.83$

B = sum of the transforms of the column totals

$= 488.666 + 225.420 + 215.405 + 551.416 = 1480.907$

C = sum of the transforms of the row totals

$= 372.189 + 1360.819 = 1733.008$

D = the transform of the grand total

$= 1920.51$

$G = 2(A - B - C + D)$

$= 0.834$

This is compared with χ^2 for $P = 0.05$ and $s - 1$ degrees of freedom (where s is the number of subdivisions); χ^2 is 7.8 for the example, so the departure from strict equality of the proportions of marked individuals among the animals seen in each area is far from significant. The marked animals appear to have mixed fully with the rest of the population.

Box 3.3. **Testing whether a population is not closed**

If a population is not closed, then the proportion of marked animals recaught or resighted on successive occasions will decline. One therefore tests the null hypothesis that there is such a decline.

List the proportions of marked animals in the sample observed on each occasion. Write down the rank order of each occasion, the occasion with the highest proportion scoring 1, the second highest scoring 2, etc.

As an example:

Occasion	Proportion marked	Rank
1	0.42	1
2	0.32	3
3	0.36	2
4	0.16	6.5
5	0.24	4
6	0.21	5
7	0.15	8
8	0.16	6.5

Note that, where two proportions are equal, they are given the appropriate average rank. In the example, 0.16 occurs twice, lying between the fifth and eighth ranks (0.21 and 0.15, respectively), so it is ranked 6.5. Such ties render the value of r_s approximate but this is unimportant unless its value is close to the critical value or there are many ties.

Calculate the usual (i.e. 'Pearson') correlation between the ranks and the occasion number. Because the data are ranked, this is Spearman's rank correlation coefficient (r_s), critical values of which are provided in Table 3.2.

In this case $r_s = 0.84$. This is compared with the critical value for a one-tailed test given in Table 3.2. (A one-tailed test is appropriate because we are concerned only with possible declines in proportions observed, not increases). For eight occasions, this is 0.62, so the result is significant: there is evidence that the proportions declined, suggesting that the population was not closed.

Table 3.2. *Critical values of Spearman's rank correlation coefficient*

n	One-tailed	Two-tailed	n	One-tailed	Two-tailed
5	0.900	1.000	18	0.402	0.474
6	0.829	0.886	19	0.392	0.460
7	0.715	0.786	20	0.381	0.447
8	0.620	0.715	21	0.371	0.437
9	0.600	0.700	22	0.361	0.426
10	0.564	0.649	23	0.353	0.417
11	0.537	0.619	24	0.345	0.407
12	0.504	0.588	25	0.337	0.399
13	0.484	0.561	26	0.331	0.391
14	0.464	0.539	27	0.325	0.383
15	0.447	0.522	28	0.319	0.376
16	0.430	0.503	29	0.312	0.369
17	0.415	0.488	30	0.307	0.363

These values are for a significance level of 5%; n is the number of data points (i.e. occasions, in Box 3.6). Note that, for values of $n > 30$, it is satisfactory to use critical values of Pearson's correlation coefficient, with $n - 2$ degrees of freedom.

Most mark–recapture methods involve more than two capture occasions. On the second and subsequent occasions, further marks are applied (see below). Multiple resighting occasions cannot be used as substitutes for multiple recaptures in the methods described below because they do not allow further marks to be applied. However, the program NOREMARK can be used to obtain population estimates from such data (see the appendix to this chapter).

Marking may be unnecessary

It may be possible to identify animals uniquely from their appearance or from DNA profiles, thus obviating the need for marking. In principle, this may be better than applying marks because it may interfere less with the animals (especially if it makes trapping unnecessary). Refer to the section 'Mark–recapture without capture' for details.

Open or closed populations?

Many of the methods described in this section assume the population to be 'closed'. That is, they assume that there are no gains (births or immigration) or losses (deaths or emigration) during the course of the study. If these methods are applied to 'open' populations (ones subject to gains or losses), the estimates of population size will generally be biased (usually upwards), though some (which will be identified below) are not affected by losses. If one uses one of the methods appropriate for a closed population, it is important to minimise the chance of losing or gaining individuals by conducting the study over a short period and at a time of year when births, deaths and movements are likely to be few. It may be possible to recognise new-born individuals and exclude them from the calculations. Immigration and emigration are more difficult to deal with, though it is sometimes possible to measure the rates and allow for them.

If one applies closed population models to open populations, the estimate of population size is biased. If one applies open models to closed populations, the estimate is not biased. Why do we not therefore apply open models routinely, both to open and to closed populations, and thus avoid the possibility of obtaining a biased estimate by applying a closed model to a population that we do not realise is actually open? The answer is that, because open population models make fewer assumptions (that is, they do not assume rates of gain and loss to be zero), the estimates of population size that they produce are less precise. Nonetheless, if one cannot be sure that a population is closed it is better to use an open model: an imprecise but unbiased estimate is generally preferable to a precise but biased estimate.

Other assumptions

In addition to the assumption that marks are not lost or overlooked and, where appropriate, that the population is closed, there are several other assumptions that apply to some or all of the methods used to analyse mark–recapture data. It is important to be aware of these in order to use methods that are least likely to give rise to problems.

The proportion of the population captured can vary from time to time depending both on uncontrollable factors such as weather and season and on the trapping effort (number of traps, time for which they are set, etc.). Although such variation sometimes produces little or no bias (unless the probability of capture tends systematically to increase or decrease over the period of the study), it does reduce the precision of the estimates. It is generally best, therefore, to keep trapping effort constant over all capture occasions.

All methods assume that all the animals in the population can be trapped. If some of the animals are too small to be caught in the traps, for example, one's estimate of population size will refer only to those animals that are large enough to be trapped. Similarly, animals that are aestivating or hibernating during the period over which the population is studied will not be included in the population estimate. More generally, the methods assume that all animals in the population are equally trappable. If they are not, population size will be seriously underestimated.

Differences in trappability are most likely, but fortunately most readily detectable, when there are identifiable subgroups in the population, such as different sexes or ages. In this case, either of two approaches is to be recommended. The simpler is to estimate the numbers in each sub-group separately. The alternative, which is statistically more efficient but less easy to implement, is mentioned under the multi-capture M_h method below, a method designed to cope with cases where there is heterogeneity in probability of capture, though only by making assumptions about the form that the heterogeneity takes. It is not appropriate to pool all the data and make a simple estimate of total population size unless the differences in trappability between subgroups are slight and sample sizes are small. (Subgroups should not be pooled without formally testing the differences between them, but we have not included such tests in the account below because pooling is so rarely to be recommended.)

It is important that differences in catchability are not created by the investigator. Recall that the basic idea (above) entails the assumption that the marked animals have had enough time between the marking and the recapture episodes to become thoroughly mixed into the rest of the population. If they have not, so that they are still closer to the trapping sites than are other members of the population, then they will have a greater than average probability of being captured. For animals that tend to stick to particular home ranges (even if only temporarily), this is a particular problem. It can be overcome only by having a sufficient density of traps that every animal's home range includes at least one. In cases in which the animals are caught (or observed) by mobile observers rather than by static traps, the same result is achieved by the observers covering all parts of the study area.

Trap responses are also important. Once caught, animals may become trap-shy, reluctant to enter a trap again, avoiding its location, or even leaving the area completely. Sometimes animals become trap-happy, especially if traps are baited. Both types of response may be short-lived or permanent. Trap-shyness results in overestimates of population size, trap-happiness in underestimates. The chances of trap-shyness developing can be reduced by employing methods that are not likely to distress the animals. If baiting is essential – either to attract the animals to the traps or to keep them alive after they have been caught – it is sometimes possible to set the traps open and baited, but fixed so that no animals entering them will be caught, for some time before the first catch is made. This allows all animals in the population to become equally trap-happy before catching starts. (Similarly, in multiple-recapture studies, one can exclude the data from the first few trapping occasions.)

Finally, the calculations of confidence intervals of the estimates of population size (though not those of the estimates themselves) assume that the probability of capturing an individual does not depend on whether others have been captured. This might not be true. For example, if only one animal can be caught in each trap and there are too few traps, some animals may encounter traps that are already occupied and thus fail to be caught; or animals may be repelled from or attracted to traps that are already occupied. Having a large number of traps relative to the number of animals reduces such problems.

Further reading

There is a huge and expanding literature on mark–recapture methods. Williams *et al.* (2002) provide an overview, as part of a massive presentation on the science of animal population management.

The two-sample method

This method, which is commonly associated with the names of the twentieth-century biologists F. C. Lincoln and C. J. G. Petersen, was first used in the eighteenth century by the mathematician Laplace to estimate the human population of France. It is the most basic method, involving just one session of catching and marking and one recapture session, and is appropriate only for closed populations. If there are losses from the population, the estimate obtained is for the size of the population at the time of the first catch; if there are gains, the estimate corresponds to the population size during the second catch; if there is turnover (gains and losses) the estimate is biased. Trap responses and heterogeneity in catchability among animals also cause the usual biases. However, if the differences in catchability among animals are not correlated between the two occasions, there is no problem. This is the basis of using the method in some studies of fisheries or hunted populations: the animals are first caught and marked by biologists; the second sample is the fisheries' catch or the hunters' bag; both the biologists' catching methods and the fishing or hunting are likely to be biased towards catching certain animals but the biases may be independent; if they are, the resultant estimate of population size will not be biased by the differences in catchability. This method may also overcome the problem of behavioural responses – even if being caught by the biologists makes the animals avoid (or seek out) the biologists' traps, this should not affect the likelihood of their subsequently being caught by fishermen or killed by hunters.

Differences in the proportions of the population caught on the two occasions have no effect on the estimate (though it is generally most cost-effective to make the same catching effort on both occasions). Capture mortality causes no statistical bias: animals found dead on the second occasion are included in the calculation; those found dead on the first occasion are left out but are usually added to the estimate made to provide an estimate of the total population before trapping commenced.

It is not possible to use the mark–recapture data themselves to check the assumptions of the two-sample method, so one must exercise particular caution when using it. If, on independent grounds, the estimate seems unlikely, it is probably wrong. The method is best regarded as a quick way to get a rough answer when there is no practical way to get a better one.

The method of estimating population size, with approximate confidence limits, is shown in Box 3.4. Note that, to eliminate bias arising for statistical reasons, the estimate of population size is not simply the common-sense value, $n_1 n_2/m_2$.

Box 3.4. **The two-sample method**

Basics

n_1 = number marked and released on the first occasion
n_2 = total number caught on the second occasion
m_2 = number of marked animals found on the second occasion

The total population size (N) is estimated as

$$\hat{N} = (n_1 + 1)(n_2 + 1)/(m_2 + 1) - 1$$

Planning

For this method, the percentage relative precision (Box 2.1) is given by

$$Q = 200\sqrt{N/(n_1 n_2)}$$

and the required sample size for a given precision is

$$n = 200\sqrt{N}/Q \qquad (\text{if } n_1 = n_2 = n)$$

Thus, the PRP attainable if one believes that the population size is about 40 000 and one is able to take samples of roughly 1500 is

$$Q = 200\sqrt{40\,000/1500^2} = 27\%$$

If one wishes to obtain a PRP of 10%, then the required sample sizes are

$$n = 200\sqrt{40\,000}/10 = 4000 \qquad (\text{for each sample})$$

Example

Climbing cutworms (larvae of various species of noctuid moths) in a field of blueberries *Vaccinium* in New Brunswick, Canada (Wood 1963).

$$n_1 = 1000, \ n_2 = 1755, \ m_2 = 41$$

Estimate of total population size

$$\hat{N} = [(1001)(1756)/42] - 1 = 41\,851 - 1 = 41\,850$$

Approximate confidence limits of the estimate

Calculate

$$P = m_2/n_2 = 0.0234$$

Calculate the two values

$$W_1, W_2 = p \pm \left[1/(2n_2) + 1.96\sqrt{p(1-p)(1-m_2/n_1)/(n_2-1)} \right]$$

$$= 0.0234 \pm \left[1/3510 + 1.96\sqrt{(0.023)(0.977)(0.959)/1754} \right]$$

$$= 0.0234 \pm 0.0072$$

$$= 0.031 \text{ and } 0.016$$

Divide W_1 and W_2 into n_1 to obtain approximate 95% confidence limits for $\hat{N}$. In the example,

$$\text{LCL} = 1000/0.031 = 33\,000 \quad \text{and} \quad \text{UCL} = 1000/0.016 = 62\,000$$

If the number of marked animals found in the second sample (m_2) is less than eight then the estimate of N is biased.

Multiple recaptures in closed populations

Various models

Several methods have been developed for obtaining population estimates from data obtained over a series of capture occasions. They differ in the ways in which the assumptions underlying mark–recapture methods are relaxed and can thus be said to correspond to a series of underlying models:

M_b	allows for behavioural responses to being trapped (trap-shyness or trap-happiness);
M_t	allows temporal differences (between capture occasions) in the probability of capture;
M_h	allows heterogeneity in catchability among individuals;
M_o	allows none of these three but, in practice, methods based on M_o have no advantage over those based on M_t;
M_{bt}, M_{bh}, and M_{th}	allow the various combinations of departures from these three assumptions; and
M_{bth}	allows all three to be breached.

CAPTURE and other programs allow one to take a coherent approach to analysing data from closed populations, by applying various likely models and comparing the results in order to discover which appears to be a better fit to the data (see the appendix to this chapter).

Here, we introduce methods that can be used for the three simple cases, M_b, M_t and M_h.

General goodness-of-fit tests for mark–recapture data

Wherever one uses a model to analyse the data, it is important to consider how closely the model fits the data. If the fit is poor, this indicates that at least one of the assumptions on which the model is based is wrong and that one should apply another model to the data. If no other model is available, then one may use the estimates from the poorly fitting model but only with due caution.

CAPTURE and other programs provide goodness-of-fit tests for the various models that can be fitted to mark–recapture data. They test not only the general assumptions but also the estimated values of the various parameters of the models.

Here we provide three simple tests that can be applied before the analysis, to test the general assumptions. Two of our tests involve calculating what the expected values of the data would be if particular assumptions were met, so that one has a set of observed values and a set of corresponding expected values, which one compares by calculating a value of the G statistic. This is then compared with the theoretical value of χ^2 for a significance level (probability) of 0.05 and the appropriate number of degrees of freedom. If the G value is greater than the critical value of χ^2, this shows that the differences between the observed and expected values are unlikely to have arisen just by chance, so we must conclude that the assumptions on which the expectations are based are wrong.

The first general test is whether the number of animals caught has been the same (within the limits of chance variation) on all occasions (Box 3.5). If the result is significant, it indicates either that the catchability (or trapping effort) was not the same on all occasions or that the behaviour of the animals changed in response to being caught. Such a behavioural response would produce a trend in total captures, with an increase (from trap-happiness) or a decline (from trap-shyness) – as would a temporal trend in catchability (which could result from animals becoming more or less active as the season advanced, from the biologist becoming more effective at catching them, etc.). Estimating the rank correlation between the numbers caught and the occasion provides a test for such a trend (Box 3.6). If there is actually such a trend, Box 3.6 provides a more sensitive test of the assumptions of no temporal changes in catchability and of no behavioural response than does Box 3.5. The example used in Boxes 3.5 and 3.6 illustrates this greater sensitivity, with a clearly significant trend even though the result from the test of equal numbers is not quite significant.

If animals have been given multiple, batch-specific, or individual marks, then a third test is available: do the numbers caught 1, 2, 3 . . . times fit what one would expect if all the assumptions were true – no behavioural response to trapping, no temporal variation in catchability, no heterogeneity in probability of being caught among animals? Box 3.7 shows this test. Note that, because the method is approximate, it is not suitable when more than about one-third of the population is caught on each occasion. It is generally a more sensitive test than those based on the numbers of total captures (because it uses more information).

Box 3.5. **Tests of the equality of total numbers caught on each occasion**

Occasion	Observed	Expected	$O \log_e(O/E)$
1	29	20.143	10.569
2	24	20.143	4.205
3	23	20.143	3.051
4	19	20.143	−1.110
5	21	20.143	0.875
6	10	20.143	−7.003
7	15	20.143	−4.422
Sum	141	141	6.165

The expected numbers are simply the average of the numbers actually caught. Note, as a check, that the sums of observed and expected must be equal.

Calculate values for the last column, with O the observed number and E the expected number.

Calculate G, which is twice the sum of the last column, 12.3 in this case.

This is to be compared with the critical value of for χ^2 for $k - 1$ degrees of freedom, where k is the number of occasions. For $k = 7$ and $P = 0.05$, this is 12.6, so the result is not quite significant.

Box 3.6. **Test of trend in numbers caught per day**

This uses the same data as in Box 3.5.

Put the numbers caught per day in rank order:

Occasion	Numbers (ranked)
1	1
2	2
3	3
4	5
5	4
6	7
7	6

Proceed, as in Box 3.3, to calculate Spearman's rank correlation and compare it with the critical value in Table 3.2. Note, however, that this is a two-tailed test, since the trend could be in either direction.

In this case, $r_s = 0.93$ and the critical value for seven occasions and $P = 0.05$ is 0.79, so the trend is clearly significant.

Note that, if one is comparing a negative r_s with the critical value, one would ignore the negative sign: that would simply indicate an increasing rather than a declining trend in catch size.

Box 3.7. **Test of goodness of fit to a zero-truncated Poisson distribution**

Basics

One must first estimate the mean of the underlying Poisson distribution ($\bar{x}$) from the observed mean number of times that each animal has been caught ($\bar{x}_T$). For observed means less than 2.0, use Table 3.3; for larger values estimate $\bar{x}$ as

$$\bar{x}_T - Z - Z^2 - 1.5Z^3 - 2.6Z^4 - 5.2Z^5, \quad \text{where } Z = x_T \exp(-\bar{x}_T)$$

As an example of this calculation, suppose that $\bar{x}_T = 2.540$. Then

$$Z = 2.54 \exp(-2.54) = 0.200$$
$$\bar{x} = 2.54 - 0.2 - (0.2)^2 - 1.5(0.2)^3 - 2.6(0.2)^4 - 5.2(0.2)^5 = 2.282$$

Having estimated $\bar{x}$ from the table or by calculation, the expected number of animals caught exactly i times is calculated from the equation for the Poisson distribution; the observed and expected numbers are used to calculate a G value in the usual way.

Table 3.3. *The relationship between the mean of a Poisson distribution ($\bar{x}$, in the body of the table) and the mean of the equivalent zero-truncated Poisson distribution ($\bar{x}_T$), for values of $\bar{x}_T = 1.01–1.99$. (for larger values, use the method of calculation shown above)*

$\bar{x}_T$	0.00	0.01	0.02	0.03	0.04	0.05	0.06	0.07	0.08	0.09
1.0	–	0.020	0.040	0.060	0.079	0.099	0.118	0.137	0.156	0.175
1.1	0.194	0.213	0.232	0.250	0.268	0.287	0.305	0.323	0.341	0.359
1.2	0.377	0.395	0.412	0.430	0.447	0.465	0.482	0.499	0.516	0.533
1.3	0.550	0.567	0.584	0.601	0.617	0.634	0.650	0.667	0.683	0.700
1.4	0.715	0.732	0.748	0.764	0.780	0.795	0.812	0.828	0.843	0.859
1.5	0.875	0.890	0.906	0.921	0.936	0.952	0.967	0.982	0.997	1.012
1.6	1.027	1.042	1.057	1.072	1.087	1.102	1.117	1.131	1.146	1.161
1.7	1.175	1.190	1.204	1.219	1.233	1.247	1.262	1.276	1.290	1.304
1.8	1.318	1.333	1.347	1.361	1.375	1.389	1.403	1.417	1.430	1.444
1.9	1.458	1.472	1.485	1.499	1.513	1.526	1.540	1.553	1.567	1.580

To find $\bar{x}$ corresponding to $\bar{x}_T = 1.23$ (for example), enter the table on the row marked 1.2 and the column marked 0.03 (which gives $\bar{x} = 0.430$).

Example

Agamid lizards *Amphibolurus barbatus* studied over 21 sampling occasions in Australia by J. A. Badham (Caughley 1977); f_k is the number of animals caught exactly k times:

k	1	2	3	4	5	6	7	8	9	10–21	Sum
f_k	23	7	3	2	1	0	1	0	1	0	38
$f_k k$	23	14	9	8	5	0	7	0	9	0	75

The observed mean is,

$$\bar{x}_T = \sum_k f_k k / \sum_k f_k = 75/38 = 1.97$$

The corresponding estimated Poisson mean ($\bar{x}$) from Table 3.3 is 1.553.

The expected number of animals caught only once is calculated from a formula and successive expectations from a simple recursive relationship:

Cumulative sum

$$E_1 = \frac{(\sum_k f_k)\exp(-x)}{1 - \exp(-x)}\bar{x} = \frac{38\exp(-1.553)}{1 - \exp(-1.553)}1.553 = 15.840 \quad 15.840$$

$$E_2 = E_1\bar{x}/2 = 15.840 \times 1.553/2 = 12.300 \qquad 28.140$$

$$E_3 = E_2\bar{x}/3 = 12.300 \times 1.553/3 = 6.367 \qquad 34.507$$

etc.

The tabulation of the cumulative sum as one works out successive values allows one to work out where to stop, which is the point at which the cumulative sum is within 5 of the total number of animals captured ($\sum_k f_k$). In this example, it occurs at $k = 3$.

The expected values are then compared with the observed, calculating G in the usual way. Note that the last expected value (9.86 in this case) corresponds to the stopping point in the preceding calculation and all greater k values. It is obtained by subtracting the sum of the other expected values from the overall sum:

	1	2	3 or more	Sum
$O_k (=f_k)$	23	7	8	38
E_k	15.84	12.30	9.86	38

As usual, calculate G as

$$2\sum_i O_i \log_e(O_i/E_i)$$

In this case

$$G = 2[23\log_e(23/15.84) + 7\log_e(7/12.30) + 8\log_e(8/9.86)] = 5.92$$

Test this against χ^2 for $P = 0.05$ and with $C - 2$ degrees of freedom (*not C* − 1), where C is the number of pairs of O and E values in the table. In this case, the critical χ^2 is 3.84, so the result is significant, indicating that the frequency of captures does not fit a zero-truncated Poisson distribution.

Caution is needed with any of these tests if the value of any of the E_i is small. Traditionally, 'small' in this context has been taken to be less than 5, but one is probably safe enough if no E_i is less than 2. (See Zar (1999, p. 470) for further detail.) If any of the E_i is too small, the data in adjacent columns may be combined: for example, the cells for all values of k greater than 2 have been combined in Box 3.7, to avoid having E_i values less than 5. This reduces the bias in the resultant G value, though at the cost of reducing the sensitivity of the test. Note that, if any $O_i = 0$, then the corresponding value of $O_i \log_e(O_i/E_i)$ is also zero.

None of these tests is particularly sensitive. For this reason, non-significant results in these tests should not be taken to mean that the mark–recapture assumptions have not been violated; the tests assist the biologist's judgement but should not be substituted for it. If the result of any of these tests is significant, at least one of the assumptions is probably violated in the data that one wishes to analyse.

The pseudo-removal method

This corresponds to model M_b; that is, it is not affected by a behavioural response to being trapped. This is because it uses data only about initial captures (which means that its precision is low).

If one traps a closed population over a period and marks the trapped animals, then the number of unmarked animals in successive catches will decline as there are fewer and fewer animals that have never been caught. If the numbers of unmarked animals caught in each trapping session are plotted on a graph, relative to the total numbers already marked in previous trapping sessions, the decline is clear (Figure 3.2). By fitting a line to the data and extrapolating it until it cuts the horizontal axis, one obtains an estimate of the total population (the arrow in Figure 3.2); at this point, one's catches would yield no more unmarked animals, however long they continued (so long as the population was closed).

We shall call this the pseudo-removal method because it is mathematically (though not biologically) equivalent to the removal method discussed later, in which animals are removed rather than being marked. Trap deaths thus present no problems for this method: if the dead animal is marked, it is of no interest; if it is unmarked, then it is counted in that day's total of unmarked animals and added to the cumulative total for the future. Trap deaths are a problem for the methods based on the models M_t and M_h: if deaths are few, the animals can be excluded from the calculations and then added to the calculated estimate of population size, as in the two-sample method; if deaths are more numerous, they prevent those other methods being used, so the pseudo-removal method or one of the models M_{nh} and M_{bht} must be applied.

Figure 3.2 refers to a study in which the same number of traps was used every day. The trapping effort can be varied from day to day: one plots catch per unit effort on the vertical axis. It is important that trapping effort is measured correctly. For example, doubling the number of

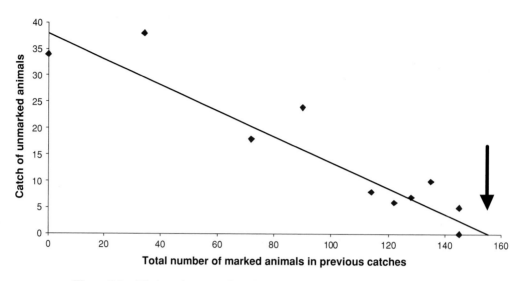

Figure 3.2   The pseudo-removal method. The line has been fitted by least-squares regression. The intersection of the fitted line with the axis for which the catch of unmarked animals is zero (marked by the arrow) is the estimate of total population size.

traps will not double true trapping effort if neighbouring traps are now so close together that they interfere with each other; or, if there are few traps relative to the number of animals and if each trap can catch just one animal at once, then the number caught may be limited by the number of traps.

Box 3.8 shows how an estimate of population size may be calculated, with confidence limits. A more refined approach is possible using CAPTURE, though this does not allow adjustment for trapping effort.

As can be seen from the example in Box 3.8, the estimates are rather imprecise even when a substantial proportion of the population has been caught. If only a small proportion of the population is caught, it can even happen that there is no decline in numbers caught over successive sampling occasions. One must then conclude that the population is large compared with the number trapped – but one cannot make any useful estimate of how large from such data. Table 3.4 shows the PRP attainable in relation to the proportion of the population trapped in total (over all trapping occasions) and to population size.

Uncontrolled variation in catchability causes problems for this method. At the very least, it increases the scatter about the line on the graph. A systematic change in catchability is worse. If catchability steadily increases, the points will lie on a right-convex curve and the fitted straight line will overestimate the population size (Figure 3.3(a)). If it steadily decreases, the converse pattern will emerge (Figure 3.3(b)). The latter pattern, with a tail of captures extending out to the right, is also seen if there is heterogeneity in catchability among animals or if new animals are entering the population. To prevent such patterns being overlooked, a graph of the data and the fitted line should always be examined and a goodness-of-fit test applied (Box 3.9). Formal tests to

Box 3.8. **The pseudo-removal and removal methods**

Basics

S = number of trapping occasions
y_i = catch on the ith occasion
x_i = total number caught before the ith occasion

Calculate

$\bar{y}$ = mean of y
$\bar{x}$ = mean of x
v_x = variance of x

For the data used in Figure 3.2,

i	1	2	3	4	5	6	7	8	9	10	Mean	Variance
y_i	34	38	18	24	8	6	7	10	0	5	15.0000	
x_i	0	34	72	90	114	122	128	135	145	145	98.5000	2439.6111

Regression analysis

Carry out a linear regression analysis in the usual way to obtain the following parameter estimates for the fitted line:

a: the intercept on the y axis (at $x = 0$);
b: the slope of the line; and
V: the variance about the regression line, also called the residual mean square in the analysis of variance of the regression.

For the example data,

$a = 39.2790, \quad b = -0.246\,487, \quad V = 23.7513$

Note that b should have a negative sign; if it does not, the graph slopes upwards and no estimate of population size is possible from the data.

The *catchability* (K) is the proportion of the population caught on each occasion, if all the assumptions of the method hold. It is estimated by taking the value of b without the minus sign.

Estimate of total population size

$\hat{N} = \bar{x} + \bar{y}/K$

$= 98.5 + 15.0/0.246\,487 = 159.355 \approx 159$

Confidence limits of the estimate

Look up Student's t for $S-2$ degrees of freedom and $P=0.025$ (2.751 53 in this example).
Calculate

$$A = k^2 - Vt^2/[(S-1)v_x]$$

$$= 0.246\,487^2 - (23.7513 \times 2.75153^2)/(9 \times 2439.6111)$$

$$= 0.052\,566\,2$$

$$B = 2\bar{y}K$$

$$= 2 \times 15 \times 0.246\,487$$

$$= 7.394\,62$$

$$C = \bar{y}^2 - Vt^2/S$$

$$= 15^2 - (23.7513 \times 2.751\,53^2)/10$$

$$= 207.020$$

$$W_1, W_2 = \left(B \pm \sqrt{B^2 - 4AC}\right)\bigg/(2A)$$

$$= \left[7.394\,62 \pm \sqrt{7.394\,62^2 - 4(0.052\,566\,2)(207.020)}\right]\bigg/2(0.052\,566\,2)$$

$$= 38.572 \text{ and } 102.100$$

The 95% confidence limits of $\hat{N}$ are

$$\bar{x} + W_1 = 137.072 \approx 137 \quad \text{and} \quad \bar{x} + W_2 = 200.600 \approx 201$$

Note that, if the data points fall for from a straight line, the calculation will provide unreal confidence limits.

Catch per unit effort

If effort varies from day to day, divide numbers caught by a measure of effort for the day, to give the catch per unit effort. Use those figures as y_i in the above calculations. Note that x_i should still be the actual numbers previously caught.

Reference

See Seber (1982).

distinguish between merely haphazard variation about the line (as in Figure 3.2 and Box 3.8) and curvature are rarely sensitive unless there are more than ten sampling occasions, but examination of the graphs can reveal a lot. If captures of new animals decline rapidly (as in Figure 3.3(a)), all one needs to do is to carry on trapping until no new animals have been caught on several successive occasions. It is then likely that all the animals in the population have been caught (unless some are uncatchable), so the total number caught can be used as a precise population estimate, with the proviso that it may be a slight underestimate.

Table 3.4. *The percentage of the population that needs to be trapped to achieve a given percentage relative precision (PRP: Box 2.1) in the pseudo-removal and removal methods, in relation to population size (N) (derived from Zippin (1956))*

	Percentage relative precision			
N	60%	40%	20%	10%
200	55	60	75	90
300	50	60	75	85
500	45	55	70	80
1 000	40	45	60	75
10 000	20	25	35	50
100 000	10	15	20	30

Percentages are rounded to the nearest 5.

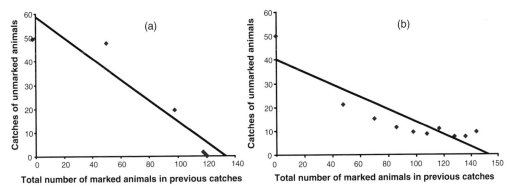

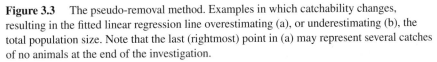

Figure 3.3 The pseudo-removal method. Examples in which catchability changes, resulting in the fitted linear regression line overestimating (a), or underestimating (b), the total population size. Note that the last (rightmost) point in (a) may represent several catches of no animals at the end of the investigation.

A right-concave pattern (Figure 3.3(b)) indicates that the estimate derived from the straight line is probably an underestimate; indeed, if any data points lie to the right of the intercept, the latter is certainly an underestimate because it is less than the total number of animals caught! In these circumstances, there are three possible approaches: find a better way to estimate the population, come to a conclusion such as 'we caught 157 animals and the shape of the graph indicates that there could be 10–40 more in the population', or give up. The second approach might not seem very respectable scientifically, but in real life it may be better than the third.

The commonest cause of a continuing tail of captures is probably that the population is not closed. In this case, the estimate from the straight line may provide a rough indication of the number present in the population at any one time, though it will always be an overestimate.

Box 3.9. **A goodness-of-fit test for the fit of a straight line to pseudo-removal or removal data**

Using the same symbols as in Box 3.8, expected values are calculated from

$$E_i = K(\hat{N} - x_i)$$

For example, for the data in Box 3.8,

$$E_4 = 0.246\,487(159.355 - 90) = 17.0951$$

Note that one uses the exact estimate of $\hat{N}$, not the value rounded to a whole number. The value of G is calculated in the normal way. It has $S - 2$ degrees of freedom.

Catch per unit effort

If one is using the catch per unit effort, the expected values are calculated from
$$E_i = Kf_i(\hat{N} - x_i), \qquad \text{where } f_i \text{ is the effort on the } i\text{th day}$$

The Schnabel method

This (see Box 3.10) corresponds to the model M_t: that is, it is not affected by temporal variation in catchability of the animals. It depends simply on observing how the proportion of marked animals in catches increases as more animals have been marked; when this proportion equals 1.0, the total number of animals previously marked must be the number in the population (Figure 3.4). Like the pseudo-removal method, it requires only that animals are marked on their first capture. Note that, for a given total effort, more precise estimates are obtained if the effort is the same throughout the study than if it varies. The method uses more information than the pseudo-removal method but its precision is also low because it allows for trappability being different on each occasion.

As with the pseudo-removal method, it is important to plot a graph such as Figure 3.4 because if the points do not fall on a straight line then the estimate of population size is likely to be biased. Curvature may be caused by a behavioural response to being trapped, by heterogeneity in trappability among individuals, or by the population not being closed. The appropriate line is that drawn from the origin to the point at which $M_i = \hat{N}$ and the proportion of animals marked is 1.0. A formal test of the model assumptions may be made by assessing the goodness of fit of the data to the regression line (Box 3.11) and by applying the test in Box 3.7.

If the result of either test is significant, indicating that the model assumptions are violated, one needs first to consider what the violations might be. As with the pseudo-removal method, if the points are simply rather widely scattered about the line rather than falling on a curve (which suggests variation in catchability between trapping sessions but neither a trend nor heterogeneity in catchability), then the population estimate will be satisfactory; if the points tend to fall on a

Box 3.10. **The Schnabel method**

Planning

For this method, provided that the size of the S samples is fairly uniform (average $= \bar{n}$), the percentage relative precision (Box 2.3) for a population of size N is given by

$$Q = 200\sqrt{N/(S\bar{n}^2)}$$

The required average sample size to achieve a given precision with a fixed number of samples is

$$\bar{n} = 200\sqrt{(N/S)}/Q$$

and the number of samples required for a given precision and sample size is

$$S = N[200/(\bar{n}Q)]^2$$

Thus, the PRP attainable if one believes the population size to be about 100 and achievable sample sizes about 50 and if one plans to take five samples is

$$Q = 200\sqrt{100/(5 \times 50^2)} = 18\%$$

If one wished to obtain a PRP of 10% with the same number of samples, then their average size would need to be

$$\bar{n} = 200\sqrt{(100/5)}/10 = 89$$

If one wished to obtain a PRP of 10% with sample sizes of about 50, then the number of samples required would be

$$S = 100[200/(50 \times 10)]^2 = 16$$

Basics

S = number of samples
n_i = number of animals in the ith sample
m_i = number of animals in the ith sample that are already carrying marks
$u_i = n_i - m_i$ = number of unmarked animals in the ith sample
$M_i = \sum\limits_{j=1}^{i-1} u_j$ = number of animals marked prior to the ith sample

Preliminary calculations:

$$A = \sum_i n_i M_i^2, \quad B = \sum_i m_i M_i, \quad C = \sum_i m_i^2 \big/ n_i$$

Example

The cricket-frog data presented in Figure 3.4:

i	n_1	m_1	u_1	M_1	$n_iM_i^2$	m_iM_i	m_i^2/n_1
1	32	0	32	0	0	0	0
2	54	18	36	32	55 296	576	6.0000
3	37	31	6	68	171 088	2108	25.9730
4	60	47	13	74	328 560	3478	36.8167
5	41	36	5	87	310 329	3132	31.6098
					$A = 865\ 273$	$B = 9294$	$C = 100.3994$

Estimate of total population size

$$\hat{N} = A/B = 865\ 273/9294 = 93.10 \approx 93$$

Confidence limits of the estimate

95% confidence limits are given by

$$A \Big/ \left[B \pm t\sqrt{(AC - B^2)/(S-2)} \right]$$

Where t is Student's t for $S-2$ degrees of freedom for a significance level of $P = 0.05$. Confidence limits for the cricket-frog estimate are

$$865\ 273 \Big/ \left[9294 \pm 3.182\sqrt{865\ 273 \times 100.3994 - 9294^2)/3} \right]$$

$$= 865\ 273/(9294 \pm 1291.82)$$

$$\approx 82 \text{ and } 108$$

curve, it will not. In this case, if one suspects that the population is open rather than closed, then one should switch to the Jolly–Seber method, but if one is convinced that the population is closed, then the pseudo-removal or Burnham and Overton methods should be used, as appropriate.

The Burnham and Overton method

This method allows animals to differ in probability of capture (the M_h model); the assumption that they do not is perhaps the commonest cause of bias in capture–recapture studies that use other methods. The information used to estimate population size in this case is the number of animals caught exactly one, two, three and four times over the entire study, which should comprise more than four sample periods. This requires that animals are individually marked or that they

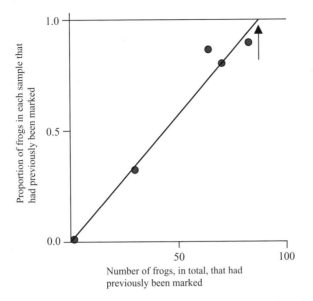

Figure 3.4   The Schnabel method applied to a population of cricket frogs *Acris gryllus* in Louisiana, USA, sampled over five successive days. The arrow indicates the estimated population size. (Data from Turner (1960).)

are marked each time they are captured (not necessarily with batch-specific marks). The critical assumption in this method is that capture probabilities do not change over time, so it is important to maintain constant trapping effort and to adopt procedures that minimise the likelihood of trap responses. The assumption of constant capture probabilities should always be tested (Box 3.5). However, simulations indicate that the method may be fairly robust in the face of violations of these assumptions, so a significant result in such a test need not be an absolute bar to using the method.

A fundamental problem with this method is that it has to make assumptions about the form that the heterogeneity of trappability takes, and different assumptions can lead to quite distinct estimates of population size (Link 2004). Furthermore, no method can allow for animals that are effectively untrappable (because, for example, they are aestivating or guarding nests). Even where the heterogeneity is less extreme, the method will perform better if one can identify characteristics, such as sex or age, that are likely to influence trappability. One can then either, as suggested above, make separate population estimates for different categories or use these characteristics as covariates in more complex models. MARK allows such an approach (see the appendix to this chapter).

Box 3.12 presents the calculations, which involve making four different estimates of population size and then testing successive values to choose the best (defined as the one which minimises the mean-square error – that is, the variance plus the square of the bias). CAPTURE conducts more refined analyses.

Box 3.11. **A test of goodness of fit to the Schnabel model**

The expected values of m_i and u_i are calculated from

$$E(m_i) = M_i n_i / \hat{N} \quad \text{and} \quad E(u_i) = n_i - E(m_i)$$

Note that one should use the precise estimate of $\hat{N}$ rather than the value rounded to a whole number.

The observed and expected values should be tabulated against each other, leaving out the first sample. In the cricket-frog example (Box 3.10)

	m_i Values				u_i Values			
Sample	2	3	4	5	2	3	4	5
Observed values	18	31	47	36	36	6	13	5
Expected values	18.56	27.02	47.69	38.31	35.44	9.98	12.31	2.69

G is then calculated in the usual way. In this case,

$$G = 2[18 \log_e(18/18.56) + \cdots + 5 \log_e(5/2.69)] = 4.20$$

G should be compared with the tabulated value of χ^2, with $S-2$ degrees of freedom. In this case, $S = 5$ and the tabulated χ^2 for three degrees of freedom and a significance level of 0.05 is 7.8. Thus the value of 4.20 is not significant, indicating that the regression line is a satisfactory fit to the data.

Box 3.12. **The Burnham and Overton method**

Basics

S = number of samples
f_k = number of animals caught exactly k times

Example

Cottontail rabbits *Sylvilagus floridanus* in an experimental enclosure in the USA (Edwards & Eberhardt 1967), sampled on 18 occasions:

k	1	2	3	4	5	6	7	8–18	Sum (T)
f_k	43	16	8	6	0	2	11	0	76

Estimate of total population

Calculate α coefficients from

$$\alpha_{11} = (S-1)/S$$
$$\alpha_{12} = (2S-3)/S$$
$$\alpha_{13} = (3S-6)/S$$
$$\alpha_{14} = (4S-10)/S$$
$$\alpha_{22} = -(S-2)^2/[S(S-1)]$$
$$\alpha_{23} = -(3S^2 - 15S + 19)/[S(S-1)]$$
$$\alpha_{24} = -(6S^2 - 36S + 55)/[S(S-1)]$$
$$\alpha_{33} = (S-3)^3/[S(S-1)(S-2)]$$
$$\alpha_{34} = (4S^3 - 42S^2 + 148S - 175)/[S(S-1)(S-2)]$$
$$\alpha_{44} = -(S-4)^4/[S(S-1)(S-2)(S-3)]$$

In this example

$\alpha_{11} = 0.9444$

$\alpha_{12} = 1.8333 \qquad \alpha_{22} = -0.8366$

$\alpha_{13} = 2.6667 \qquad \alpha_{23} = -2.3562 \qquad \alpha_{33} = 0.6893$

$\alpha_{14} = 3.4444 \qquad \alpha_{24} = -4.4150 \qquad \alpha_{34} = 2.4937 \qquad \alpha_{44} = -0.5231$

Calculate β values as the difference between each α value and the one above it in the display ($\beta_{jk} = \alpha_{jk} - \alpha_{(j-1)k}$). Note that β_{11} is not required, although all other values of β_{jk} for which $j = k$ are required; they are taken to be the same as the corresponding α values. In this example

$\beta_{12} = 0.8889 \qquad \beta_{22} = -0.8366$

$\beta_{13} = 0.8333 \qquad \beta_{23} = -1.5196 \qquad \beta_{33} = 0.6893$

$\beta_{14} = 0.7778 \qquad \beta_{24} = -2.0588 \qquad \beta_{34} = 1.8044 \qquad \beta_{44} = -0.5231$

Note that, if Excel® is used for these calculations, the six β values for which $j \neq k$ can be calculated using the same cell formula. Similar easings of the calculations apply in much of the rest of this box.

Calculate a table of values of $\alpha_{jk}f_k$. Calculate the sum of each row. The row sum plus the sum of the total number of animals caught provides an estimate of N. (In our example, $\alpha_{11}f_1 = 0.9444 \times 43 = 40.611$; this is the only value in the first row, so $\hat{N}_1 = 116.611$.)

Finally, calculate the difference between each N and the one above it ($D_j = \hat{N}_j - \hat{N}_{j-1}$). In our example, the results of these calculations are

		k					
j	1	2	3	4	Sum	$\hat{N}_j$	D_j
1	40.611				40.611	116.611	
2	78.833	−13.386			65.448	141.448	24.837
3	114.667	−37.699	5.515		82.482	158.482	17.034
4	148.111	−70.640	19.949	−3.139	94.281	170.281	11.799

To decide which of the estimates is the best, begin by calculating a table of values of $(\beta_{jk})^2 f_k$.

Calculate the sum of each row (B_j) and the three values of $t_j = D_j \sqrt{(T-1)/(B_j - D_j^2)}$:

		k				
j	1	2	3	4	B_j	t_j
2	33.975	11.198			45.174	4.05
3	29.861	36.947	3.801		70.610	2.07
4	26.012	67.820	26.045	1.642	121.519	1.07

Inspect the list of t_j values to find the first one for which $t_j < 1.96$. A value less than 1.96 means that the corresponding N_j is not significantly better than the one above, so the one above is taken to be the best estimate of the population. In the example, t_4 is the first t_j less than 1.96, so $\hat{N}_3$ is taken as the best estimate of N.

If none of the t_j is less than 1.96, $\hat{N}_4$ is taken to be the best estimate of N.

Approximate confidence limits of the estimate

Approximate 95% confidence limits of the estimate $\hat{N}_j$ are given by

$$\hat{N}_j \pm 1.96\sqrt{\sum_k [f_k(\alpha_{jk} + 1)^2] - \hat{N}_j}, \qquad \text{summation being over all } k$$

In this case, for which $\hat{N}_3$ is taken as the best estimate, these are

$$158.5 \pm 1.96\sqrt{43(3.667)^2 + 16(1.3562)^2 + 8(1.6893)^2 - 158.5}$$

$$= 158.5 \pm 42.6$$

$$\approx 116 \text{ and } 201$$

References

See Burnham & Overton (1978, 1979).

This method appears to need particularly high capture rates to achieve precise estimates: the wide confidence limits obtained in Box 3.12 are a consequence of only about 10% of the population being caught each day. Unfortunately, there is no general equation for calculating the percentage relative precision for this method.

Choosing among methods

How does one choose among the various methods, particularly when using programs such as CAPTURE, which allow the more complex models to be fitted? The most important criterion is one's knowledge of the animal being studied, for this tells one which of the various assumptions are likely not to apply. It is, of course, often the case that none of them applies – one knows that some members of the population are particularly difficult to trap and that the capture rate varied and one suspects that the animals learn to avoid the traps. It is then a matter of considering which of the assumptions is likely to be least violated in one's data.

A pilot study has the advantage of giving one insights about the assumptions. It also reveals practical problems. Also, it allows the percentage relative precision to be estimated, so that one can choose a precise method – or abandon the study if it appears that the required precision is unobtainable given the resources available.

Once some results have been obtained, so that one has an idea of the size of the population and the proportion being caught, it is possible to build simulation models. Modern computing power makes it easy to run models many times and to use the various methods to estimate the 'known' population; this allows one to discover how well the various methods perform. One can make judgements about the likely magnitude of heterogeneity, temporal variation in catchability and behavioural responses, to test the robustness of the methods in the face of violations of the assumptions. Simulations can be constructed in Excel® and are available in WISP.

One might naïvely think that the more complex models (M_{bt} etc.) are better than simpler ones (M_b, M_t and M_h) because they make fewer assumptions. It is true that they are less likely to be biased but they are also likely to deliver less precise estimates. This is because making fewer assumptions means estimating more parameters: for example, model M_b requires three parameters to be estimated (N and the different probabilities of capture of previously caught and naïve animals); M_t requires $K + 1$ (N and the different probabilities of capture on the K occasions); and M_{bt} requires, at the minimum, $K + 3$. The problem is that more parameters means less precision. Thus, one should choose the method that is simplest and most precise (the latter being discovered empirically, by applying various models and calculating confidence limits for each) – always provided that it seems biologically reasonable.

More refined statistical methods than we have presented here allow various formal comparisons of the models (using likelihood ratios, Akaike's Information Criterion, or discriminant functions). However, simulations using model populations show that these may often choose the wrong model, so these methods, objective as they are, should not be allowed to over-rule the investigator's judgement. (Johnson & Omland (2004) provide a clear introduction to the topic of the assessment of alternative models in ecology.)

One may sometimes conclude that there is no way of getting an unbiased result. The question is then as follows: is a biased result better or worse than no idea at all? Each case has to be judged on its merits.

Multiple recaptures in open populations

The general approach

In an open population, the number of animals carrying marks is unknown because some of those that have been marked may have died or emigrated. However, if there are at least three capture occasions, the number of animals carrying marks, and hence the total population size, on each occasion (except the first and the last) can be estimated. The rationale of the calculations is explained by McCallum (2000).

The method also involves estimating rates of loss from and recruitment to the population. Because so many parameters are estimated, the precision of each estimate is poor – as can be seen for the example in Box 3.15 later, despite that study having involved 13 capture occasions with about 120–240 animals being caught on each occasion. To improve precision, it is important to catch a substantial proportion of the population on each occasion and to space the occasions close enough together that there is a high recapture rate.

Animals must, at a minimum, be marked batch-specifically. Individual marking is better, in that it may allow more complex analyses to be undertaken if necessary.

The Jolly–Seber method

This is the usual method used for analysing multiple-recapture data from open populations. Like all methods for such data, it requires one to count the numbers of animals with various 'capture histories' and, from these, to generate sets of key numbers to go into the calculations (Box 3.13).

Box 3.13. **Preliminary calculation for the Jolly–Seber method**

The following four sets of numbers are needed before calculations can begin:

n_i = total number of animals caught in the ith sample
R_i = number of animals that are released after the ith sample
m_i = number of animals in the ith sample that carry marks from previous captures
m_{ij} = number of animals in the ith sample that were most recently caught in the jth sample

To get these, the data should be recorded in the form of a 'capture history', as in the table on the left, which relates to a study with three capture occasions. A 1 is written when the animal is caught, a 0 when it is not. In the data tabulated below, the first animal was caught on all three occasions, the second only on the first and the second, etc. Data for animals that were caught but not released (as sometimes happens) are marked with an asterisk. The numbers caught on each occasion (n_i) and the numbers not released are obtained by adding down the columns; the numbers released (R_i) are calculated by subtraction.

If you have entered the data on a spreadsheet, such as in Excel®, it is easy to sort them into a table as on the right in the table below, such that all the animals with the same capture history are brought together.

Note that each part of the table is here arranged in two sets of columns, merely to save space.

	Original data								Sorted data						
	Occasion				Occasion				Occasion				Occasion		
Animal	1	2	3	Animal	1	2	3	Animal	1	2	3	Animal	1	2	3
1	1	1	1	20	1	1	1	1	1	1	1	19	1	0	0
2	1	1	0	21	0	1	0	6	1	1	1	22	0	1	1
3	1	0	1	22	0	1	1	8	1	1	1	23	0	1	1
4	1	0	0	23	0	1	1	13	1	1	1	25	0	1	1
5	1	1	0	24	0	1	0	16	1	1	1	27	0	1	1
6	1	1	1	25	0	1	1	20	1	1	1	31	0	1	1
7	1	0	1	26	0	1	0	2	1	1	0	24	0	1	0
8	1	1	1	27	0	1	1	5	1	1	0	26	0	1	0
9	1	1*	0	28	0	1	0	9	1	1	0	28	0	1	0
10	1	0	1	29	0	1*	0	11	1	1	0	29	0	1	0
11	1	1	0	30	0	1	0	3	1	0	1	30	0	1	0
12	1	0	1	31	0	1	1	7	1	0	1	21	0	1	0
13	1	1	1	32	0	0	1	10	1	0	1	32	0	0	1
14	1*	0	0	33	0	0	1	12	1	0	1	33	0	0	1
15	1	0	0	34	0	0	1	17	1	0	1	34	0	0	1
16	1	1	1	35	0	0	1	4	1	0	0	35	0	0	1
17	1	0	1	36	0	0	1	14	1	0	0	36	0	0	1
18	1	0	0	37	0	0	1	15	1	0	0	37	0	0	1
19	1	0	0	38	0	0	1	18	1	0	0	38	0	0	1
	n_i				20	21	23						20	21	23
	Not released				1	2									
	R_i				19	19									

From the sorted data, it is easy to summarise the numbers of animals with different capture histories (below). From these, one can then pick out the animals contributing to the other two sets of totals, m_{ii} and m_{ij} (defined above):

Occasion									
1	2	3	No.		m_2	m_3	m_{21}	m_{31}	m_{32}
1	1	1	6		6	6	6		6
1	1	0	4		4		4		
1	0	1	5			5		5	
1	0	0	5						
0	1	1	5			5			5
0	1	0	6						
0	0	1	7						
			Sums		10	16	10	5	11

Box 3.14. **Working out capture histories when marks are batch-specific**

On each occasion, the numbers caught with various histories are known. The numbers of those not caught on that occasion but which have been caught before can be worked out by subtraction. The number of those that have never been caught is unknown (but can be estimated using the Jolly–Seber and related methods).

	History	Number
First capture		
Captured	1	N_i
Not captured	0	$?_i$
Second capture		
Captured	1 1	N_{12}
	0 1	N_{02}
Not captured	1 0	$N_1 - N_{12}$
	0 0	$?_2$
Third capture		
Captured	1 1 1	N_{123}
	1 0 1	N_{103}
	0 1 1	N_{023}
Not captured	1 1 0	$N_{12} - N_{123}$
	0 1 0	$N_{02} - N_{023}$
	1 0 0	$N_1 - N_{12} - N_{023}$
	0 0 0	$?_3$

When animals are individually marked, the accumulation and counting of capture histories is straightforward (Box 3.13). When marks are merely batch-specific, histories need to be enumerated on each sampling occasion in order that the numbers of animals with various histories that are not caught on the last occasion can be worked out (Box 3.14).

Having worked out the numbers with various capture histories and used these to produce the four sets of key numbers, one proceeds as in Box 3.15 to estimate first the numbers of marked animals in the population at the time of each capture and then, from these, population sizes, survival rates and recruitment rates. (Note that 'survival' is the complement of mortality plus emigration; 'recruitment' is the sum of births and immigration, net of deaths of those recruited since the previous capture).

Box 3.15. **The Jolly–Seber method: main calculations**

To illustrate the calculations needed to derive population estimates from the data, we use data for black-kneed capsids *Blepharidopterus angulatus* caught at three- or four-day intervals in a British apple orchard (Jolly 1965).

The data are first organised following Box 3.13, and are then laid out as in the table below. Note that in this there appear two columns and two rows of r and z values, which are obtained from the rest of the table. The two columns on the left are

> r_i = the number of animals that were released from the ith sample and were subsequently recaptured (these are simply the row sums); and
>
> z_i = the number of animals caught both before and after the ith sample but not in the ith sample itself (z_i is the sum of all the m_{ij} that fall in all columns to the right of column i and all rows above row i, thus the dashed lines in the table delimit the m_{ij} values that must be summed to obtain z_4, for example).

The r_i and z_i rows are just copies of the columns, laid out under the main table for ease of further calculation. (In Excel®, use Paste Special, with Values and Transpose checked.)

	1	2	3	4	5	6	7	8	9	10	11	12	13		r_i	z_i
n_i	54	146	169	209	220	209	250	176	172	127	123	120	142			
R_i	54	143	164	202	214	207	243	175	169	126	120	120	0			
i																
1		10	3	5	2	2	1	0	0	0	1	0	0		24	0
2			34	18	8	4	6	4	2	0	2	1	1		80	14
3				33	13	8	5	0	4	1	3	3	0		70	57
4					30	20	10	3	2	2	1	1	2		71	71
5						43	34	14	11	3	0	1	3		109	89
6							56	19	12	5	4	2	3		101	121
7								46	28	17	8	7	2		108	110
8									51	22	12	4	10		99	132
9										34	16	11	9		70	121
10											30	16	12		58	107
11												26	18		44	88
12													35		35	60
m_i	0	10	37	56	53	77	112	86	110	84	77	72	95			
z_i	0	14	57	71	89	121	110	132	121	107	88	60				
r_i	24	80	70	71	109	101	108	99	70	58	44	35				

The following rows of the table (which are explained below it) are four sets of parameter estimates plus (with four intermediate steps) lower and upper confidence limits for the population estimates

$\hat{M}_i$	0.0	34.9	169.5	256.2	227.0	323.7	358.2	318.3	399.7	314.3	313.6	273.7
$\hat{N}_i$		466.2	758.1	943.8	928.8	871.6	795.7	647.6	623.0	473.3	498.6	453.6
Φ_i	0.646	1.009	0.864	0.564	0.834	0.790	0.651	0.981	0.685	0.880	0.767	
$\hat{B}_i$		290.5	293.0	400.2	101.5	109.2	134.4	−11.6	48.5	82.8	73.3	
T_i		5.966	6.509	6.729	6.703	6.638	6.500	6.321	6.279	6.010	6.075	5.970
s_{Ti}		0.331	0.186	0.164	0.147	0.123	0.109	0.107	0.112	0.119	0.144	0.168
W_{Li}		229.6	498.6	643.3	644.4	627.4	558.8	468.9	445.7	336.6	345.6	298.9
W_{Ui}		862.6	1047.4	1238.7	1159.6	1025.0	862.8	718.3	698.8	542.4	613.9	586.2
LCL		308	587	752	759	736	691	561	536	403	410	362
UCL		937	1134	1345	1272	1132	992	809	787	608	677	648

Parameter estimates

$\hat{M}_i$ = number of marked animals in the population when the ith sample is taken (but not including animals newly marked in the ith sample).

$= m_1 + (R_i + 1)z_i/(r_i + 1)$

$\hat{N}_i$ = population size at the time of the ith sample

$= \hat{M}_i(n_i + 1)/(m_i + 1)$

Φ_i = proportion of the population surviving (and remaining in the study area) from the ith sampling occasion to the $(i + 1)$th

$= \hat{M}_{i+1}/(\hat{M}_i - m_i + R_1)$

$\hat{B}_i$ = number of animals that enter the population between the ith and $(i + 1)$th samples and survive until the $(i + 1)$th sampling occasion

$= \hat{N}_{i+1} - \Phi_i(\hat{N}_i - n_i + R_i)$

Note that one cannot calculate $\hat{M}$ for the last sample, $\hat{N}$ for the first or last, Φ for the last two and $\hat{B}$ for the first or last two. $\hat{M}_1$ is always zero.

Example calculations for the four parameter estimates follow:

$$\hat{M}_2 = 10 + (143 + 1)14/(80 + 1) = 34.89$$

$$\hat{N}_2 = 34.89(146 + 1)/(10 + 1) = 466.1$$

$$\Phi_2 = 169.46/(34.89 - 10 + 143) = 1.009$$

$$\hat{B}_2 = 758.1 - 1.009(466.2 - 146 + 143) = 290.7$$

Confidence limits for $\hat{N}_i$

Methods usually presented for calculating confidence limits of Jolly–Seber estimates are inadequate for the commonly encountered sample sizes. The method presented here, which is due to Manly (1984), provides better limits.

Calculate a transformation of each $\hat{N}_i$ and the standard error of the transformation:

$$T_i = \log_e \left\{ \hat{N}_i \left[1 + \sqrt{1 - n_i/\hat{N}_i - n_i/(2\hat{N}_i)} \right] \Big/ 2 \right\}$$

$$s_{T_i} = \sqrt{ \left(\frac{\hat{M}_i - m_i + R_i + 1}{\hat{M}_i + 1} \right) \left(\frac{1}{r_i + 1} - \frac{1}{R_i + 1} \right) + \frac{1}{m_1 + 1} - \frac{1}{n_i + 1} }$$

For example, the transformation for $\hat{N}_2$ is

$$T_2 = \log_e \left\{ 466.2 \left[1 + \sqrt{1 - 146/466.2 - 146/(2 \times 466.2)} \right] \Big/ 2 \right\} = 5.966$$

$$s_{T_2} = \sqrt{ \left(\frac{34.89 - 10 + 143 + 1}{34.89 + 1} \right) \left(\frac{1}{80 + 1} - \frac{1}{143 + 1} \right) + \frac{1}{10 + 1} - \frac{1}{146 + 1} }$$

$$= 0.3309$$

Calculate the W values from

$$W_{Li} = \exp(T_i - 1.6s_{T_i}) \qquad W_{Ui} = \exp(T_i + 2.4s_{T_i})$$

In the example,

$$W_{L2} = \exp[5.966 - 1.6(0.331)] \qquad W_{U2} = \exp[5.966 + 2.2\,(0.331)]$$
$$= 229.6 \qquad\qquad\qquad = 862.6$$

Finally, calculate 95% confidence limits for the $\hat{N}_i$ as

$$\text{LCL} = (4W_{Li} + n_i)^2/(16W_{Li}) \qquad \text{UCL} = (4W_{Ui} + n_i)^2/(16W_{Ui})$$

In the example,

$$\text{LCL} = [4(229.6) + 146]^2/[16(229.6)] \qquad \text{UCL} = [4(862.6) + 146]^2/[16(862.6)]$$
$$= 308 \qquad\qquad\qquad\qquad = 937$$

Assumptions and biases

All the assumptions for closed populations, except that of closure, apply to open populations. Further assumptions are the following.

1. All animals have the same probability of survival between capture occasions. (It is important to minimise the likelihood of trapping causing animals to die or emigrate.)
2. Capture and release are 'instantaneous', that is, short in relation to survival rate.
3. Emigration is permanent.
4. The fate of each animal (whether it is captured again, whether it dies, whether it emigrates) is independent of the fate of others.

The last of these assumptions generally affects only one's estimates of precision. Violation of the others results in biased estimates of the parameters. As with closed populations, heterogeneity in catchability among animals is probably the commonest cause of severe bias in the estimates of population size, though it only slightly biases the estimates of survival. Box 3.16 provides a goodness-of-fit test for the Jolly–Seber method.

Box 3.16. **The goodness-of-fit test for Jolly–Seber data**

The table below provides a capture history for a study in which animals were captured four times. The goodness-of-fit test is applied to all occasions except the first and the last. For each occasion, the history table is inspected to discover animals with each of the following histories:

f_1 = first captured before this sample, subsequently recaptured
f_2 = first captured before this sample, not subsequently recaptured
f_3 = first captured in this sample, subsequently recaptured
f_4 = first captured in this sample, not subsequently recaptured

The numbers of these animals are laid out to the right of the history table. Note that more than one row of the history table may contribute to an f value; and some rows contribute to none. Columns are summed to produce the four final f values for each occasion.

Occasion 1 2 3 4	No.	Occasion 2 f_1	f_2	f_3	f_4	Occasion 3 f_1	f_2	f_3	f_4
1 1 1 1	3	3				3			
1 1 1 0	3	3					3		
1 1 0 1	2	2							
1 1 0 0	6		6						
1 0 1 1	2					2			
1 0 1 0	3						3		
1 0 0 1	5								
1 0 0 0	8								
0 1 1 1	9			9		9			
0 1 1 0	6			6			6		
0 1 0 1	5			5					
0 1 0 0	8				8				
0 0 1 1	5					5		5	
0 0 1 0	8								8
0 0 0 1	7								
Total f		8	6	20	8	19	12	5	8
n					42				44
		a_1	a_2	a_3	a_4	a_1	a_2	a_3	a_4
		14	28	28	14	31	13	24	20
$f \log_e(f)$		16.636	10.751	59.915	16.636	55.944	29.819	8.047	16.636
g_1					103.936				110.446
$a \log_e(a)$		36.947	93.302	93.302	36.947	106.454	33.344	76.273	59.915
g_2					260.497				275.986
$n \log_e(n)$		156.982				166.504			
G		0.84				1.93			
Total G						2.77			

The table is extended as follows.

From the *f* values are calculated the following:

$$a_1 = f_1 + f_2, \qquad a_2 = f_3 + f_4, \qquad a_3 = f_1 + f_3, \qquad a_4 = f_2 + f_4$$
$$n = f_1 + f_2 + f_3 + f_4$$

The four values of $f\log_e(f)$ and of $a\log_e(a)$ are calculated, their sums being g_1 and g_2, respectively. The test statistic is then

$$G = 2[g_1 - g_2 + n\log_e(n)]$$

Each *G* is compared with χ^2 for $P = 0.05$ and one degree of freedom (3.84). Any *G* greater than this indicates that the assumptions of the method have been violated, at least on that occasion.

The sum of *G* values provides a test of the overall goodness of fit, with a number of degrees of freedom equal to the number of samples providing individual *G* values. In this example the sum is 2.77, with four degrees of freedom, which is far from significant.

Samples for which any of the expected frequencies of the four groups of animals are less than 2 should be left out of this test.

Modified models

Various modifications of the basic model have been developed, many of which are implemented in MARK and CAPTURE (see the appendix to this chapter). Some involve placing constraints on one of the parameters, such as no recruitment, no losses, constant rates over all occasions, or time-specific covariates, which reduces the number of separate parameters to be estimated, so improving precision – but at the cost of making the additional assumptions. For example, by modelling the effect of weather on trappability, one can reduce the number of parameters needed to describe variation in trappability, which would otherwise be one per day. (Another technique is simply to ignore the data from days of extreme weather and assume constant trappability for the remainder.)

Other variants, in contrast, involve the modelling of additional parameters, such as the effects of being trapped on behaviour or survival. Temporary emigration has also been modelled.

Choosing among alternative models follows the same principles as for closed populations, which we discussed above.

Using ancillary information

In some studies, ancillary information that can be used to increase the precision of estimates is available. One of the chief means implemented in MARK is to use information about animals found dead – such as tagged animals killed by hunters and fishermen or ringed birds found by members of the public. Such recoveries provide extra information about survival, thus indirectly increasing the precision of estimates of population size. They differ from the recaptures used in the basic method in that they provide no parallel information about unmarked animals. The same is true of information derived from resightings of marked animals or from radiotelemetry.

The robust model

Suppose that animals can be caught most readily during some particular season of the year. One could then make a catch every year at that time and then over a period of several years generate annual Jolly–Seber estimates. To increase precision, one could make several catches each year, over a sufficiently short period that the population could be regarded as closed at that time. Thus one would have a series of 'primary periods' (years, in our example), over which the population was open and within each of which there were 'secondary periods' over which the population was closed. The secondary periods provide more precise information about population size than would otherwise be obtained, because of the closure assumption. This is the basis of the robust model, in which closed-population estimates from the secondary periods are combined with Jolly–Seber estimates over the primary periods.

Annual population, survival and recruitment estimates based on a certain number of catching occasions per year are always more robust and precise if the catches are concentrated into a period when the population can be considered closed than if they are spread throughout the year. That may also be the logistically more efficient arrangement. However, seasonal changes in numbers, survival and recruitment can be measured only by spreading the catches over the year. The objective of the study must determine the methods.

What area does a trapping grid cover?

Some populations live in circumscribed habitat islands that are small enough to be surveyed in their entirety by a single mark–recapture study. More commonly, one is faced with setting a grid of traps in a small area that is representative of a much larger area occupied by the species of interest. It is a basic assumption of the mark–recapture idea that the animals are sufficiently mobile for the marked ones to mix thoroughly with the others; if this is true, then it will also be true that animals from the surrounding area will sometimes enter the study area and may be caught. Thus the population being studied includes animals from an ill-defined strip around the grid of traps, as well as those living on the grid itself. Other things being equal, it is therefore better to use a square grid rather than an oblong, since the square has relatively less edge.

Various methods have been used to estimate the width of this strip. Most are ad hoc and difficult to justify, except insofar as they are likely to give an answer that is closer to reality than if one simply assumes that there is no problem of marginal animals. The best method, in contrast, is based on a simple geometrical model; it requires that the grid of traps is square, with the same number of traps in both directions and all traps the same distance apart (though it could be adapted to rectangular, hexagonal, or other shapes). It assumes that the density and mobility of the animals is constant not just over the whole grid but also in the immediately surrounding area. Since the capture probability of animals that are resident in the study area is likely to be greater than that of animals that originate from the surrounding area, the Burnham and Overton method is likely to be the best method to use in this situation.

Simple geometry dictates the relationship between the number of animals estimated to be present in the study area and the surrounding strip (N), the population density (D) and the size of

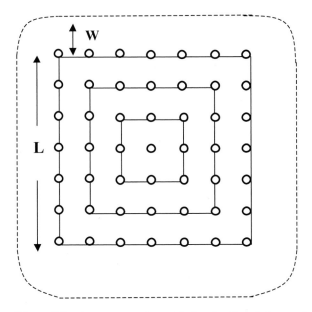

Figure 3.5 A square trapping grid, showing the whole area from which animals caught on the grid may originate, including both the grid itself (which has sides of length *L*) and a surrounding strip (of width *W*).

the grid (plus the surrounding strip). If the grid measures $L \times L$ and the width of the strip is W (Figure 3.5) then it will be true that

$$N = D(L^2 + 4LW + \pi W^2)$$

N is estimated in the mark–recapture study and L is fixed by the investigator; D and W are unknown. Hence, if two grids of different sizes are used, D and W can be estimated. In practice, one uses just the single grid but treats it as a series of nested grids: in Figure 3.5, there is not just a 7×7 grid but a 5×5 grid nested within it and a 3×3 grid within that. With more than two values of N and L thus available, estimates of W and D can be obtained using non-linear least-squares regression. The method is implemented in CAPTURE.

This method assumes that, apart from the animals effectively drawn from the strip of width W, the population is closed. It also assumes that the density of animals is constant across the grid and the surrounding area. CAPTURE provides tests of these assumptions.

Sample counts (2): some other methods based on trapping

The removal method

This is formally equivalent to the pseudo-removal method, differing only in the animals being removed from the population rather than marked and returned. Figure 3.6 shows a classic application, in which the animals were removed because they were killed in the traps.

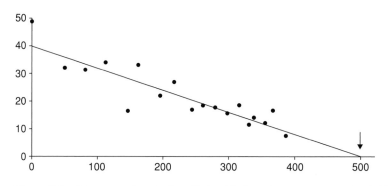

Figure 3.6   The removal method applied to black rat *Rattus rattus* catches in 70 houses in Freetown, Sierra Leone, over 18 days. The arrow indicates the estimated population size. (Data from Leslie & Davis (1939).)

The assumptions and limitations of this method are identical to those of the pseudo-removal method. If animals from the surrounding area move in to fill gaps left by those that have been removed (thus violating the assumption of closure), the points will lie on a right-concave curve rather than a straight line, so a graph such as Figure 3.6 should always be drawn and departure from linearity should be tested (Box 3.9).

The change-in-ratio method

Some populations consist of two or more classes that can be readily distinguished in the field, such as two sexes. The two classes may be selectively harvested, as when hunters target males. This causes a change in the proportions of the classes in the population, from which, so long as one knows the numbers removed, one can estimate the original population size (Box 3.17).

There are three key assumptions of this method: first, that the population is closed (except for those deliberately removed); second, that the numbers removed are precisely known; and third, that the proportions of the various classes are estimated without bias. The last is unlikely: if the two sexes (for example) are so different that humans can readily distinguish them in the field, the species is likely to have social behaviour that results in non-random mixing of the two sexes, making it very difficult to estimate the overall sex ratio accurately.

Even if all the assumptions are met, the removals need to be great enough and selective enough to make the proportions change by at least 0.1 if the estimate of population size is to be at all precise. (Note the very wide confidence limits in the example in Box 3.17, with the lower calculated limit even being less than the number of animals directly observed.)

Because it provides inaccurate and imprecise estimates, the method is scarcely used, except for hunted populations for which bag statistics are available and a rough estimate of population size is considered to be better than no estimate at all.

Box 3.17. **The change-in-ratio method**

The details shown here relate to a population in which there are two classes and a single period of removal. It has been generalised to more than two classes and more than one period of removal (Williams *et al.* 2002).

P_1, P_2 = estimated proportions of class A in the initial population and after removals, respectively.

R_A, R_B = numbers of classes A and B, respectively, that were removed.

The estimates of the initial population size and of the numbers remaining after the removals are

$$\hat{N}_1 = [R_A - P_2(R_A + R_B)]/(P_1 - P_2)$$
$$\hat{N}_2 = \hat{N}_1 - (R_A + R_B)$$

Thus, if (in unbiased samples) we observed 40 males and 60 females initially and 29 and 78 after the removals, and if 137 males and 2 females were removed, then

$$P_1 = 40/100 = 0.4000, \quad P_2 = 29/107 = 0.2710$$
$$\hat{N}_1 = [137 - 0.271(137 + 2)]/(0.4 - 0.271) = 770.14 \approx 770$$
$$\hat{N}_2 = 770.14 - 137 - 2 = 631.14 \approx 631$$

To calculate confidence limits for $\hat{N}_1$, calculate the variances of P_1 and P_2:

$$s_1^2 = P_1(1 - P_1)(\hat{N}_1 - n_1)/[\hat{N}_1(n_1 - 1)]$$
$$s_2^2 = P_2(1 - P_2)(\hat{N}_2 - n_2)/[\hat{N}_2(n_2 - 1)]$$

where n_1 and n_2 are the sizes of the samples on which P_1 and P_2 were based. The variance of $\hat{N}_1$ is

$$s_N^2 = \left(\hat{N}_1^2 s_1^2 + \hat{N}_2^2 s_2^2\right)/(P_1 - P_2)^2$$

Approximate confidence limits for $\hat{N}_1$ are

$$\hat{N}_1 \pm 2s_N$$

(These are too narrow if either n_1 or n_2 is less than 30.)

In our example,

$$s_1^2 = 0.400\,(1 - 0.400)(770.14 - 100)/[770.14(100 - 1)] = 0.002\,109$$
$$s_2^2 = 0.271(1 - 0.271)(631.14 - 107)/[631.14(107 - 1)] = 0.001\,548$$
$$s_N^2 = [(770.14^2 \times 0.002\,109) + (631.14^2 \times 0.001\,548)]/(0.4 - 0.271)^2$$
$$= 112\,287$$

The approximate lower and upper confidence limits of $\hat{N}_1$, calculated from the expression above, are 100 and 1440. Since we know that at least 246 animals were initially present (because 107 were observed after 139 had been removed), it is sensible to take that number as the lower confidence limit, rather than 100.

Simultaneous marking and recapture: the method of Wileyto et al.

Suppose that one uses traps of two different sorts for a closed population. One is a normal trap, which catches animals permanently; the other is identical, except that animals that enter it are automatically marked and are able to escape. Such automatic marking is typically carried out by having the inside of the trap coated with a dye that colours the animals that enter. The traps from which the animals cannot escape catch both marked animals (which have been through the other traps) and unmarked animals (which have not). The relative numbers of the two provide an estimate of total population size.

The method assumes that equal numbers of the two sorts of traps are employed and that they are equally efficient at catching animals (but that the animals then escape from the marker traps). It works best if the traps are operated in pairs, with one trap of each type in each part of the study area. Differences in trappability among animals do not bias this method badly – but serious bias arises if marked animals are less (or more) likely to enter the permanent traps than are unmarked animals. Although the method applies to closed populations, turnover of the population (through births, deaths, immigration and emigration) has no serious effect so long as the number of animals entering and leaving the population is less than 10% of the numbers being trapped. It is not possible to check the assumptions from the data themselves.

Box 3.18 shows how to calculate the estimate of population size, with approximate confidence limits. Note that the calculations break down if $R \geq U$. In this case $R + U$ provides a reasonable estimate of the population size: it is obvious that this is a minimum estimate; what is less obvious, but true, is that R is likely to equal or exceed U only when a very high proportion of the population has been marked. The precision to be expected for this method can be calculated from the expression given for confidence limits in Box 3.18. If one empties the traps periodically, then one may make population estimates as the study progresses and, from their confidence limits, assess for how much longer the work needs to continue.

Continuous captures and recaptures: the Craig and du Feu method

The previous method involved simultaneous marking and capture, with the relative numbers of marked and unmarked animals being assessed at the end point of the study. Studies in which animals are caught one at a time and quickly released (having been marked if not already carrying a tag) are quite different, but the relative numbers of captures of unmarked and of already marked animals once again provide an estimate of total population size.

Again, the population must be closed – and the estimate cannot be corrected to allow for immigrants.

In practice, one may make estimates of the population size at intervals (or even continuously) during the course of the study. At first, successive estimates fluctuate wildly, but they then settle down to more stable values (Figure 3.7) – unless the population is not closed, in which case they tend to drift upwards (or downwards, if animals lost from the population are not replaced by new recruits).

Box 3.18. **The method of Wileyto *et al.***

Basics

> U = number of unmarked animals caught in the permanent traps
> R = number of marked animals caught in the permanent traps

Example

Indian mealmoths *Plodia interpunctella* in a warehouse in the USA (Wileyto *et al.* 1994),

$$U = 74, \qquad R = 7$$

Estimate of total population size

Estimate of population size,

$$\hat{N} = (U + R)^2/[2(R + 1)]$$
$$= 81^2/[2(8)] = 410$$

Confidence limits of the estimate

Very approximate 95% confidence limits are

$$\hat{N} \pm 2\sqrt{\hat{N}[U(U + R) + R(U - R)]/R}$$
$$= 410 \pm \sqrt{2(410)[74(81) + 7(67)]/7} = 410 \pm 465$$
$$= -55 \text{ and } 875$$

Since 81 animals were found in the permanent traps, this number (rather than -55) should be taken as the lower confidence limit.

If the animals are individually marked, it is possible to count the number of times that each has been caught and thus use the goodness of fit to the zero-truncated Poisson distribution (Box 3.7) as a test for the assumption of equal trappability. (The accuracy of the estimate is probably not much reduced by modest differences in trappability.)

Box 3.19 shows how to estimate population size at any stage during the study, with approximate confidence limits. The expression given for the latter can be used to estimate the precision attainable using this method. Note that it performs better than partial trapping-out, in terms of the precision achieved for a given number of captures. Because estimates and confidence limits may be calculated as the study proceeds, one can (in principle) carry on until a desired level of precision is achieved.

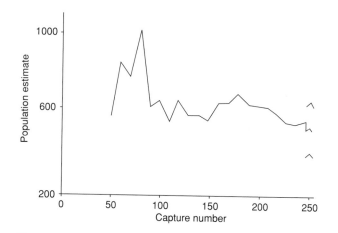

Figure 3.7   The Craig and du Feu method applied to malachite sunbirds *Nectarinia famosa* trapped over three days at an isolated food source in South Africa. The line shows successive population estimates in relation to the total number of captures made up to that point, with confidence limits applied to the later estimates. (Data supplied by M. W. Fraser and L. G. Underhill, see Underhill & Fraser (1989).)

Box 3.19. **The Craig and du Feu method**

To illustrate this method, we use the final figures obtained in the malachite sunbird study shown in Figure 3.7.

Basics

Tabulate the number of birds caught exactly k times (f_k).

At this point, it is sensible to check that that the data fit a zero-truncated Poisson distribution (Box 3.7), since the estimate will be biased if they do not. (For the sunbird data, $G = 1.78$, with two degrees of freedom, which is not significant.)

The sum of the f_k column is the total number of birds caught (U).

Calculate a column of kf_k values. The sum of this column is the total number of captures made (C).

Calculate a column of $k^2 f_k$ values. The sum of this column (A) is needed later.

k	f_k	kf_k	$k^2 f_k$
1	159	159	159
2	34	68	136
3	10	30	90
4	0	0	0
Sums	$U = 203$	$C = 257$	$A = 385$

First approximate estimate of population size

$$\hat{N}_0 = C^2/(A - C) = 257^2/(385 - 257) = 516.01$$

Assessment of estimate

Calculate the indicator H_0: it will be zero if $\hat{N}_0$ equals the best estimate of N that the data can provide:

$$H_0 = \log(\hat{N}_0 - U) + (C - 1)\log \hat{N}_0 - C\log(\hat{N}_0 - 1)$$
$$= \log 313.01 + 256 \log 516.01 - 257 \log 515.01 = -0.0006$$

Thus in this case H_0 is negative, meaning that $\hat{N}_0$ is too small.

Refinement of estimate

One now makes a guess at a potentially better estimate ($\hat{N}_1$) and uses this to calculate H_1 – using the above equation for H_0 and substituting $\hat{N}_1$ for $\hat{N}_0$. Our guess is $\hat{N}_1 = 530$, resulting in $H_1 = +0.0011$ (meaning, because it is positive, that $\hat{N}_1$ is too large).

The next estimate ($\hat{N}_2$) is calculated from

$$\hat{N}_2 = \hat{N}_1 - (\hat{N}_1 - \hat{N}_0)H_1/(H_1 - H_0)$$
$$= 530 - (530 - 516.01)0.0011/(0.0011 + 0.0006) = 521.00$$

Using the equation for H_0 but now substituting $\hat{N}_2$ for $\hat{N}_0$, one obtains $H_2 = +0.000\,02$; thus $\hat{N}_2$ is also too large.

$\hat{N}_3$ is calculated from the equation for $\hat{N}_2$, substituting $\hat{N}_2$ for $\hat{N}_1$, $\hat{N}_1$ for $\hat{N}_0$, H_2 for H_1 and H_1 for H_0; and H_3 is calculated in the now familiar way.

If necessary, one continues further. In this case, however, we get $\hat{N}_3 = 520.78$ and $H_3 = -0.000\,000\,1$, meaning that $\hat{N}_3$ is too small, which tells us that the best estimate of N is that it lies between $\hat{N}_3$ (520.78) and $\hat{N}_2$ (521.00). To the nearest whole number, we can take 521 as the best estimate.

Note that this iterative method can be effectively implemented in Excel®.

Confidence limits of the estimate ($\hat{N}$)

Approximate 95% confidence limits are

$$\hat{N} \pm 2\sqrt{[\hat{N}/(\exp(C/\hat{N}) - C/\hat{N} - 1)]}$$
$$= 521 \pm 2\sqrt{[521/(\exp(0.4933) - 0.4933 - 1]} = 521 \pm 120$$
$$= 401 \text{ and } 641$$

References

See Craig (1953) and du Feu *et al.* (1983).

Passive distance sampling

These methods (trapping webs and trapping line transects) are discussed under 'Distance sampling' (below).

Sampling from the whole area

In all of the above methods, it is important that all animals have an equal chance of being captured. This means that a sufficient number of traps, well distributed over the study area, has to be used.

Sample counts (3): 'mark–recapture' without capture

Marking without capture

Several methods that have the same statistical base as mark–recapture methods but which do not require animals to be captured are available. The assumptions on which they are based are the same as for mark–recapture. The assumption that the detectability of all individuals is identical applies to all of them. As with true mark–recapture methods, one needs to think carefully about what area one's observations are effectively covering and whether the population is open rather than closed.

Individual recognition without capture

Individuals of some species can be recognised through differences in appearance that can be observed directly or from photographs, obviating the need to catch them. The recovery of DNA from shed hair, faeces and even urine can also provide unique identification without the animals being caught and has proved particularly useful for species that are difficult either to observe directly or to capture, such as bears, wolves, forest elephants and dolphins. This allows one, by making repeated surveys over time, to apply the mark–recapture methods discussed above – subject, of course, to the assumptions inherent in them being valid for the study one has carried out. It is important to divide one's observations into sessions equivalent to capture sessions; each may extend over a period but this should be short enough that the population is closed within each period (and, of course, between periods if closed-population methods are to be used).

Since both trapping and marking may affect the animal's behaviour (and even its survival), which can lead to bias in estimating population size, methods that depend on natural distinguishing features and that avoid the need for trapping are to be preferred. Recognising animals without having to capture them is also useful because, since it is less time-consuming, it may be possible to obtain much greater sample sizes. This is important because mark–recapture methods (like most indirect methods) require large sample sizes if they are to give precise estimates.

Critical to these methods is accuracy of identification. If two or more individuals look the same, the population will be underestimated. This has been a problem with DNA recognition, though improved methods are enabling finer identification. For some purposes, in any case, estimating the minimum number of animals known to be in the population may be useful, so long as one remembers that improved techniques will push this estimate up even when the actual population is constant.

The occasional misidentification of one animal as another will not seriously bias population estimates, but the misidentification of an animal as an individual not previously recorded will inflate the estimates. There may be a simple error in recording an individual's characteristics: for example, DNA from hair or faeces may be contaminated with DNA from other organisms. If this issue is ignored, it can cause significant overestimation of the size of the population, because each of the false identifications will enter the calculations as individuals. However, by using the excess of individuals recorded only once, it is possible to correct for this (Lukacs & Burnham 2005). Alternatively, individuals may permanently change their appearance so that, over time, one individual appears to drop out of the population and another to enter it. This will have little or no effect on the estimate of population size, though it will inflate estimates of rates of loss and recruitment.

The double-observer method

Mark–recapture theory can be used even when the organisms are not individually identifiable. The double-observer method depends on an extension of the theory of the pseudo-removal method. Two observers operate simultaneously, covering the same ground. The 'primary' observer identifies to the 'secondary' observer all the individuals that he or she detects; the secondary observer records these and also all those that he or she detects that the primary observer has overlooked. The two observers alternate the primary and secondary roles, each being primary for half the time. It is not necessary for them to achieve the same probabilities of detecting what they are counting. It is, however, important that they count the same population; if conducting point counts, for example, they should each count only up to a fixed distance; otherwise, the one who can detect animals at a greater distance is counting a larger population than the other. It is also important that the probability of detecting an animal should be the same for each observer, irrespective of whether he or she is operating in the primary or secondary role. Observers sometimes feel that their efficiency is reduced in the secondary role because of the time they spend writing; having a third person as a scribe circumvents this problem but raises the difficulty of how the secondary observer communicates with the scribe without alerting the primary observer to animals that he or she would otherwise have missed. Conversely, the primary observer may pick up cues as to the presence of animals he or she has not yet detected from seeing the secondary observer looking in a particular direction or writing things down; circumventing this by having the secondary observer stand behind the first limits the ability of each to make observations.

Box 3.20 presents the basic calculations of the method.

Nichols *et al.* (2000) have specifically developed this method in relation to point counts, showing how SURVIV can be used to fit various particular models to the data.

Box 3.20. **The double-observer method**

Example

Two observers carried out a set of random strip transects to determine the population density of a plant. Time was too short to allow them to conduct independent surveys, mapping the plants on each strip (which would have allowed them to apply the mark–recapture analysis appropriate for a double survey). Instead, they applied the double-observer method, each acting as primary observer for half the strips and as secondary observer for the other half.

Data

Numbers of plants observed (in total, over all strips):

$X_{11} = 210$: by observer 1, when acting as primary observer;

$X_{22} = 125$: by observer 2, when acting as primary observer;

$X_{12} = 30$: by observer 1, additional to those observed by observer 2, when observer 1 was acting as secondary observer; and

$X_{21} = 23$: by observer 2, additional to those observed by observer 1, when observer 2 was acting as secondary observer.

Some sums

$$X_{\cdot 1} = X_{11} + X_{21} = 233$$
$$= \text{total number observed when observer 1 was primary}$$
$$X_{\cdot 2} = X_{12} + X_{22} = 155$$
$$= \text{total number observed when observer 2 was primary}$$
$$X_{\cdot \cdot} = X_{\cdot 1} + X_{\cdot 2} = 388$$
$$= \text{grand total number observed}$$

Some products

$a = X_{11} X_{22}$	$b = X_{12} X_{21}$	$c = X_{22} X_{21}$	$d = X_{11} X_{12}$
$= 26\,250$	$= 690$	$= 2875$	$= 6300$

Estimated proportions observed

By observer 1: $\hat{p}_1 = (a - b)/(a + c) = 0.8776$
By observer 2: $\hat{p}_2 = (a - b)/(a + d) = 0.7853$
By either: $\hat{p} \quad = 1 - b/a \quad = 0.9737$

(That is, observer 1 detected 88% of the plants present; observer 2 independently detected 79%; and together they detected 97%.)

Estimate of population size

$$\hat{N} = X../\hat{p} = 398.47$$

Note that this is the number in the sampled strips. If the whole study area is G times greater than the total area of the sampled strips, the estimate of its total populations is $G\hat{N}$.

Confidence limits

Calculate

$$\beta_1 = X_{.1}/X.. \qquad \beta_2 = X_{.2}/X..$$
$$= 0.6005 \qquad\qquad = 0.3995$$

The variance of $\hat{p}$ is

$$\text{var}(\hat{p}) = \hat{p}(1-\hat{p})^2 \{[1/(\hat{p}_1\beta_1) + 1/(\hat{p}_2\beta_2) + 1/\hat{p}_2(1-p_1)\beta_1]$$
$$+ 1/[p_1(1-p_2)\beta_2]\}/X.. = 0.000\,061\,89$$

in this case.

The variance of $\hat{N}$ is

$$\text{var}(\hat{N}) = \hat{N}[X..\text{var}(\hat{p})/\hat{p}^3 + (1-\hat{p})/\hat{p}]$$
$$= 398.47[(388 \times 0.000\,061\,89/0.9737^3) + 0.0263/0.9737]$$
$$= 21.12$$

The estimated number of plants not detected by either observer is

$$\hat{X}_0 = \hat{N} - X.. = 10.47$$

Calculate

$$C = \exp\left(1.96\sqrt{\log_e\left(1 + \text{var}(\hat{N})/\hat{X}_0^2\right)}\right)$$
$$= \exp\left(1.96\sqrt{\log_e(1 + 21.12/10.47^2)}\right) = 2.276$$

Confidence limits are then

$$X.. + \hat{X}_0/C = 392.6, \qquad X.. + C\hat{X}_0 = 411.8$$

(Multiply by G to get confidence limits for the estimate in the whole study area, $G\hat{N}$.)

The double-survey method

This method is applicable to things whose positions can be mapped, such as plants, nests or large and relatively immobile animals. Two surveys are carried out independently (that is, by different observers, neither person or team knowing the other's results). From the mapped positions, it is subsequently possible to identify the objects detected in both surveys and those detected in each that were not detected in the other. Substituting these numbers appropriately into the equations for the two sample mark–recapture method (Box 3.3), with the two observers substituting for the two trapping sessions, allows the population to be estimated. This method is better than the double-observer method because there are fewer problems in ensuring independence of the two surveys and because it produces more precise estimates (since it uses more information).

Subdivided point counts

Consider an observer recording animals detected within a fixed distance of the point at which the observer is standing for two successive periods of equal length. Suppose that the observer can distinguish individuals sufficiently to come up with two figures:

x_1 is the number of individuals recorded in the first time period;
x_2 is the number recorded in the second period that had not already been recorded in the first.

Following the logic of the (pseudo-)removal method, an estimate of the number of individuals present within the circle centred on the observer is

$$\hat{N} = x_1^2/(x_1 - x_2)$$

Farnsworth *et al.* (2002) have shown how this method can be extended to more than two time periods, which need not be of equal length, and point out that CAPTURE may be used to analyse the data. Models in which detection probabilities in the population are heterogeneous are possible, though (as always) their efficacy depends on how well the assumed frequency distribution of detection probabilities fits the data. The usual assumptions of the population being closed and individuals' detectability being constant also apply.

   If densities are high, it may be difficult to distinguish individuals, so the data may suffer from either overcounting or undercounting. More often, densities are so low that useful population estimates are possible only if the data from several points are combined; this raises no problems so long as detectabilities are the same at all points.

Sample counts (4): N-mixture models

Provided that certain assumptions are met, one can estimate total population size from incomplete counts, if one counts each population more than once.

   Suppose that one visits a population more than once and counts the number of individuals that one detects on each occasion. Suppose that the probability of detection, p, is the same for all individuals in the population and on every visit, and that whether an individual is detected on one visit does not influence its probability of detection on other visits. In this case, the counts will follow a binomial distribution and this can be used to provide an estimate of the population size. For a series of study plots within an area in which animals are randomly distributed (so the numbers in the plots follow a Poisson distribution), if two counts are made in each plot, then a good estimate of p is simply the correlation coefficient between the successive counts on each plot (r); if more than two counts are made, it is the multiple correlation coefficient (R). An estimate of the population size in each plot is then the average count divided by r (or R) (J. A. Royle, personal communication).

A more refined method is to make maximum-likelihood estimates of p and of the average density (and of numbers of individuals on each study plot if needed), using information about the spatial variation in numbers as well as the within-plot correlations (Royle 2004). This allows one to put standard errors on the estimates and to model the effects of covariates (such as weather and habitat) on both detectability and population density. The method assumes that the spatial variation in numbers follows a negative-binomial or a Poisson distribution. The estimates have to be obtained by iterative numerical methods: see Royle (2004) for details. Kéry *et al.* (2005) and Dodd & Dorazio (2004) have applied the method to birds and salamanders, respectively.

As always, the robustness of the method depends on how closely the assumptions fit reality. As with mark–recapture and similar methods, population size will be underestimated if individuals differ in detectability.

Sample counts (5): distance sampling

General

Field methods and principles

Suppose that one is observing big game from a low-flying aircraft, whales from a survey ship, or birds from a position in the middle of a forest. Even an experienced observer will not detect animals that are far away from the survey line. Line and point transect methodology may be applied to such situations. It requires that the distances from the observation point or line to each animal are recorded, so it is generally known as 'distance sampling'; one then uses a model of the way in which detectability drops off with increasing distance to estimate the number of animals that one has missed. It is particularly suitable for animals that are difficult to catch, especially those, such as birds, that can be observed directly in the field but detected with any certainty only if they are close. Basic distance sampling is fully covered by Buckland *et al.* (2001), advanced ideas by Buckland *et al.* (2004).

Distance sampling assumes that the density of animals is constant in the area around the transect lines or points and it provides an estimate of that density. Because distance-sampling methods provide density estimates over rather ill-defined areas, the best way of estimating numbers in a defined area is to conduct a number of replicate distance samplings, randomly or systematically located within it, and calculate a mean. As for any sampling method, a balance has to be struck between the convenience of making a few long line transects and the greater information that comes from making more short ones. Small variations in density along a single line transect do not cause much bias in the estimates of mean density but, if the habitat is clearly heterogeneous, then the various habitat divisions should be studied separately. This is true for any census method but is particularly important in distance sampling because, as well as the density of animals being different in different habitats, their detectability is likely to be different.

It is important that the points and lines are placed randomly (or systematically, if appropriate) over the study area. If, for example, they follow tracks, one's estimates are likely to be biased because animals may avoid or be attracted to tracks. If one has physically to deviate around an obstacle, one should inspect the correct line from as close as one can get to it, sufficiently carefully to ensure that all animals on it are detected.

The theory of distance sampling assumes that all animals at the observation point or on the line are detected: it is the comparison of the number of animals detected at greater distances with those detected at the point or line that allows detectability to be modelled. Thus, if animals flee away from (or are attracted to) the observer, they must be detected before they are disturbed. If the field methodology cannot guarantee that all the animals on the point or line are detected, then ancillary observations may be made to provide an estimate of detectability there. Typically, two observers are used, as in the double-observer method (above), so the correction factor is sensitive to violations of the same assumptions as that method. In aerial surveys, the aircraft itself may obscure the ground immediately below it; one then ignores this strip of ground and treats the immediately adjacent strip as the centre. This assumes that detectability for that adjacent ground is perfect (Buckland *et al.* (2001), Section 8.4.3).

If detections of individuals are not independent, for example if neighbouring birds tend to sing at the same time, the variability of the results is increased. Another example of non-independence would be when animals occur in flocks. In this case, one can treat the flock, rather than the individual, as the sample unit; having estimated the density of flocks, one multiplies by average flock size to estimate the density of the population; unfortunately, this simple approach will be upset if detectability is greater for larger flocks, as it usually will be.

It is best if each animal is counted once only, which requires careful fieldcraft if the animals are mobile and in a habitat in which they may disappear and reappear sporadically.

The method assumes that the detectability of all individuals decreases with distance in the same way. This is often not true. For example, a female bird sitting on a nest may be as certainly detectable as a singing male if her nest is exactly on the transect line, but she may be almost undetectable if the nest is only 5 m from the line whereas the songster may still be almost certainly detectable at 50 m. If possible, one should treat such categories separately, making separate density estimates for each.

Distance-sampling methods also depend for their success on the accuracy with which the distance to each animal is measured. Wherever possible, a tape measure or range-finder should be used. If the distances have to be estimated, it may be better simply to record whether the animal is within or beyond some fixed distance, for it is easier to train observers to estimate one fixed distance than to estimate all distances accurately. This may more than compensate for the fact that methods based on only two recording zones (within and beyond the fixed distance)

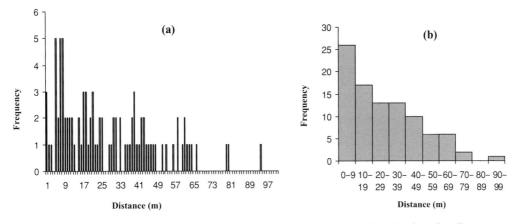

Figure 3.8 Frequency distributions of simulated line-transect data: (a) detection distances as originally measured, to the nearest metre; (b) detection distances grouped into 10-m classes, to show the overall shape of the distribution more clearly.

are inherently less precise than those based on full distance measurements (because they use much less information): for point counts, the variances are 25%–50% greater than when recording exact distances. A compromise method is to record the number of animals seen within several bands on each side of the line, such as 0–10 m, 10–25 m, 25–100 m and beyond 100 m.

Analysts typically ignore the most distant 5%–10% of observations because detectability is so low at extreme distances that their inclusion actually decreases precision. To ignore a certain percentage is generally a better approach than deciding in advance not to record beyond a certain distance (especially in multi-species studies, because the optimal cut-off distance is species-specific). However, it may be useful to use a cut-off in the field, to ensure that all observations come from the same habitat.

Apart from the use of trapping webs and trapping line transects, distance sampling appears at first sight to be simpler and less demanding than making observations sufficiently intensively that one can guarantee finding every organism in the study area or than conducting a mark–recapture study. But beware! If distance-based estimates are to be accurate, the assumptions must be met; if the estimates are to be precise, large numbers of animals need to be observed – 60–80 is the recommended minimum.

Analysis

The first step in the analysis of distance-sampling data that are in the form of exact measurements should be to plot a histogram of the data, grouping the data into 10–20 distance classes (Figure 3.8). This allows one to check that the data have a 'shoulder' around zero, rather than falling off sharply. If there is no shoulder (the data in Figure 3.8 are a little marginal in this respect), this implies that the detectability at and close to the line might not be perfect, which will reduce the reliability of the estimate. The best solution is to repeat the survey using better field techniques. Alternatively,

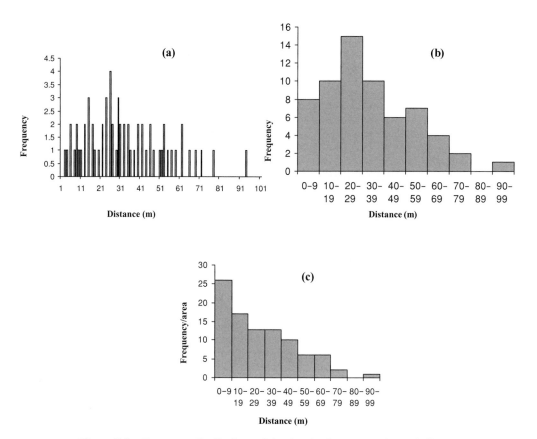

Figure 3.9   Frequency distributions of simulated point-transect data: (a) distances as originally measured, to the nearest metre; (b) distances grouped into 10-m classes; (c) with frequencies corrected by area, showing how detectability decreases with distance once the increasing area of observation at greater distances is accounted for.

one can use the data with particular care (see below). One should also check that the data are not 'heaped' around values such as 0, 5, 10 etc. (as often happens if distances are estimated rather than measured precisely). If they are, one can circumvent the problem by grouping the data into classes, with the heap-points falling roughly in the middle of the classes, and analysing the grouped data. The histogram also enables observations of unusually large distances to be identified and excluded from the analysis, following the general principle of truncating the outlying 5%–10% of observations.

Histograms for point-transect data should be adjusted according to the area of ground covered at each distance (Figure 3.9).

It is possible to estimate densities both for line and for point transects using simple formulae. However, the DISTANCE software (see the appendix to this chapter) provides a much better alternative because it provides much more than the estimates (such as confidence limits,

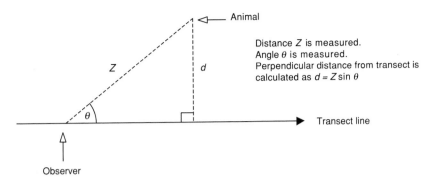

Figure 3.10 Calculating the perpendicular distance from an animal to the transect line.

goodness-of-fit tests, etc.); it also allows various analyses to be employed. It is easy to use. On opening a file, one is asked to tick boxes to indicate whether the survey was a line transect, a point transect or a cue count, whether the data are perpendicular distances or radial distances with angles (Figure 3.10), whether the observations were of singletons or clusters and whether the observations were made on both sides of the line or just one. A Data Explorer utility allows one to enter and manipulate data and to specify the sampling distribution and intensity.

The Analysis Browser utility allows one to create and run analyses, setting zone widths and truncation distances, choosing a model type, and specifying the required outputs. The model types allow a variety of detectability functions to be used (uniform, half-normal and hazard rate), each adjustable with cosine, simple-polynomial or Hermite-polynomial expansions, to provide a closer fit to the data. The uniform and half-normal models are considered good starting points, though the hazard rate is recommended if the data do not exhibit a clear 'shoulder'. It is sensible to try a variety of models, using AIC and biological insight to choose between them. DISTANCE also allows one to simulate data, which assists in assessing the robustness of one's models.

The use of ancillary information

The precision of distance-sampling estimates may be improved by building into the analysis covariates likely to affect detectability – habitat conditions, weather, type of cue being used to detect animals, observer, etc.

Line transects

General

A line transect involves an observer moving along a predetermined route through the study area, recording the distances at which each individual is seen (either exactly or in bands). Line transects are often used for birds (especially in open habitats, where it is relatively easy to record birds while walking), for aerial surveys of big game and for ship surveys of whales. The observer should move sufficiently slowly to detect all the animals on the transect line and most of those nearby. However, if one moves too slowly, animals ahead of one on the line have a greater chance of fleeing before one sees them and individuals that move are more likely to be counted twice; slow movement also uses time that could otherwise be used in surveying a greater transect length. The distances must be measured perpendicular to the transect line; if animals flee as the observer approaches one should measure (or estimate) both the distance from the observer to the animal when it is first detected and the angle between the animal and the transect line, so that the perpendicular distance may be estimated (Figure 3.10). Animals are usually recorded on both sides of the line but may be recorded on one side only.

Note that line transects are not the same as strip transects, in which one counts all the organisms in a strip across the study area, assuming that every individual in the strip is detected. Neither are they the same as line intercepts, in which one counts the number of sessile organisms crossed by a line laid down across the study area.

Transect lines should ideally be straight. It may be necessary to depart from this ideal for practical reasons but particular care is then needed to ensure that the line followed is truly representative of the habitat: following an existing track through a forest will result in fewer encounters with species that prefer dense cover than will following a randomly placed line. To cover an area properly, a number of independent transects should be walked. As with most line-based methods, it is generally better to avoid overlap of the lines by having them all parallel. It is usual to space them sufficiently far apart that the observations of neighbouring lines do not overlap. Setting the lines at fixed distances apart is practically convenient and if there is a gradient in density across the study area one can substantially reduce the variance of one's estimate of mean density. (Since the data for all lines are typically combined for analysis, so that they form a single sample, such a non-random distribution is no problem.) As usual, lines placed across a known (or suspected) gradient in density produce estimates with smaller variances than do lines placed parallel to the contours of density. If one walks a triangular route to get one back to one's starting point (see Chapter 2), one should omit enough of the corners of the triangle to avoid overlap of observations of adjacent sides; similarly for a square route (with the alternative of making no observations at all on two of the sides). Note that the sections of such routes are not statistically independent, so their data should be pooled for analysis. The same is true of the segments into which observers may divide a long transect (for example, five 200-m sections of a 1-km transect). Such division may be useful for practical recording purposes and it allows habitat differences between the sections to be taken into account, but the data from the various sections are not statistically independent.

While the use of DISTANCE software is preferable, the methods of Box 3.21 can be used to estimate densities from line-transect data.

Mobile animals, stationary observer

There are situations in which an observer can count animals passing in one direction, such as at some migration-watch points. Sometimes, the closer animals can be detected with certainty but the more distant ones cannot, making distance-sampling analysis appropriate. However, in many such cases the density of animals is not uniform across the span that is observed, which violates a key assumption.

Box 3.21. **Simple estimates of density from line-transect data**

The data represent 94 detections along a 1-km transect (with observations being made on both sides). The frequency distribution of detection distances is that shown in Figure 3.8.

Analysis based on exact distances

This analysis assumes a half-normal detectability function.

n = total number of animals detected
x_i = perpendicular distance of the ith animal detected from the transect line
L = length of transect

Density around the transect is estimated as

$$\hat{D} = n \sqrt{\left[2n \bigg/ \pi \sum_i (x_i^2) \right] \bigg/ (2L)}$$

For the example data, $\sum_i (x_i^2) = 109\,146\ \text{m}^2$.
The estimate is thus

$$\hat{D} = 94\sqrt{(188/109\,146\pi)}/2000$$
$$= 0.00110 \text{ animals per m}^2 = 11.0 \text{ per hectare}$$

Analysis based on two recording zones

This analysis assumes a negative-exponential detectability function.
The first zone extends from the transect line to Z units on either side of the line; the second extends from Z to infinity.

n_1 = number of animals detected in first zone
n_2 = number of animals detected in second zone

Density around the transect line is estimated as

$$\hat{D} = \{(n_1 + n_2) \log_e [(n_1 + n_2)/n_2]\}/(2\,L\,Z)$$

For the example data, $Z = 25\,\text{m}$, $n_1 = 52$, $n_2 = 42$.
The estimate is thus

$$\hat{D} = 94 \log_e (94/42)/(2000 \times 25)$$
$$= 0.001\,51 \text{ animals per m}^2 = 15.1 \text{ per hectare}$$

Point transects

A point transect entails the observer remaining at one point for a fixed time and recording the animals he or she sees (or otherwise detects). As with line transects, distances may be recorded in terms of concentric zones around the point or the exact distance to each animal may be measured or estimated. Because the analysis of point-transect data involves calculating areas from the distances measured, they are especially sensitive to errors in the distances, compared with line transects, since the error is squared in calculating the area.

Point-transect methodology assumes that there is no immigration into the area during the observation period; otherwise density will be overestimated. The recording period must, however, be long enough for all animals close to the observer to be detected. Indeed, because birds, for example, may fall silent as an observer approaches, it is common to allow a brief 'settling-down' period after arriving at the observation point before beginning the observations. (The assumption of no immigration is irrelevant to line transects, where observations at each point on the line are effectively instantaneous.)

The number of animals recorded at a single point is often small. This problem may be overcome by combining the data from a number of replicate counts, to provide an estimate of mean density in the area – though one must remember to allow for the number of replicates in the analysis. Even then, the number of animals detected at or very close to the points may be small (because so little of the study area falls at or very close to the points) and, since this number is crucial to the estimation of detectability, the variance of density estimates from point transects can be greater than is desirable. Since the data from the various points are combined, the points do not need to be randomly distributed. Indeed, if there is any variation in true density across the study area, the variance in one's estimate of mean density is reduced if the points are spaced uniformly.

If any of the points falls close enough to the edge of the study area for some animals beyond the edge to be detected, one should include those observations only if the edge of the area is arbitrary. If it represents a real ecological boundary, beyond which one expects the density to be different, a different approach is needed (Figure 3.11). If W is the radius beyond which observations are ignored, extend the sampling area to include both the actual study area and a buffer zone of width W. Place the points over the whole area but ignore any detections that do not fall within the core study area.

Calculations to estimate densities from point-transect data are presented in Box 3.22 (though use of the DISTANCE software is preferable).

Because line transects usually produce more data than point transects and are less sensitive to inaccuracies in distance measurement and to animals moving into the area, point quadrats are used only where line transects are impractical. For example, it may be impossible to make observations while one is moving through dense vegetation or over steep terrain, especially if one is attempting to stick to a straight line regardless of obstacles.

Passive distance sampling

Trapping webs

In this technique, a gradient in detectability is imposed not by Nature but by the investigator. A trapping web is a set of concentric circles of traps; each circle has the same number of traps in it

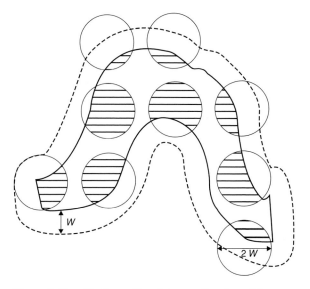

Figure 3.11 Dealing with points near the edge of the survey area. For details, see the text. (From Buckland *et al.* (2004), with permission.)

(usually arranged for convenience so that the traps in the various circles lie in lines, like the spokes on a wheel – see Figure 3.12). Ideally, if the innermost circle has a radius R, succeeding circles have radii of $3R$, $5R$ etc. – i.e. the circles lie a constant distance apart ($2R$) and the inner one has a radius equal to half that distance. The estimation of population density from trapping webs rests on the fact that animals that are near the centre of a web are more likely to be caught than those near (though within) its outer margins – simply because the traps are more crowded in the centre. It might appear possible to work out how the probability of capture varies with distance from the centre, from the areas of the successive circles. The central traps are so close together, however, that they interfere with each other (to an unknown degree), upsetting such simple calculations. Since the size of the interference effect is unknown, we are in the same position with these data as with those from a point transect and we use the same methods to analyse them (i.e. DISTANCE). The familiar assumptions apply: all animals at the centre are caught, the trapping of one animal is independent of the trapping of its neighbours, all individuals are equally catchable and density is constant across the web. As with point transects, the precision of this method tends to be low unless population densities are high enough to produce large numbers of captures at the centre.

Animals are caught over a series of trapping sessions, until no more new animals are being caught in the central traps. It is likely that new animals will still be getting caught in the outer traps since, although there is the same number of traps in the outer circles, the area of ground from which they draw animals is larger. For the purpose of obtaining a density estimate, only the position at which an animal is first captured is used. We need to know the number of first captures in each circle. The total numbers of animals first caught in the few circles that are furthest out may be clearly greater than the numbers first caught in slightly less marginal circles (Figure 3.13), as a result of catching animals originating from beyond the outermost circle. If such

---

Box 3.22. **Simple estimates of density from point-transect data**

The data represent 63 detections made on 40 points. The frequency distribution of detection distances was that shown in Figure 3.9.

Analysis based on exact distances

This analysis assumes a half-normal detectability function.

n = total number of animals detected
x_i = distance of the ith animal detected from the point
k = number of points

Density in the area in which the points fell is estimated as

$$\hat{D} = n^2 \Big/ \Big[\pi k \sum_i (x_i^2)\Big]$$

For the example data, $\sum_i (x_i^2) = 91\,118\,\text{m}^2$, so

$$\hat{D} = 63^2/[\pi \times 40 \times 91\,118]$$
$$= 0.000\,35 \text{ animals per m}^2 = 3.5 \text{ per hectare}$$

Analysis based on two recording zones

This analysis assumes a negative-exponential detectability function.

The first zone is a circle of radius Z around the point; the second is outside that, to infinity.

n_1 = number of animals detected in the first zone
n_2 = number of animals detected in the second zone

Density in the area in which the points fell is estimated as

$$\hat{D} = \{(n_1 + n_2)\log_e[(n_1 + n_2)/n_2]\}/(\pi k Z^2)$$

For the example data, $Z = 30$ m, $n_1 = 33$ and $n_2 = 30$.
The estimate is thus

$$\hat{D} = \{63\log_e(63/30)\}/(\pi \times 40 \times 900)$$
$$= 0.000\,41 \text{ animals per m}^2 = 4.1 \text{ per hectare}$$

---

immigration is obvious from the data, the affected circles should be ignored: for the data in Figure 3.13, for example, it would be appropriate to truncate the data beyond the 18th circle. Undetected immigrants will result in the density of animals being overestimated. For this reason, it is good practice to release trapped animals (having marked them, to ensure that they are not scored again if they are retrapped later); removing animals leaves a population vacuum into which immigrants may move.

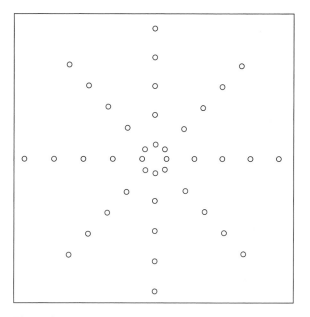

Figure 3.12   A representative layout of a trapping web.

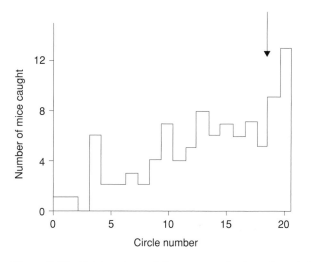

Figure 3.13   Total captures of *Peromyscus* mice in various circles of a trapping web. Circle number 1 is the central circle; circle number 20 is the outermost. The arrow marks the point at which the data were truncated for analysis. (Data from Anderson *et al.* (1983).)

A study comparing trapping webs with rectangular trapping grids, analysed using DISTANCE and mark–recapture estimates (CAPTURE models M_b and M_{bh}), respectively, on populations of rodents in open scrubland in New Mexico, showed the former to be generally superior (Parmenter *et al.* 2003). The web-based estimates generally had narrower confidence limits than did the

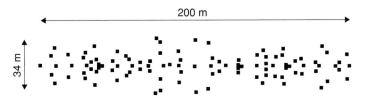

Figure 3.14   A trapping line transect. (From Buckland *et al.* (2004), with permission.)

mark–recapture estimates and the latter seemed more prone to give biased estimates at particularly low or high densities. (Actual densities were assessed by intensive pseudo-removal, the animals being in immigrant-proof enclosures.)

Trapping line transects

This method is similar to the trapping web approach but the traps are laid around a line rather than around a point. On the line itself, they are placed at high density at regular intervals; on either side of the line they are placed at random, with their density decreasing away from the line according to a hazard rate or similar function (Figure 3.14). The analysis is similar to that for line-transect data.

This method was first used in 2002 and needs further testing to establish its precision relative to, say, mark–recapture. Laying out the traps takes longer than placing them on a regular grid (as one would do for a mark–recapture study) and finding them again is more difficult.

Sample counts (6): interception methods

Point quadrats

Some plants and other sessile organisms may vary considerably in size. Indeed, they need not even occur as obviously distinct individuals but may rather form more-or-less-continuous patches. It is then often more appropriate to measure the extent to which they occupy an area than to count them. An easy way to do so is to estimate the extent of cover by eye, but this may be inaccurate, especially if there are small gaps in the patches. It is more accurate (but more time-consuming) to lay down a set of points over the study area and count the proportion of points that lie over the species in question. The 'points' are typically narrow wires laid down in sets of ten in a 'point frame'. Each point frame provides an estimate of cover for only the area that it encompasses, so a series distributed according to the principles in Chapter 2, is required in order to represent a wider area. Note that, whatever system is used, the 'points' should be as narrow as possible. If they are too wide relative to the organism, its extent of cover will be overestimated. Chapter 4 provides practical details.

Point quadrats are particularly useful when the cover of a species within a patch is not continuous, since a point falling in a small gap does not register positive, whereas the gap might not be noticed when one is estimating cover by eye.

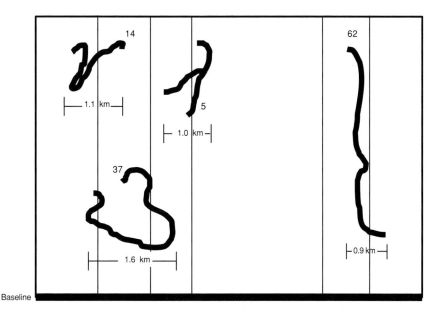

Figure 3.15 Line intercepts of tracks. The thin straight lines are the intercept lines flown; the thick rambling lines are the tracks intercepted. The number by each track is the number of animals in the herd associated with that track. The length of the projection of each track on the baseline is also shown.

Line intercepts (cover)

A thin line laid down over an area is another way of estimating cover, if one measures the lengths of the sections of the line that do and the lengths of those that do not fall across the species in question. The method may be more efficient than point quadrats when cover is continuous within patches.

Line intercepts should be laid down according to the same protocols as strip transects.

Line intercepts (counts)

In theory, line intercepts can be used to count the number of patches of a species that are intercepted (which may but need not be the same as the number of individuals). Allowance may need to be made for the different probabilities of interception of patches of various shapes and sizes. Strip transects are generally more effective than line intercepts.

Another use of line intercepts is in respect of aerial surveys of group-living animals in the winter. A set of lines is flown (at random positions along a baseline) and all the tracks made in the snow by the species in question that are encountered are explored in their entirety. The tracks are mapped and the number of animals associated with each is determined, as in Figure 3.15. From the map, the projection of each track on the baseline from which the flight lines are extended is measured. This allows one to make a simple estimate of the number of animals in the entire area over which the sample lines were flown (Box 3.23).

Box 3.23. **The line-intercept method for animals associated with definite tracks**

$B =$ length of the baseline

$n =$ number of lines surveyed

$N_i =$ number of animals associated with the ith track

$W_i =$ projected width of the ith track on the baseline

$p_i = W_i B$

   $=$ proportional width of the ith track

$\pi_i = 1 - (1 - p_i)^n$

   $=$ probability that the ith track is intercepted by at least one line in a random set of n lines

The estimated population in the whole study area is then

$$\hat{N} = \sum_i (N_i / \pi_i), \qquad \text{summation being over all tracks.}$$

Using the example in Fig. 3.15 which has a baseline of 8 km and $n = 5$,

Track	N_i	W_i	p_i	π_i	N_i/π_i
1	14	1.1	0.1375	0.5227	26.78
2	37	1.6	0.2000	0.6723	55.03
3	5	1.0	0.1250	0.4871	10.27
4	62	0.9	0.1125	0.4494	137.96
					$\hat{N} = 230$

The assumptions of this method are that all the animals associated with each track are detected and that the detectability of each track depends only on its projection on the baseline (in particular, that its detectability does not depend on the number of animals associated with it). Its precision is low, unless the number of animals per track is fairly uniform and most tracks are detected. Thompson (2002) provides more detail.

Sample counts (7): migrating animals

Continuous migration

Suppose that migrants can be counted as they pass. Sometimes constant observation is possible, as with automatic counts of fishes passing up fish-ladders. More often, counts are interrupted by darkness, bad weather or lack of manpower. If the latter is the only problem, counts may be taken at random intervals through the migration season and the total numbers passing during the season estimated in the usual way. (It is most efficient to stratify the season, taking relatively more counts at the height of the passage.) However, interruptions in the counts caused by light, weather

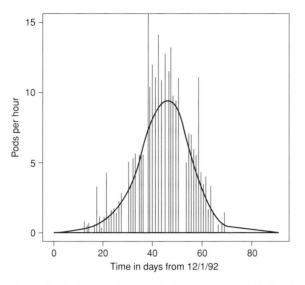

Figure 3.16 Observed counts during watched periods (bars), with fitted model (line) from the 1992 survey of the migration of grey whales past Monterey, California. (From Borchers *et al.* (2002), with permission.)

and similar factors may result in sampling being non-random, which renders this sample-based approach suspect.

An alternative is to use the observations that are made to model the relationship between numbers passing and date. The model is a curve, fitted either by eye or by statistical calculation, as in Figure 3.16. Calculation is more objective and can provide confidence limits. Borchers *et al.* (2002) provide further details.

Stop-over sites

A migratory stageing site may be used by many more animals in the course of a season than are present at any one time, because early arrivals move on before late-comers arrive. It is often important to know the total using the site but, unless all the animals are individually recognisable, this cannot be determined directly. However, methods to produce estimates based on combining the counts with estimates of the survival-rate of marked individuals have been developed, 'survival' in this case meaning that the animals remain both alive and on the site (Frederiksen *et al.* 2001; Schaub *et al.* 2001).

Population indices

The idea of an index

The ideal index

An index is a measurement that is related to the actual total number of animals or plants. Ideally, the relationship should be such that the ratio of the index I to the number N is constant:

$$I/N = K$$

In this case, even if the index ratio (K) is unknown, one can compare populations in various places or at various times by comparing the indices. In reality, the index ratio tends to vary, as can be appreciated by considering some examples of indices.

Examples of indices

The numbers of animals seen during a certain time period while standing at a point or the numbers seen while walking a set distance (that is, point and line counts rather than point and line transects) are not estimates of population size. They are merely indices. Although these counts are likely to be larger when the number of animals in the population is larger, they also vary according to the competence of the observer, the behaviour of the animals (they may be less detectable when resting, for example) and all those environmental factors that influence detectability, such as the density of the vegetation, the lie of the ground and weather conditions. The number of calls heard per minute can be an index of the numbers of frogs in an area but calling rate varies with season, time of day (or night) and weather. The number of animals shot by hunters is commonly used to index the numbers of game animals; it is influenced by the number and enthusiasm of hunters. The number of animals caught in pitfall traps depends not only on the density of animals in the area but also on their activity (which varies with season, weather, etc.), on the surrounding vegetation, on the colour of the trap, on the size and shape of the aperture, on the material, depth and shape of the trap (which influence whether animals can escape), on whether or not there is a preservative fluid in the traps, what that fluid is and how deep it is (which influence the attractability of the traps and the ease of escape from them) and on the spacing and arrangement of the traps. The population of small seeds or micro-organisms in soil may be assessed by germinating or culturing them; the numbers observed will depend on the culture conditions. Tracks in mud provide an index of deer populations, which is influenced both by the animals' activity and by soil conditions; tracks left by small mammals on smoked paper left in tunnels provide more standardised recording conditions but still depend on activity as well as numbers. The number or weight of droppings in an area depends both on the size and on the metabolic activity of animals, as well as on their number. (This can be turned to advantage by ecologists studying, for example, the availability of caterpillar food to birds, since actively feeding caterpillars are easier for birds to find than inactive ones and large ones provide more food than small ones; thus the weight of caterpillar droppings falling on the forest floor may be a better index of available food supply than is a simple count of caterpillar numbers.)

Counts of a part of the population

It may be possible to make an accurate count or estimate of a part of the population. This may be used as an index of the whole, so long as one remembers that the proportion of the whole that it comprises may vary. Thus a count of males holding territories is an index of the breeding population but one must not forget that under many social systems more (sometimes fewer) females may breed than males and that variable proportions of males may hold territory but not breed or, conversely, breed without holding territory. If one wishes to go further and use the count of territorial males as an index of the whole population, one must also remember that a variable proportion of adults might not breed (especially for longer-lived species) and that the adult-to-juvenile ratio probably also varies.

Counts and density estimates as indices

The first example of an index listed above was an incomplete direct count. Earlier in this chapter, we have considered various indirect methods of estimating population numbers and density, noting that it is often difficult to avoid bias. Such biased estimates can be taken as indices, recognising that the factors leading to the bias should not vary if the index ratio itself is to be constant, as is ideal. For example, if we have to assume that a population being studied using mark–recapture methods is closed (because the Jolly–Seber method cannot be applied, for practical reasons) but we know that it experiences some immigration and emigration, we can use our estimate as an index, so long as we bear in mind that it will change not only if true population density changes but also if the animals become more or less mobile.

We have also seen that it might not be possible to estimate the true density of a population using many of the techniques based on trapping, because one is not sure from how large an area the trapped animals have been drawn; the estimated size of the trappable population is then an index of population density in the area, which will also vary according to the mobility of the animals.

The minimum number known to be alive

If a population of distinguishable individuals is studied for a period, the total number of individuals observed (the minimum known to be alive, MNA) can be used as an index.

Beware, however, a common error in studies extending over a number of sampling occasions: individuals not recorded on a particular occasion are added to its MNA total if they are recorded before and after that occasion. In the absence of resurrection after death, this is clearly logical: those individuals must have been alive on that occasion, even though they were not recorded. However, there can be no such additions for the first or last of the series (and fewer for other early and late occasions than for occasions in the middle of the series), so the MNA will appear to change even if the true population size does not.

Cue counting

Suppose that one wishes to use the number of droppings found in an area as an index of population size. The validity of the index would depend not only on the relationship between the number of animals and the number of droppings in the area but also on the effectiveness of one's searching for the droppings. One may regulate the latter by walking definite transect lines and applying distance sampling to estimate the density of droppings, thus reducing one level of variation. The method has been used, for example, in studying whale numbers from distance sampling of the numbers of 'blows' observed, supplemented with separate surveys of the frequency with which whales blow. Buckland *et al.* (2001) provide further details.

Comparing index values: take care!

Particular care is needed when comparing index values obtained in various places or at various times in the same place. It is all too easy to assume that differences between index values reflect differences in the populations and to ignore the possibility that they are the result of differences in detectability. These may arise from only minor differences in circumstances: for example, skylarks *Alauda arvensis* are 50% more detectable in the stubble left after a crop has been harvested than they are in managed grassland (and the exact height of the stubble also affects their detectability). In monitoring programmes that use indices, it is particularly important to be aware that long-term changes in detectability may mimic population trends. So, even if it is more costly to mount a monitoring programme based on direct population estimates rather than one based on indices, this cost needs to be weighed against the cost of coming to the wrong conclusion about the population's status. Moore & Kendall (2004) consider such costs.

Overcoming variation in the index ratio

The value of standardisation

Standardisation of methods is valuable in reducing the variation in one's measurements. It is essential if one wishes, as is usual, to compare populations in various places or to study the variation in numbers over time. Much of the variation likely to affect index ratios can be substantially reduced by making the observations under similar weather conditions, at the same time of day and in the same season; using the same type, number and layout of traps (containing the same attractants or foods); having the observations made by the same person, using the same binoculars, etc. If various places are to be compared, simultaneous trapping sessions at each place ensure standardisation of factors such as weather, time of day and season – though subtle microclimatic differences among the sites can cause different levels of activity in their populations. Comparisons between years demand careful attention to the species' natural history. It may, for example, be inadequate to try to standardise by surveying on the same date every year if the date of species' peak of activity varies as a result of differences in weather conditions. While it is never possible to standardise all of the conditions that are likely to affect the relationship between the index and the population size, it is useful to standardise as much as possible.

The value of multiple, randomised observations

Suppose that an observer wished to compare the number of animals in two areas, knowing that the bias in her counts was strongly weather-dependent. If she made counts in the two areas on two different days, she would have no way of knowing whether the differences between the counts were the result of differences in the populations or of differences in the weather on the two days. To overcome this problem, she could extend the study over several days; on half the days, chosen at random, she would count in one area and on the other days she would count in the second area. The standard techniques of analysis of variance could then be used to compare the mean counts in the two areas, the effects of weather being subsumed in the error variance.

Allowing for variation

Better than simply including the effects of weather (or any other factor that influences the population index) in the error variance, one can take them into account in the analysis. An imaginary example is shown in Box 3.24, in which the variable to be allowed for is the time of day. The method assumes that the effect of time of day is the same in the various study areas; if it is not, it is more difficult to allow for it. In the example in Box 3.24, the 95% confidence limits of the difference between the mean indices for the two areas are $+0.6$ and $+35.2$ (in favour of area B). If time of day had been ignored in this analysis (so that the data for all three times were lumped into a single set for each area), the variation associated with time would have been subsumed in the residual variance, so the mean squares would have been 2412 (1 d.f.) (place) and 2245 (24 d.f.) (residual). As a result, the confidence limits of the difference between the mean indices of the two areas (-17.6 and $+53.4$) would have given a much wider interval. This shows how taking extraneous factors into account increases the precision with which means and differences are estimated.

Box 3.24. **Using two-way analysis of variance to allow for extraneous sources of variation in index studies**

Example

An index that is greatly affected by time of day. Observations to compare populations in two areas were made; because not all observations could be made at a standard time of day, this factor has to be allowed for in the analysis.

The observations were as follows:

| Area | Observations | | |
	Morning	Mid-day	Afternoon
A	40, 49, 66, 69, 90	100, 107, 130, 140, 174,	140, 150, 165, 170, 180
B	59, 69, 84, 87, 105	119, 125, 145, 163, 195	150, 170, 170, 195, 203

Analysis of variance

Conduct a two-way analysis of variance, with time and place as the two factors. For the example data, the results are

Source of variation	Mean squares	Degrees of freedom
Time	25 000.53	2
Place	2 412.03	1
Interaction	4.22	2
Residual (or 'within')	535.87	24

Are the effects of time the same in the two areas? This hypothesis is tested by calculating the variance ratio for the interaction:

$$F = 4.22/535.87 = 0.008, \qquad \text{with 2 and 24 degrees of freedom}$$

This value is not significant, so we can conclude that the effects of time of day are the same in the two areas.

Differences between the areas

Mean indices, averaged over all times of day, are

$$(40 + 49 + \cdots + 180)/115 = 118.0 \qquad \text{for area } A$$
$$(59 + 69 + \cdots + 203)/115 = 135.9 \qquad \text{for area } B$$

The mean difference between the two areas is thus

$$d = 135.9 - 118.0 = 17.9$$

The standard error of this difference is

$$s_d = \sqrt{2s^2/n}$$

where

$$s^2 = \text{residual mean square from the analysis of variance}$$
$$n = \text{total number of observations in each of the samples being compared}$$

In this case,

$$s_d = \sqrt{2 \times 535.9/15} = 8.45$$

The 95% confidence limits are

$$\text{difference} \pm t \times s_d$$

where t is Student's t for $2(n-1)$ degrees of freedom and $P = 0.05$.
The limits in this case are

$$17.9 \pm 2.05 \times 8.45 = +0.6 \text{ and } +35.2$$

Unequal sample sizes

Cases in which the sample sizes are not the same in all combinations of area and time employ the same principles, though they are not currently provided for in Excel® and are more complex to calculate.

Had the example in Box 3.24 involved more than two areas, the confidence limits of the difference between any two of them would have been worked out in the same way, using the residual mean square from an overall analysis of all the areas.

Note that, despite the power of analysis of variance to separate out the influence of factors such as time of day from the differences between places, it is still better to standardise. Had the 30 observations in Box 3.24 all been made at the same time of day, then the difference between the sites would have been estimated more precisely.

Correction factors

The mean indices for the three times of day in Box 3.24 are 71.8 (morning), 139.8 (mid-day) and 169.3 (afternoon). Since these observations have established the differences among these times, one could, in principle, take observations made at only one time of day from other areas and correct them to some standard time of day. For example, suppose that one surveyed areas C, D and E in the morning, around mid-day and in the afternoon respectively. To convert the indices for C and E to values 'standardised on mid-day', one would add 68.0 (139.8 − 71.8) to the index for C and subtract 29.5 (169.3 − 139.8) from that for D. This practice is, however, best avoided, since it assumes that the differences between times of day are the same in all the other areas, which might not be true.

Regression techniques

Suppose that one was trying to index frog populations by counting the number of calls heard in standard ten-minute periods, at a standard time of night. It is likely that calling rate varies with temperature, so it would be best to make observations only on nights on which the temperature fell within a certain range. Were this to be impossible, one could classify the nights into categories such as 'cool', 'warm' and 'hot' and then apply the analysis-of-variance method of Box 3.24 to the data. Greater precision would be obtained if one measured the temperature at the time of each observation and used the usual statistical techniques of regression analysis to discover the relationship between call rate and temperature. The extent to which the call rate at an individual location departed from the regression line would then provide an index corrected for the effect of temperature (Figure 3.17).

Greater precision is obtained by making several observations at each location, on different nights, so that separate regression lines can be fitted to each (using the standard techniques of analysis of covariance). The differences in elevation of the lines then provide measures of differences among their populations (Figure 3.18).

When the index ratio varies with population density

The ratio of an index to the number of animals present may vary not only in relation to extrinsic factors but also in relation to population density itself. For example, the frequency with which an animal calls may increase if others are calling nearby; if so, a doubling in numbers may lead to the call rate increasing more than twofold (Figure 3.19, dashed line). Conversely, an observer

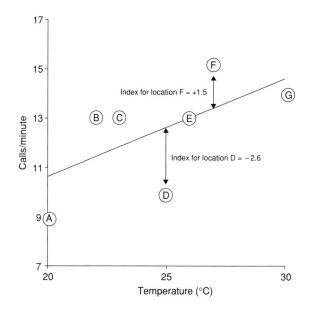

Figure 3.17   An imaginary example showing total call rates of frogs at various locations (A–G) in relation to temperature at the time the observations were made. The deviation of individual points from the regression line provides an index of numbers at that location, corrected for the effect of temperature on call rate.

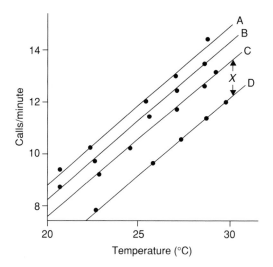

Figure 3.18   An imaginary example showing total call rates of frogs at four locations (A–D) in relation to temperature at the time at which the observations were made, there being several observations at each location. The differences in elevation (e.g. *x*) between the parallel regression lines provide indices of differences in numbers, corrected for the effect of temperature on call rate.

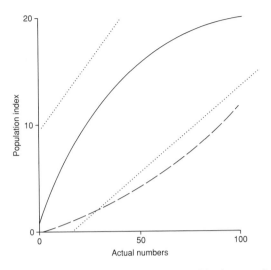

Figure 3.19 Various possible relationships between index values and number of animals present (imaginary examples). See the text for further explanation.

may be overwhelmed by the number of calls at high densities, so that changes in the counts are less than proportional to changes in numbers present (Figure 3.19, continuous line).

Indices may, indeed, have maximum possible values. If an index comprises the number of animals caught in a session and if only one animal can be caught per trap, then the index has a maximum value set by the number of traps. Another example is when the index is based on frequency of occurrence (see below): a common species will occur everywhere and it cannot occur more than everywhere even if its actual numbers increase.

Indices that are not linearly related to numbers can be useful so long as one bears in mind that a difference of, say, 10 points in the index has different implications if it is the difference between 20 and 30 from the implications it has if it is the difference between 120 and 130. This means, in particular, that average values have to be treated with caution. Box 3.25 shows how samples of observations can have identical mean values of the index but different mean population sizes (compare samples A and B) and conversely (compare B and C); indeed, one sample may have a smaller mean population size than another but a larger mean index (compare C and D). When the relationship between the index and numbers is convex (like the continuous line in Figure 3.19), samples that are more variable (C compared with B, in Box 3.25, for example) have lower mean indices for a given mean population size. The converse is true for concave relationships (Figure 3.19, dashed line). Note that indices that are linearly related to abundance but which do not pass through the origin (such as the dotted lines in Figure 3.19) present similar problems to those exhibiting curvilinear relationships.

Calibration

One way of dealing with the problem caused (for calculation of means) by non-linear relationships between population indices and actual numbers is to convert the index values to numbers of

Box 3.25. **Showing how a non-linear relationship between an index of population size and actual numbers can lead to means of indices not reflecting means of numbers**

The relationship between index and numbers used is that shown by the continuous line in Figure 3.19:

		Index values		Actual numbers	
Sample	Mean	Individual values		Individual values	Mean
A	18.0	16.0 18.0 18.0 20.0		50 67 67 100	71
B	18.0	18.0 18.0 18.0 18.0		67 67 67 67	67
C	17.6	15.2 17.1 18.6 19.4		47 60 74 87	67
D	17.0	14.0 14.0 20.0 20.0		40 40 100 100	70

Note that, in terms of means,
(1) although index(A) = index(B), numbers(A) ≠ numbers(B);
(2) although index(B) ≠ index(C), numbers(B) = numbers(C); and
(3) although index(C) > index(D), numbers(C) < numbers(D).

animals before calculating means. This requires that the relationship is known. It can be known only if one has studied a representative sample of study sites and has simultaneously measured both the index and the actual numbers. Such calibration then allows numbers to be estimated from studies in which only indices have been obtained.

Box 3.26 presents an example of such a calibration for a case in which the relationship between the index and numbers follows a straight line that does not pass through the origin.

Box 3.26. **The calibration of an index**

Example

Over a number of winters and in a number of areas of northern Wisconsin, USA, the number of trails of white-tailed deer *Odocoileus virginianus* encountered on strictly standardised transects was counted; the population densities of the deer in these areas were known through more direct methods (McCaffery 1976). Provided that the temperature had dropped to $-2\,^\circ$C by 20 October, the number of trails was linearly related to the density of deer:

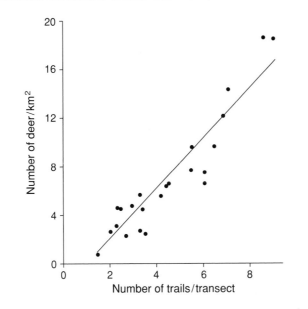

Methods

Calculate the least-squares regression of density (or numbers) on the index in the usual way to obtain the equation

$$N = a + bL$$

For the deer, the relationship between numbers per km^2 (N) and trails per transect (L) (the line plotted on the above graph) was

$$N = 2.16L - 2.83$$

(Note that, although in a biological sense the number of trails depends on the number of deer, which would make it appropriate to regress number of trails on number of deer if one wished to describe the relationship, it is appropriate to regress the population on the index when one wishes to be able to predict an unknown population number from an index value.)

From this equation, one can estimate the density of deer in an area simply from the number of trails per transect. Suppose, for example, that one encountered 5.0 trails per transect; then the estimated number of deer would be

$$\hat{N} = 2.16 \times 5.0 - 2.83 = 7.97 \text{ deer/km}^2$$

Confidence limits

From the original data set used to calculate the regression line,

m = sample size
s_N^2 = variance of numbers (densities)
s_L^2 = variance of index values
$\bar{L}$ = mean index value
r = correlation coefficient between numbers and index values

Then the 95% confidence limits of $\hat{N}$ are

$$\hat{N} \pm t \sqrt{s_N^2(1-r^2)\left(\frac{m-1}{m} + \frac{(L'-\bar{L})^2}{s_L^2}\right) \bigg/ (m-2)}$$

where

t = Student's t for $P = 0.05$ and $m - 2$ degrees of freedom
L' = the index value corresponding to $\hat{N}$

For the deer example,

$s_N^2 = 23.86,$ $s_L^2 = 4.47$
$\bar{L} = 4.54,$ $r = 0.937$

Hence, confidence limits of $\hat{N}$ are

$$7.97 \pm 2.08 \sqrt{23.86(1 - 0.937^2)\left(\frac{22}{23} + \frac{(5-4.54)^2}{4.47}\right) \bigg/ 21}$$

$$= 7.97 \pm 0.78$$

$$= 7.19 \text{ and } 8.75 \text{ deer/km}^2$$

Warning

Estimates made from calibration equations should be applied only under the same range of conditions and over the same range of densities as the original calibration data.

Territory mapping (Chapter 4) provides a form of calibrated index, in which the calibration is informal, being based on expert knowledge of the relationship between such data and the true numbers of animals present. This means that it is important to have trained analysts who work to common standards to produce the territory estimates.

Ratio estimators for lines passing through the origin

A better approach is available if it is both biologically reasonable to suppose that the index will be zero when there are no animals present and the value of the intercept (a) obtained from the regression is not significantly different from zero. One simply takes the sum of the numbers of animals found on all the calibration sites (ΣN) and the sum of all the indices from those sites (ΣI) to provide the ratio estimator

$$R = \sum N / \sum I$$

Then, for another site where an index I' has been observed but numbers have not been estimated directly, an estimate of numbers is then

$$\hat{N} = R I'$$

Non-linear relationships

If the relationship between numbers and the index is non-linear, one may fit a relationship either by using non-linear regression techniques or by using a suitable statistical manipulation to transform the relationship to a linear one (transforming both numbers and indices to logs often works).

Biologically based calibrations

It is sometimes possible to produce an analytical model of the relationship between an index and the true population, rather than having to rely on a simple mathematical relationship. For example, counts of dung have successfully been used to produce population estimates of mammals ranging from rabbits to elephants, and even of lizards. From measurements of rates of defaecation and of decay of dung, one can model the quantity of dung to be expected per animal in an area. (See Buckland *et al.* (2004), Section 11.10, for modelling methods.) It is essential to measure these rates accurately and under normal field conditions: calculations of decay rates that ignore the effect of dung beetles, for example, will be misleading. The effects of weather and of seasonal differences must also be taken into account. Because these fundamental rates are difficult to measure accurately, it is tempting to use rates already obtained in another area; due caution should be exercised.

Extrapolation: be cautious!

Obtaining data on true population sizes may be costly. For this reason, it is often tempting to extrapolate from an original calibration to rather different circumstances – higher or lower densities, other habitats, different seasons, etc. This should be done only with extreme caution because the relationship between true numbers and the index may be different under the various conditions.

For rare species, it may be impossible to obtain enough data for direct calibration. Apart from giving up an attempt to estimate true numbers for a rare species (which may be a sensible decision), there are two approaches to this problem. The first is simply to use a calibration obtained for a common species to which the rare species appears to be sufficiently similar in crucial respects. The other, appropriate to multi-species studies, is to combine data for several similar species (Bart & Earnst 2002).

Double sampling

Double sampling refers to a formal way of using calibration based on a sample to obtain an estimate for the whole area of interest. The idea is that one takes a random sample of the actual or notional plots into which that whole area is divided and a population index is obtained for each. (If sufficient resources are available, this sample can include the whole population of interest.) A subsample of the indexed plots is also subject to intensive studies that provide accurate population estimates for each of them. The index is calibrated on the information from the subsample and that calibration allows the populations to be estimated for the plots included in the main sample, the population of the whole area then being estimated from this sample. Box 3.27 provides details of the calculations. It assumes that all sample plots are of equal size and uses a ratio estimator for calibration; the method can be extended to plots of unequal size and may employ calibration statistics other than the ratio estimator (see Cochran (1977), for statistical details, and Bart & Earnst (2002), for a modern application).

This method is useful when it is cheaper or easier to obtain indices rather than accurate direct estimates of the population on sample plots. In the example in Box 3.27 (where there is a 40-fold difference in cost between indexing and true censusing), if the workers had relied simply on making enough true censuses, without the support of the index survey, it would have cost over ten times as much to obtain an estimate for the whole population of interest as precise as the one obtained by using the double-sampling method.

Frequency of occurrence

Basics

Theory

If observations are made in a number of locations, then the proportion of localities in which a species is observed is likely to be greater if the species is common than if it is rare. The curve relating frequency of occurrence to abundance will have the general form shown in Figure 3.20(a). Indeed, if the species has the same average density in all localities and individuals are randomly distributed then the curve in Figure 3.20(a) and the linear transformation in Figure 3.20(b) will apply exactly, with the coefficient a depending on how detectable the species is. The same principles apply if observations are made on a number of occasions at the same place.

If the species is more uniformly distributed than random, the Poisson equation shown in Figure 3.20 is still likely to apply because the random element will generally arise from whether or not an individual present at a locality is detected by the observer. If the species' distribution is clumped but the size of the clumps is fairly constant, the Poisson equation will still be a reasonable fit to the observations. If, however, the size of the clumps rises with the overall density of the population, it will not; the logarithmic transformation will then produce a right-convex curve rather than a straight line.

Box 3.27. **Double sampling**

Example

The total area of interest comprised 50 601 km² of the North Slope of Alaska. Sample plots were each 0.16 km², so the total area could be considered to comprise 316 256 plots ($=N$). Dunlin (*Calidris alpina*) were counted systematically but rapidly on 338 ($=n'$) sample plots. A detail of this particular study was that repeated counts were used to ensure representative coverage over the breeding season. This analysis uses the mean values (the x_i below). In nine ($=n$) of these, intensive studies were also conducted, to census dunlin numbers accurately.

Results of rapid counts

The mean number of dunlin encountered in rapid counts of 338 plots was

$$\bar{x}' = 4.139\,053$$

Results of the intensive counts

The data obtained from the nine intensively studied sites were as follows (ignore the right-hand columns at present):

Plot i	Rapid survey x_i	Intensive Study y_i	Rx_i	$(y_i - Rx_i)^2$
1	3.25	4	4.290 323	0.084 287
2	2.00	2	2.640 199	0.409 854
3	1.14	1	1.508 685	0.258 760
4	0.75	1	0.990 074	0.000 098
5	0.75	1	0.990 074	0.000 098
6	1.50	3	1.980 149	1.040 096
7	2.50	4	3.300 248	0.489 653
8	0.88	1	1.155 087	0.024 052
9	1.63	2	2.145 161	0.021 072
			Sum $=$	2.327 971

If $\bar{x}$ and $\bar{y}$ are the means of the above counts, then the ratio estimator is given by

$$R = \bar{y}/\bar{x}$$
$$= 2.111\,111/1.599\,206 = 1.320\,099$$

Estimation of total population

The estimate of the mean true number present in each of the 338 rapidly surveyed plots is

$$\bar{y}_N = R\bar{x}'$$
$$= 1.320\,099 \times 4.139\,053 = 5.463\,961$$

This is the best estimate of the mean number present in each of the 316 256 (N) notional plots in the whole area of interest ($\bar{y}_N$). Thus the estimate of the total population is

$$\hat{y} = N\bar{y}_N$$
$$= 316\,256 \times 5.463\,961 = 1\,728\,010$$

Confidence limits

Calculate var(y), the sample variance of the y_i values, in the usual way. In this case it is 1.611 111.

For each of the i intensive plots, calculate Rx_i and then $(y_i - Rx_i)^2$: these values are tabulated on the right in the table above.

Calculate

$$\text{var}(g) = \sum_i (y_i - Rx_i)^2/(n-1)$$
$$= 2.327\,971/8 = 0.290\,996$$

The variance of $\bar{y}_N$ is then

$$\text{var}(\bar{y}_N) = (N - n')[\text{var}(y)]/(Nn') + (n' - n)[\text{var}(g)]/(n'n)$$
$$= (999\,662 \times 1.611\,111)/(316\,256 \times 338) + (329 \times 0.290\,996)/(338 \times 9)$$
$$= 0.036\,222$$

Approximate confidence limits for $\hat{Y}$ are

$$\hat{Y} \pm tN\sqrt{\text{var}(\bar{y}_N)}$$

where t is Student's t for $n - 1$ degrees of freedom and $P = 0.05$.

In this case, the limits are

$$1\,728\,010 \pm 2.306\,006 \times 316\,256 \times \sqrt{0.036\,222}$$
$$= 1\,728\,010 \pm 138\,821 \approx 1.59 \text{ and } 1.87 \text{ million}$$

Optimal allocation of sampling effort

If

C = total resources available

c' = cost of conducting the cheap index survey per plot

c = cost of conducting the accurate census on one plot

then, for optimal allocation of sampling effort, the sample sizes should be

$$n'_{opt} = C/(c' + c\alpha) \quad \text{and} \quad n_{opt} = (C - c'n'_{opt})/c$$

where

$$\alpha = \sqrt{c'\text{var}(g)/\{c[\text{var}(y) - \text{var}(g)]\}}$$

Note that, if n'_{opt} is greater than N, the whole of the study area of interest should be surveyed and N should be substituted into the formula for n_{opt}.

In the current example, with $c' = 2$, $c = 80$ and $C = 1400$ man-hours,

$$\alpha = \sqrt{2 \times 0.290\,996 \Big/ \left[80 \times (1.611\,111 - 2.290\,996)\right]}$$

$$= 0.074\,235$$

$$n'_{opt} = 1400/(2 + 80 \times 0.074\,235) = 176 \text{ plots}$$

$$n_{opt} = (1400 - 2 \times 176)/80 = 13 \text{ plots}$$

Reference

The data are a small subset from a study covering the whole of arctic North America, with a complex study design involving stratification both by region and by habitat (Bart & Earnst 2002; Bart *et al.* 2006).

Practice

Frequency of occurrence ignores the numbers of individuals of a species seen at each location and may therefore seem to be inefficient. There are, however, many practical situations in which it is easy to determine whether one has encountered a species but difficult to make any sort of quantitative assessment of its abundance in a locality. This is especially true if the species' distribution is highly clumped, with the species being very numerous in most places where it does occur. The counting of individuals rather than the presence or apparent absence of each species is often particularly difficult in multi-species surveys. For this reason, frequency of occurrence has been used widely in bird surveys but is also applicable to other groups that can be identified without trapping and in which many species may occur in the same place.

Frequency of occurrence is particularly applicable when observations are gathered by volunteers, a significant proportion of whom may be happy enough to report whether or not they have seen a species but not prepared to spend their leisure time making exhaustive counts.

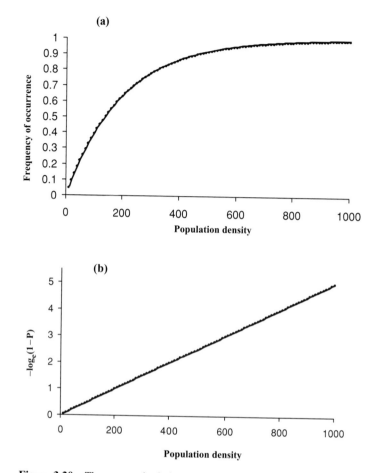

Figure 3.20 The expected relationship between frequency of occurrence and abundance when individuals are randomly distributed: (a) linear scale; (b) transformed scale. In the equations, N = population density, d = detectability and P = frequency of occurrence.

The method is not suitable for the most widespread species because they are usually seen almost everywhere, except when at their lowest densities. Thus the yellowhammer data in Figure 3.21 show that in over half the locations surveyed the frequency of occurrence gave no useful information about numbers – there was little difference in frequency of occurrence between localities in which the mean count was 0.75 per ten minutes and those in which it was 1.5 per ten minutes. The method was much more useful for the stock dove, which was less widespread.

Calibration

Figures 3.20 and 3.21 raise the possibility of calibrating frequency of occurrence against density (or, at least, against actual counts). The fact that the logarithmic transformation is expected to produce a straight line if the organisms are distributed approximately randomly and that this expectation is commonly realised makes the idea of such calibration attractive, since the line can

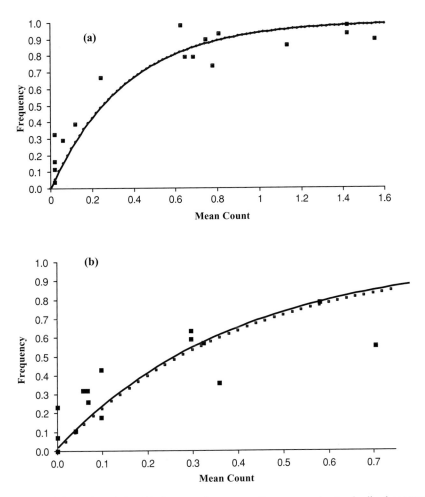

Figure 3.21 The relationship between frequency of occurrence in standardised surveys of all tetrads (2-km × 2-km squares) within each of 32 hectads (10-km × 10-km squares) in Britain and mean numbers of individuals observed in two 10-min point counts in each tetrad: (a) yellowhammer *Emberiza citronella*; (b) stock dove *Columba oenas*. (Data from Gibbons (1987).)

be fitted using familiar regression methods for lines that pass through the origin of the graph (available in Excel®). For statistical reasons, it is probably better to take logs again before fitting the regression (so that one ends up fitting $\log_e[-\log_e(1-P)] = \log_e(dN)$). Both methods run into trouble if any of the frequencies of occurrence is 100% because the log of zero is minus infinity; a simple ad-hoc way of circumventing this is to add 1 to the total sample size for each data point before calculating the 'frequencies'. Advanced statistical packages provide ways of fitting the untransformed equation directly, using general linear models, which is both statistically superior in principle and does not require this ad-hoc adjustment. (When applying such methods, one specifies the 'complementary log–log link function', to correspond to the double log transformation.)

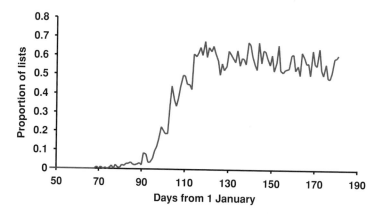

Figure 3.22   The proportion of birdwatchers' lists that included swallows *Hirundo rustica* in Great Britain in spring 2004. (Data from the British Trust for Ornithology; model-fitting by S. R. Baillie, personal communication.)

Managing the methodology

Unstandardised observations can be useful

Figure 3.22 shows the temporal pattern of arrival of the swallow *Hirundo rustica* in Great Britain in 2004, on the basis of lists of birds seen each day by birdwatchers; the metric is the proportion of lists on which swallows occurred. The only standardisation was that people had to record whether or not they had recorded each species on the list of species supplied to them. They could collect their records anywhere and at any time, in any habitat; they could spend as much time collecting records as they wished; people with any level of expertise could participate. Thus the increase in positive records during the spring could be a result of observers completely ignoring places where swallows occurred in the early spring but visiting them frequently later, or of people spending more time per day birdwatching later in the season, etc. However, the magnitude of the increase is such that it is more reasonable to suppose that it reflects a real change in numbers. It is simply too big to be an artefact.

Standardisation

The last example notwithstanding, standardisation is to be preferred. Surveys using frequency of occurrence as an index of abundance usually standardise the time over which observations are made. Those that require the observer to move about, rather than record from one spot, also usually standardise the area searched or the length of the route. It is usually also important to standardise time of day, time of year and weather, since all of these are likely to affect detectability.

Differences in observer ability may also cause bias. Thus, if one wishes to compare two areas and two observers are available, it is not good to have one observer make all the observations in one area, the other making them all in the other. Rather, each should make half the observations in each area. It has been suggested that another way of overcoming differences in ability between observers is to standardise by the number of species observed, e.g. carrying on observations at

(a)

Figure 3.23 Apparent patterns of abundance in Britain and Ireland of (a) meadow pipit *Anthus pratensis*; (b) robin *Erithacus rubecula*. Abundance was indexed as the proportion of tetrads (2-km × 2-km squares) within each hectad (10-km × 10-km square) in which the species was recorded in standardised surveys. Shades of grey indicate proportions from zero to one, in ten equal intervals. (Redrawn from Gibbons *et al.* (1993).)

each point until 20 species have been detected. This is not a good procedure: see remarks on McKinnon lists, below.

Comparisons across habitats

Detectability varies with habitat, so this method is not suitable for comparing densities in different habitats. In studies of geographical patterns of abundance, in which all habitats are included, interpretation has to be done carefully. Thus the apparent pattern for the meadow pipit (Figure 3.23(a)) could be partly the result of habitats in the north and west of Britain and Ireland being generally more open than those in the south and east (though it agrees with independent evidence that is less biased by detectability). The map for the robin (Figure 3.23(b)) may be biased for the same reason but in this case the bias would have generally diminished rather than accentuated the true pattern.

(b)

Figure 3.23   *(cont.)*

Correction, in the absence of standardisation?

If one doubles the length of time and area of search at a location one is likely, on average, to double the number of individuals observed; but one will not double the proportion of locations in which any one species is observed. Thus correcting for differences in effort between frequency-of-occurrence surveys is difficult. It is much better to standardise. Using the number of species observed as a correcting factor is as unsatisfactory as is using it for standardisation.

Sampling strategy and statistical analysis for frequency of occurrence

The size of samples

In principle the size of locations (or length of time spent at each, if standardisation is by time) should be small, in order to reduce the likelihood of encountering a species in every sample. Moving to a new location, however, takes time, so there is, as usual, an optimum – short enough that the probability of detecting a common species is well below 100% but not so short that most of one's time is taken up in moving between locations rather than making observations. Since frequency of occurrence is usually used as an index in multi-species surveys and the optimum will differ among species, the best approach is to conduct a pilot study and make a judgement based on that.

Distribution of sample locations

The preferred distribution of the locations for frequency-of-occurrence measurements follows the usual principles discussed in Chapter 2: it should be random or stratified random (unless there are good reasons for using a systematic distribution). Two-stage sampling may be used to increase efficiency. Note that a set of surveys to determine frequency of occurrence in a location or at a particular time is effectively only a single observation, for it provides just one measure of abundance in that location or at that time.

Statistical manipulation of frequencies

Suppose that one wishes to test the null hypothesis that the frequency of occurrence of a species is the same in a number of localities (or at a number of different times), in each of which one has made a set of surveys recording the species' presence or apparent absence. The appropriate test uses the usual $2 \times N$ contingency table described in standard statistical textbooks, the N columns being the N localities (time) and the two rows being the numbers of presences and of apparent absences, respectively.

Suppose that one measures the frequency of occurrence of a species in samples in each of a large number of grasslands and one wishes to test whether the mean frequency of occurrence differs among acidic, neutral and calcareous grasslands. Given that one has measurements from each of several localities of each type, the obvious approach is to carry out an analysis of variance. The results will be more robust if the analysis is carried out not on the frequencies of occurrence themselves but on the figures after taking logs twice – that is, taking $\log_e[-\log_e(1 - P)]$, where P is the frequency of occurrence (the complementary log–log transformation again). The same applies to any statistical analysis of frequency of occurrence that assumes the sampling error of the data to be normally distributed – t-tests, regression, correlation, etc. If parameters and confidence limits are estimated, they can be converted to the original frequency-of-occurrence scale by taking antilogs twice, though with the usual caveat that these back-transformed estimates will be only approximate.

Subdivision of samples

Frequency in subsamples

An easy way of increasing the precision of frequency-of-occurrence indices is to subdivide the samples: instead of recording whether a species is detected along a 1-km route, in a 1-ha area, or during a 30-min observation period, one records its occurrence in each of the ten 100-m sections of the route, in each of the sixteen 25-m × 25-m subdivisions of the hectare, or in the six 5-min divisions of the observation period. The benefit is similar to that of taking many small samples rather than a few large ones, but it is easier to do because it does not involve moving to new sample sites. Note that this is a form of cluster sampling because the subsamples are not independent.

Timed species counts

In some situations, it is easier simply to record which of the subsamples each species is first recorded in, rather than recording every subsample in which each occurs. The method was originally used for 1-h surveys divided into six 10-min subsamples but could be extended to any number of subsamples or also to spatial subsampling.

In the original formulation, species detected in the first subsample are given a score of 6, those not detected until the second subsample score 5, and so on. Replicate 1-h samplings are conducted, to accumulate values of n_0, n_1, n_2, n_3, n_4, n_5 and n_6 for each species: that is the number of samples in which a species gained the score 0, 1, 2, 3, 4, 5 and 6. The data are processed as in Box 3.28, to provide either simple indices or estimates of λ, the average number of individuals of that species encountered per 10-min period. (Divide by ten to estimate λ on a per-minute basis.) Periods other than 10 min can, of course, be used. Indeed, it would be possible to generalise the method for observations divided into fewer or more than six subsamples).

Either the simple index or λ can be taken as an index of density, though λ is preferable because it is more likely to be linearly related to density, whereas values of the timed-species-counts index reach a plateau at high densities.

McKinnon lists

The detectability of a species varies according to locality and the ability of the observer. McKinnon proposed an approximate method for accumulating measures of species richness that was less sensitive to such differences than more formal approaches are. The idea is to list all species until one has a certain number (say 20) on the list; then start again, to make a second list of 20 (including again those species detected that were recorded on the first list); and so on. The increasing total number of species detected as one accumulates lists tends to reach a plateau, the height of the plateau indicating total species richness.

It has been suggested that the frequency with which a species is found on the lists of such a survey can be used an index of its abundance. So it can, as long as it is recognised that it is an index of abundance relative to those of other species. As such, it is not generally to be recommended because it indexes abundance of each species not in absolute terms but relative to those of the others and so reveals no differences between places in which relative abundances of all species are the same but between which overall abundance differs; equally, it may suggest the existence of differences in abundance of an individual species just because the abundance of other species has changed.

Box 3.28. **Frequency of occurrence**

Indices from timed species counts

Bateleurs (*Terathopius ecaudatus*) were recorded at Pakuba airstrip, Uganda, during eight standard TSC surveys (Freeman *et al.* 2003). The numbers of times that the species was scored 0, 1, 2, . . ., 6 were

n_0	n_1	n_2	n_3	n_4	n_5	n_6	Total
4	0	1	0	1	2	0	8

The TSC index

This is simply the mean score, averaged over all eight surveys and all 10-min subdivisions:

$$(4 \times 0 + 0 \times 1 + 1 \times 2 + 0 \times 3 + 1 \times 4 + 2 \times 5 + 0 \times 6)/8 = 2.0$$

Estimating λ

Calculate

$$a = n_5 + 2n_4 + 3n_3 + 4n_2 + 5n_1 + 6n_0$$
$$= 2 + 2 + 0 + 4 + 0 + 24 = 32$$
$$b = N - n_0 = 8 - 4 = 4$$

Then estimate λ as

$$\hat{\lambda} = -\log_e[a/(a + b)]$$
$$= -\log_e(32/36)$$
$$= 0.117\,783 \text{ encounters per 10-min period}$$

Confidence limits for $\hat{\lambda}$

$\hat{\lambda}$ is the 'maximum-likelihood estimate'. The log-likelihood corresponding to any particular value of λ is

$$l = b\log_e[1 - \exp(-\lambda)] - a\lambda$$

So, the maximum value of log-likelihood (l_{max}) is obtained by substituting $\hat{\lambda}$ into this equation. In the current example,

$$l_{max} = 4 \times \log_e[1 - \exp(-0.117\,783)] - 32 \times 0.117\,783$$
$$= -12.558$$

The 95% confidence limits for $\hat{\lambda}$ are the values of λ for which the likelihood is given by

$$l_{CL} = l_{max} - 1.921$$
$$= -14.479 \text{ in the current example}$$

These values are easily found using a spreadsheet facility such as Excel®. Label two adjacent columns 'lambda' and 'l', respectively. In the top cell of the lambda column insert the value of $\hat{\lambda}$; in the (adjacent) top cell of the l column insert the above formula for l, using the actual values for a and b, and referring to the adjacent cell for the value of λ. The previously calculated value of l_{max} will appear in this cell, provided that you have entered the formula correctly.

Enter a series of numbers into the lambda column, starting close to the value of $\hat{\lambda}$ and decreasing in steps appropriate to the level of precision with which you wish to obtain confidence limits. It is easy to do this using 'drag-and-fill' in Excel®. Next, drag the formula from the top cell of the l column down the rest of the column, to get the values of l corresponding to the values of λ that you have entered. Carry on until you reach a value of l less than l_{CL}.

Here is such a spreadsheet result for the current example, though with most of the rows omitted for brevity:

Lambda	l
0.117 83	−12.558
0.116	−12.763
0.115	−12.777
0.114	−12.560
.	.
.	.
.	.
0.038	−14.372
0.037	−14.445
0.036	−14.521

In this case, of the last two values of l, it is the penultimate that is closest to l_{CL}; thus 0.037 is one of the approximate confidence limits of $\hat{\lambda}$.

Repeat the process for values of λ greater than $\hat{\lambda}$ to obtain the other confidence limit (0.273 in this example).

Reference

Freeman *et al.* (2003) present both the method and the example.

Estimating the true frequency of occurrence

In all of the above, we have been concerned only with the apparent frequency of occurrence: because the detectability of a species cannot be assumed to be perfect, failure to see a species does not mean that it is absent. However, just as it is possible to estimate true population size from counts of samples (as in mark–recapture and distance sampling), so it is possible to estimate the true frequency of occurrence from observations made with imperfect detectability (MacKenzie & Nichols 2004). As with the estimation of numbers, the method involves making particular assumptions that must be satisfied if the estimate of the true frequency of occurrence is to be accurate.

Appendix: software packages for population estimation

For those with good programming skills, many of the models described in this chapter can be fitted in general statistical packages. A flexible, powerful and freely available statistics program is R (http://www.r-project.org/).

However, various software packages are available to do the more complex calculations that we describe, most being freely available on the web. This section lists the most general packages, together with some indications of what they do and how easy they are to use. A comprehensive discussion of these and other packages can be found in Lebreton *et al.* (1993) and Williams *et al.* (2002). The following websites provide links to most of the packages available (including all those mentioned here):

Phidot: http://www.phidot.org/software/
Patuxent software archive: http://www.mbr-pwrc.usgs.gov/software.html
Colorado State University: http://www.colostate.edu/depts/coopunit/
CNRS, Montpellier: http://www.cefe.cnrs.fr/BIOM/En/Software-BIOM.htm

Packages are generally designed to run on IBM PCs or compatibles, either under MS-DOS or Windows. DOS-based programs tend to require little in the way of computer memory, so they will probably run on all but the oldest computers; on computers with 286, 386 or 486 chips, having a maths co-processor will help enormously. They use a command line interface and some require simple modifications to the program files each time they are run, so they require some confidence to use, though good instructions are usually provided. Windows-based programs are much more resource-hungry, particularly the latest generation. Hence, although they are much easier to use than DOS-based programs, they do require a powerful computer – find the best you can. As a general guideline, using packages like MARK on any computer without a Pentium processor and with less than 64 MB of RAM will lead to many frustrating computer crashes on all but the simplest of problems. Closing down other programs and running just a few models at a time can help if you are running into memory problems.

Some of these programs may run under UNIX (the Phidot website is a good place to look).

Capture–recapture: closed populations

The first, and most comprehensive, program to estimate density and abundances, when a closed population can be assumed, is CAPTURE (Otis *et al.* 1978; Rexstad & Burnham 1991). This DOS-based program has eight different ways of modelling capture probabilities and provides estimates for modelling time and heterogeneity effects. Goodness-of-fit tests are provided, as are model-selection capabilities using a discriminate function procedure.

Most of the models fitted in CAPTURE can also be fitted in MARK, which also includes the possibility of modelling capture probabilities as individual covariates, or SURVIV (see below).

Capture–recapture: open populations

Perhaps the longest-standing piece of software for implementing Jolly–Seber models is POPAN, which is now in its sixth version (Schwartz & Arnason 1996; Arnason & Schwartz 1999). POPAN is a Windows-based package (though there are also versions for UNIX and Linux), which runs analyses in batches, rather than interactively as in some packages, thus ensuring that a record of analyses is kept. POPAN allows fitting of 32 different versions of the Jolly–Seber model, and provides model-selection facilities based on using maximum-likelihood tests and Akaike's Information Criterion (AIC). It also provides a goodness-of-fit test using chi-squared statistics and the 'standard' set of tests implemented originally in RELEASE (Burnham *et al*. 1987).

SURGE (Lebreton & Clobert 1986; Pradel & Lebreton 1991) was developed for modelling survival rates (rather than abundance) in open populations, within a linear model framework; consequently, it is also very flexible. Although it was originally a DOS-based program, the latest version can be run through Windows and comes with a user-friendly guide. SURGE provides a powerful way of fitting Cormack–Jolly–Seber-type models, and, although it provides for model selection (by likelihood ratio and AIC), it does not include goodness-of-fit testing. This should be done using a package like RELEASE (Burnham *et al*. 1987).

As with the closed-population situations, many of the models fitted by POPAN and SURGE can be fitted in MARK and SURVIV (see below).

Ring-recovery models

For analyses of ring-recovery data and, indeed, capture–recapture data, most people will want to use MARK (White & Burnham 1999). MARK has been developed as a Windows package and presents an almost bewildering array of model types, with many of the latest models incorporated. It provides extensive goodness-of-fit and model-selection (likelihood ratio and AIC) tests and provides some simulation capabilities. It will provide both population and survival estimates (together with recapture/recovery probabilities) and also includes facilities for modelling individual covariates and multi-state models (in which individuals may move from one class to another).

In some cases, even the range of models implemented in MARK does not provide exactly the situation you are studying. If the recovery probabilities can be written in multinomial form (i.e. entirely as a product of each of the parameters), then use SURVIV (White 1983). This is an extremely flexible, DOS-based program, with which the user can specify the individual probabilities associated with each capture event, allowing almost complete flexibility in model choice. (An additional version, TM-SURVIV, is available for cases in which transient individuals are present in the population.) SURVIV is very flexible but requires a certain amount of programming ability.

Multi-state models

The latest generation of models allows the modelling not just of survival rates and recapture/ recovery probabilities, but also of transition probabilities between states (e.g. site or age/weight class). MS-SURVIV (Hines 1994) provides a multi-state capability for SURVIV, saving the user from having to write the individual cell probabilities. More recently, MS-SURGE (Choquet *et al.* 2004) has been developed in a Windows environment using an extremely powerful model-specification language, which makes fitting of such models relatively easy. Although it provides model-selection facilities, it does not provide goodness-of-fit testing; the program U-CARE is recommended for this (Pradel *et al.* 2003).

Observation-based methods

For modelling population numbers with just observational count data from line or point transects, the software to use is DISTANCE (Buckland *et al.* 2001), which is available in DOS- and Windows-based formats (from http://www.ruwpa.st-and.ac.uk/distance/). It enables a range of models to be fitted, under a basic detection-function approach. DISTANCE provides several different approaches to model selection and includes goodness-of-fit tests and bootstrap estimation of errors.

Acknowledgements

Many colleagues, especially at the British Trust for Ornithology, have been generous in answering queries and supplying copies of papers. We are especially grateful for the advice of J. Bart, S. T. Buckland, M. J. Conroy, S. N. Freeman, P. M. Lukacs, J. D. Nichols, J. A. Royle and S. K. Thompson.

References

Anderson, D. R., Burnham, K. P., White, G. C. & Otis, D. L. (1983). Density estimation of small mammal populations using a trapping web and distance sampling methods. *Ecology* **64**, 674–680.

Arnason, A. N. & Schwartz, C. J. (1999). Using POPAN-5 to analyse banding data. *Bird Study* **46** (Suppl.), S157–S168.

Bart, J. & Earnst, S. (2002). Double sampling to estimate density and population trends in birds. *Auk* **119**, 36–45.

Borchers, D. L., Buckland, S. T. and Zucchini, W. (2002). *Estimating Animal Abundance: Closed Populations*. London, Springer-Verlag.

Buckland, S. T., Anderson, D. R., Burnham, K. P. *et al.* (2001). *Introduction to Distance Sampling: Estimating Abundance of Biological Populations*. Oxford, Oxford University Press.

(2004). *Advanced Distance Sampling – Estimating Abundance of Biological Populations*. Oxford, Oxford University Press.

Burnham, K. P. & Overton, W. S. (1978). Estimation of the size of a closed population when capture probabilities vary among animals. *Biometrika* **65**, 625–633.

(1979). Robust estimation of population size when capture probabilities vary amongst animals. *Ecology* **60**, 927–936.

Burnham, K. P., Anderson, D. R., White, G. C., Brownie, C. & Pollock, K. P. (1987). *Design and Analysis of Methods for Fish Survival Experiments Based on Release–Recapture*. Bethesda, Maryland, American Fisheries Society.

Caughley, G. (1977). *Analysis of Vertebrate Populations*. London, John Wiley & Sons.

Cochran, W. G. (1977). *Sampling Techniques*. 3rd edn. New York, John Wiley & Sons.

Choquet, R., Reboulet, A.-M., Pradel, R., Gimenez, O. & Lebreton, J.-D. (2004). M-SURGE: new software specifically designed for multistate recapture models. *Animal Biodiversity & Conservation* **27**, 207–215

Craig, G. C. (1953). On the utilisation of marked specimens in estimating populations of insects. *Biometrika* **40**, 170–176.

Diggle, P. J. (1983). *Statistical Analysis of Spatial Point Patterns*. London, Academic Press.

Dodd, C. K. & Dorazio, R. M. (2004). Using counts to simultaneously estimate abundance and detection probabilities in a salamander community. *Herpetologia* **60**, 468–478.

du Feu, C., Hounsome, M. & Spence, I. (1983). A single session mark/recapture method of population estimation. *Ringing & Migration* **4**, 211–226.

Edwards, W. R. & Eberhardt, L. L. (1967). Estimating Cottontail abundance from live-trapping data. *Journal of Wildlife Management* **31**, 87–96.

Farnsworth, G. L., Pollock, K. H., Nichols, J. D. *et al.* (2002). A removal model for estimating detection probabilities from point-count surveys. *Auk* **119**, 414–425.

Frederiksen, M., Fox, A. D., Madsen, J. & Colhoun, K. (2001). Estimating the total number of birds using a staging site. *Journal of Wildlife Management* **65**, 282–289.

Freeman, S. N., Pomeroy, D. E. & Tushabe, H. (2003). On the use of times species counts to estimate avian abundance indices in species-rich communities. *African Journal of Ecology* **41**, 337–348.

Gibbons, D. W. (1987). *The 'New Atlas' Pilot Fieldwork*. BTO Research Report No. 30. Tring, British Trust for Ornithology.

Gibbons, D. W., Reid, J. B. & Chapman, R. A. (1993). *The New Atlas of Breeding Birds in Britain and Ireland: 1988–1991*. London, T. & A. D. Poyser.

Hines, J. E. (1994). *MSSURVIV User's Manual*. Laurel, MD, National Biological Survey.

Johnson, J. B. & Omland, K. S. (2004). Model selection in ecology and evolution. *Trends in Ecology & Evolution* **19**, 101–108.

Jolly, G. M. (1965). Explicit estimates from capture–recapture data with both death and immigration – stochastic model. *Biometrika* **52**, 225–247.

Kéry, M., Royle, J. A. & Schmid, H. (2005). Modeling avian abundance from replicated counts using binomial mixture models. *Ecological Applications* **15**, 1450–1461.

Lebreton, J.-D., & Clobert, J. (1986). *User's Manual for Program SURGE. Version 2.0*. Montpellier, C.E.F.E., C.N.R.S.

Lebreton, J.-D., Reboulet, S. M. & Banco, G. (1993). A review of software for terrestrial vertebrate population dynamics. In *The Study of Bird Population Dynamics Using Marked Individuals*, eds. Lebreton, J.-D and North, P. M., Berlin, Birkhäuser Verlag, pp. 357–372.

Leslie, P. H. & Davis, D. H. S. (1939). An attempt to determine the absolute numbers of rats on a given area. *Journal of Animal Ecology* **8**, 94–113.

Link, W. A. (2004). Individual heterogeneity and identifiability in capture–recapture models. *Animal Biodiversity & Conservation* **27**, 87–91.

Lukacs, P. M. & Burnham, K. P. (2005). Estimating population size from DNA-based closed capture–recapture data incorporating genotyping error. *Journal of Wildlife Management* **69**, 396–403.

MacKenzie, D. I. & Nichols, J. D. (2004). Occupancy as a surrogate for abundance estimation. *Animal Biodiversity & Conservation* **27**, 461–467.

Manly, B. F. J. (1984). Obtaining confidence limits on parameters of the Jolly–Seber model for capture–recapture data. *Biometrics* **40**, 749–758.

McCaffery, K. R. (1976). Deer trail counts as an index to populations and habitat use. *Journal of Wildlife Management* **40**, 308–316.

McCallum, H. (2000). *Population Parameters: Estimation for Ecological Models.* Oxford, Blackwell Science.

Moore, C. T. & Kendall, W. L. (2004). Costs of detection bias in index-based population monitoring. *Animal Biodiversity & Conservation* **27**, 287–296.

Nichols, J. D., Hines, J. E., Sauer, J. R. *et al.* (2000). A double-observer approach for estimating detection probability and abundance from point counts. *Auk* **117**, 393–408.

Numata, M. (1961). Forest vegetation in the vicinity of Chosi. Coastal flora and vegetation at Chosi, Chiba Prefecture IV. *Bulletin of the Chosi Marine Laboratory of Chiba University* **3**, 28–48 [in Japanese].

Otis, D. L., Burnham, K. P., White, G. C. & Anderson, D. R. (1978). Statistical inference from capture data on closed animal populations. *Wildlife Monographs* **62**, 1–135.

Parmenter, R. R., Yates, T. L., Anderson, D. R. *et al.* (2003). Small-mammal density estimation: a field comparison of grid-based vs. web-based density estimators. *Ecological Monographs* **73**, 1–26.

Pradel, R. & Lebreton, J.-D. (1991). *User's Manual for Program SURGE. Version 4.1.* Montpellier, C.E.F.E., C.N.R.S.

Pradel, R., Wintrebert, C. M. A. & Gimenez, O. (2003). A proposal for a goodness-of-fit test to the Arnason–Schwartz multisite capture–recapture model. *Biometrics* **59**, 43–53.

Rexstad, E. A. and Burnham, K. P. (1991). *User's Guide for Interactive Programme CAPTURE. Abundance Estimation of Closed Animal Populations.* Fort Collins, CO, Colorado State University.

Royle, J. A. (2004). N-mixture models for estimating population size from spatially replicated counts. *Biometrics* **60**, 108–115.

Schaub, M., Pradel, R., Jenni, L. & Lebreton, J.-D. (2001). Migrating birds stop over longer than usually thought: an improved capture–recapture analysis. *Ecology* **82**, 852–859.

Schwartz, C. J. & Arnason, A. N. (1996). A general methodology for the analysis of capture–recapture experiments in open populations. *Biometrics* **52**, 860–873.

Seber, G. A. F. (1982). *The Estimation of Animal Abundance and Related Parameters.* London, Charles Griffin.

Southwood, T. R. E. (1978). *Ecological Methods.* 2nd edn. London, Chapman & Hall.

Thompson, S. K. (2002). *Sampling.* 2nd edn. New York, John Wiley & Sons.

Turner, F. B. (1960). Size and dispersion of a Louisiana population of the cricket frog *Acris gryllus. Ecology* **41**, 258–268.

Underhill, L. G. & Fraser, M. W. (1989). Bayesian estimate of the number of malachite sunbirds feeding at an isolated and transient nectar resource. *Journal of Field Ornithology* **60**, 381–387.

White, G. C. (1983). Numerical estimation of survival rates from band recovery and biotelemetry data. *Journal of Wildlife Management* **47**, 716–728.

White, G. C. & Burnham, K. P. (1999). Program MARK; survival rate estimation from both live and dead encounters. *Bird Study.* **46** (Suppl.), S120–S139.

Wileyto, E. P., Ewens, W. J. & Mullen, M. A. (1994). Mark–recapture population estimates: a tool for improving interpretation of trapping experiments. *Ecology* **75**, 1109–1117.

Williams, B. K., Nichols, J. D. and Conroy, M. J. (2002). *Analysis and Management of Animal Populations.* San Diego, TX, Academic Press.

Wood, G. W. (1963). The capture–recapture technique as a means of estimating populations of climbing cutworms. *Canadian Journal of Zoology* **41**, 47–50.

Zar, J. H. (1999). *Biostatistical Analysis.* New Jersey, PA, Prentice-Hall, Inc.

Zippin, C. (1956). An evaluation of the removal method of estimating animal populations. *Biometrics* **12**, 163–169.

4 Plants

James M. Bullock

NERC Centre for Ecology and Hydrology, Winfrith Technology Centre, Dorchester,
Dorset DT2 8ZD, UK

Introduction

Seeds and phytoplankton can be extremely mobile, but usually we are interested in surveying plants that cannot move. Sessile plants are usually arranged over a substrate (soil, sediment, etc.) and can be found, identified and examined at leisure. This characteristic means that, in many ways, it is much easier to census plants than it is to census other organisms and estimates of, e.g., density, species number and composition and distribution of a species are more accurate for plants. A second characteristic of plants, however, causes problems in deciding how best to characterise the abundance of species. Plant species, and even individuals within a species, in a community can differ enormously in size. An English wood may contain oak trees 30 m tall and with a canopy diameter of 40 m, in contrast to herbs, grasses and oak seedlings in the understorey, which are only a few centimetres in height. Even in a grassland where all plants are a few centimetres tall, there will be huge differences in the horizontal spread of individuals, from a few millimetres to several metres. While the standard measure of abundance of animals, a count of individuals, can be used for plants, this variety in plant size will mean that counts ignore a large amount of information about the community. For instance, there may be equal numbers of individuals of two species in your study area but the species with a larger average size will have a greater importance for the ecological processes. Clonal plants present another problem. They grow as a set of connected shoots or ramets (e.g. grass tillers). The connections may be buried and ramets of various plants may be intermingled. So it is impossible to distinguish the complete individual (the genet). It is sensible and usual in this case to count ramets rather than genets and this has the added benefit that ramets will exhibit less size variation.

Because of these issues, various alternative approaches to assessing plant abundance have been developed. The censusing methods described in this chapter (Table 4.1) are often amenable to more than one approach.

Density is the standard count of the number of organisms in a prescribed area. For sessile plants this is a straightforward measure in comparison with that for some animals, which can move in and out of the area during the census period.

Frequency is a measure of the chance of finding an individual of a species in the sample area. It is related to density, but is a rather odd measure. Frequency is most commonly used in quadrats and is discussed in that section.

Ecological Census Techniques: A Handbook, ed. William J. Sutherland.
Published by Cambridge University Press. © Cambridge University Press 1996, 2006.

Table 4.1. *A summary of the methods suitable for various groups*

Method	Trees	Shrubs	Herbs and grasses	Aquatic macrophytes	Algae	Bryophytes, fungi and lichens	Seeds	Page
Counts	*	*	*	*	?	+		188
DAFOR	*	*	*	*	?	+		189
Quadrats	*	*	*	*	+	*		189
Point quadrats			*	?	?	+		194
Transects	*	*	*	*	?	+		196
Mapping terrestrial vegetation	*	*	*			+		197
Mapping aquatic vegetation				*	+			200
Seed traps							*	201
Sampling seedbanks							*	204
Phytoplankton					*			207
Benthic algae					*			209
Mapping and marking individuals	*	*	*	*		+		210

* Method usually applicable, + method often applicable, ? method sometimes applicable. The page number for each method is given.

Cover is a size-based measure of the area covered by the above-ground parts of plants of a species when viewed from directly above; i.e. it is the proportion of ground (within the prescribed area) occupied by the vertical projection onto it of the parts of all individuals of a species. Because the vegetation may be layered the cover of all species often sums to >100%.

Biomass is another size-based measure and is usually the above-ground weight of the plants of a species in a prescribed area.

Plant censusing tends to concentrate on the visible above-ground parts. Roots are usually ignored. This is generally accepted as a necessity by plant ecologists because measuring roots is very difficult and most likely to be destructive, resulting in damage to, or death of, the plants. Density and frequency count individual plants, so including roots is irrelevant, but, because plants differ greatly in the allocation of biomass to above- and below-ground parts, size-based measures of cover or biomass could give a much different picture of relative abundances if roots were included.

The sessile nature of plants leads to a clear and slowly changing spatial pattern in the distribution of species. Almost invariably, patchiness in environmental variables, restricted dispersal of propagules and clonal growth all bring about a clumped distribution of plants of a species.

The scale of this patchiness will depend on the type of plant community and the mechanism causing the patchiness. For example, in a fertilised lawn where management has removed most spatial variation in environmental factors, patchiness will be mostly at a small scale (decimetres), being driven by clonal growth and local dispersal. In an unimproved grassland, patchiness in soil nutrients, hydrology, etc. will impose an extra scale of patchiness (metres to tens of metres). Your sampling strategy (see Chapter 2) must be designed to compensate for this patchiness and to give an accurate representation of the abundances of the species in the whole study area.

Species also undergo changes in relative abundance through the year, particularly in response to seasons. This is caused by 'phenological' differences among species in the time of year in which they germinate, begin growth, peak in biomass or flowering and die back for the winter or summer. The latter is seen most spectacularly for plants that die back to, for instance, a bulb for part of the year. These patterns are clearest in temperate regions, but species-specific phenologies are also seen in the tropics. If you are comparing sites or studying change over years, censusing should be at the same time of year (usually within the same month) for all sites or years. Climatic variation among years, e.g. late frosts, early spring, or summer drought, will lead to differences in phenology among years and you should consider this variation when interpreting results. If time permits, repeating censuses within a year can lead to a huge improvement in data quality. In temperate regions repetition in late spring and late summer is useful.

Although plants are usually categorised into species in surveys, you could use a coarser system if you require a different type of information. This can be based on higher taxonomic groupings such as genera, families or even orders. Loose classifications based on taxonomy can be used: e.g. mosses, grasses, herbs, trees, etc. 'Functional types' are popular and involve classification based on a mix of taxonomy and ecological function. Angiosperms are often divided into grasses (or graminoids, which includes the Cyperaceae), legumes and other herbs. Lower plants such as ferns and mosses may be included as other functional groups. Algae can be classified by their morphologies (e.g. single cell, colony or filament) and cell type rather than by individual species. For many of the methods described in this chapter these classification systems can be used instead of species.

The methods described in this chapter can be used to assess the abundances of individual species in the study area or of all species in the community. They can also be used on other organisms than plants: fungi, lichens and sessile animals such as corals or encrusting bryozoans. They are possible in many benthic aquatic plant communities. Phytoplankton and the propagules of adult sessile plants do not stay still and wait to be counted, so different measures must be used for them.

Counts

Counts are used for assessing densities of large or obvious plants that are present at low density.

Method

Every individual of a species or a number of species in the whole study area is counted. It can be useful to label counted individuals to avoid counting them twice.

Advantages and disadvantages

Because the study area is usually several orders of magnitude larger than the plants this technique is often much too time-consuming (imagine counting every plant in a 1-ha grassland). However, you should be able to use this technique if a species has a low enough density and is easily spotted (e.g. trees in a prairie) and the whole of the study area can be covered.

Biases

Because no subsampling of the study area is used, this method measures the true density of species and therefore has no biases.

DAFOR

This method is used for assessing abundances of plants over large areas.

Method

It is a density or cover measure. This involves simply assigning each species as dominant, abundant, frequent, occasional or rare ('DAFOR'). These classes have no strict definition and you must decide on your own interpretation, including whether your score is based on relative cover or density of species.

Advantages and disadvantages

This is a quick and simple technique that can be used to characterise large areas of vegetation. It is often used to give background data prior to more detailed censuses, e.g. by quadrat. It is, however, very coarse and has poor power in detecting vegetation differences either over time or between sites.

Biases

More obvious (bigger or more clumped) plants may be given higher scores incorrectly. Because there is no rigorous definition of the categories there may be large variation between observers and even with the same observer at different times. This can be controlled by ensuring that the categories are defined carefully, observers are trained and the observers cross-check definitions throughout the census.

Quadrats

Quadrats are for measuring abundances of sessile species in any vegetation, including aquatic macrophytes.

Method

Quadrats can be used to measure density, frequency, cover or biomass. They are used to define sample areas within the study area and are usually four strips of wood, metal or rigid plastic, which are fixed together to form a square. It can be useful to use bolts so that the quadrat can be dismantled for storage or transport. For aquatic macrophytes a wood or plastic frame will float and can be used to sample floating or emergent vegetation on the water surface. For large quadrats, over 4 m², a frame will be unwieldy and as an alternative you could measure out the quadrat using tape measures, folding rulers or string. Corners are marked by posts and it is important to keep a constant quadrat shape, for example by using a set-square to measure out right-angles. Although a square is often used, the quadrat shape is unimportant as long as you keep the shape constant and know its area. For certain purposes the quadrat can be divided into a grid of equal-sized squares using regularly spaced lengths of string or wire.

Different vegetation types require different quadrat sizes. Vegetation with smaller plants, greater plant density or greater species diversity should require smaller quadrats. The sizes most often used are 0.01–0.25 m² for bryophyte, lichen and algal communities (for instance, on rocks or tree bark), 0.25–16 m² for grassland, tall-herb, short-shrub and aquatic-macrophyte communities, 25–100 m² for tall-shrub communities and 400–2500 m² for trees in woods and forests. Different quadrat sizes can be used to survey different vegetation types within a study area, such as the understorey layer and canopy layer in a forest.

To get a good estimate of species' abundances multiple quadrats should be used in each study area according to your sampling design and various measures can be used to survey the vegetation (see Figure 4.1).

Density is measured by counting the number of individuals of each species within the quadrat. Many plants will lie on the edge of the quadrat and for this and other measures you must decide which to classify as inside the quadrat. Often only the plants rooted within the quadrat are counted.

Cover can be measured by estimating visually the proportion of the quadrat occupied by each species (i.e. the vertical projection of each plant; see above). Various measures can be used. You can estimate cover to the nearest per cent (or less), but this might give a spurious accuracy, given the problems of estimating cover (see below). It may be more sensible to use percentage classes, e.g. in 10% or 25% steps, or use those given in the Domin or Braun–Blanquet scales (Table 4.2). You may find it useful to divide the vegetation into layers, e.g. a bryophyte layer, a herb layer and a shrub layer, and make cover estimates separately for each layer.

Frequency can be measured in two ways. One is to use the quadrat as the sampling unit. A large number of quadrats is placed in the study area and the proportion of quadrats containing the species is counted. The abundance of the species in the quadrat is ignored. A more local measure of frequency can be derived if the quadrat is subdivided into a grid and the percentage of grid squares containing the species is calculated. Local frequency is often used at points on a transect or where only a few quadrats can be used.

Biomass of species can be measured by cutting all the above-ground parts of the plants in the quadrat at a certain height from the surface of the substrate, usually at or close to ground-level. Root harvesting (below-ground biomass) is too difficult and error-ridden to consider seriously.

Table 4.2. *The Domin and Braun–Blanquet scales for visual estimates of cover*

Value	Braun–Blanquet	Domin
+	<1% Cover	1 Individual, with no measurable cover
1	1%–5% Cover	<4% Cover with few individuals
2	6%–25% Cover	<4% Cover with several individuals
3	26%–50% Cover	<4% Cover with many individuals
4	51%–75% Cover	4%–10% Cover
5	76%–100% Cover	11%–25% Cover
6		26%–33% Cover
7		34%–50% Cover
8		51%–75% Cover
9		76%–90% Cover
10		91%–100% Cover

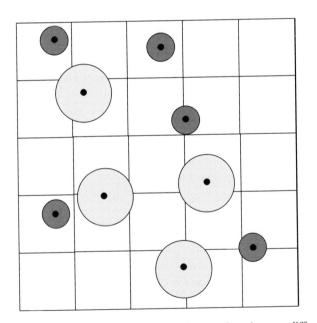

Figure 4.1   The various measures in a quadrat give very different information about the relative abundances of two species; a large, pale species and a small, dark species. Density counts suggest relative abundances of the large and small species of 1 : 1.25; frequency among the 15 subdivisions (rooted frequency, where the small inner circle represents the stem – note that this sometimes spans two subdivisions) 1 : 1.17; cover 1 : 0.31; and biomass (assuming that plants are hemispheres and have the same tissue density) 1 : 0.16.

A knife, scissors, shears, saw or chainsaw may be used, depending on the vegetation type. The plant material should be taken to the laboratory, kept separately in bags, and then sorted into species. Each species is either weighed as it is, giving a measure of 'fresh weight', or it is dried first, giving the 'dry weight'. The scales used will depend on the size of plants, but for any plants below the size of tall shrubs scales with an accuracy of at least 0.01 g should be used. Drying should be carried out in an oven at about 100 °C for 1–2 days, but if an oven is not available natural drying for several days will work. You must sort the species before drying or else the plants will be unidentifiable. It is important to remove or wash off any soil or detritus on the plant material before weighing. Aquatic macrophytes may be harvested with a corer, which is effectively a quadrat frame that doubles as the harvesting tool. It is a sheet metal cylinder with one rim sharpened and it is placed over the sample area and pushed into the substratum far enough to sever stems and roots. You could also use a grab to harvest an area of macrophytes.

Advantages and disadvantages

Quadrats are the basic tool of many plant ecologists. They are very easy to use and can be employed in a wide range of studies. It can be difficult and time-consuming to measure out very large quadrats. If, as is common, species have a non-random distribution over the study area then estimates of abundance from a single quadrat will be changed by the size of the quadrat. This is because larger quadrats will even out the patchiness in the vegetation more than smaller ones will. Single quadrats are never used and carefully planned sampling (Chapter 2) will eliminate these size effects.

The measures have different advantages and disadvantages. Density- and size-based measures correlate poorly and even cover and biomass (size-based) are not very compatible (Figure 4.1). At intermediate densities frequency has a loose logarithmic relationship to density, but saturation of quadrats or quadrat subdivisions at high densities will lessen any correlation. The use of smaller (and more) quadrats or subdivisions maintains the relation with density better as the area sampled approaches that of a single plant. Frequency is a very quick and easy method to use but the estimate of frequency will always be influenced by quadrat or subdivision size. This is because the quadrat, rather than being used to select a sample area, is used as the dimensionless unit of measurement. Therefore, larger quadrats or subdivisions will usually be more likely to find the study species and will give higher frequency estimates. When using quadrats as the measurement unit, patchiness in species distribution will reduce the likelihood of a randomly placed quadrat finding the species and will therefore also reduce the frequency estimate. For these reasons you should take great care when interpreting frequency measures, especially when comparing study areas. Local frequency has the same problems.

'Nested quadrats' have been suggested to improve frequency measures, especially to improve the correlation with density (Morrison *et al.* 1995). The most straightforward of a number of nesting approaches involves concentric nesting of a series of gradually larger sub-quadrats (e.g. using a geometrical relationship) within the largest quadrat. The observer notes the species in the smallest nested sub-quadrat, then moves on to the next smallest quadrat and notes species found only in the area of this sub-quadrat not occupied by the smallest sub-quadrat. This goes on, surveying only the area newly encompassed by each progressively bigger quadrat, and the frequency is the proportion of sub-quadrats that contain the species. This approach is thought to give a closer approximation of density and is less affected by the spatial distribution of species. However, nested quadrats are rarely used because of the difficulty of actually nesting a series of quadrats in the field. Despite the conceptual problems with frequency, the speed, accuracy and reproducibility among observers of this approach mean it is widely used.

Counting individuals for density can be very time-consuming and difficult unless the plants have a low density or you use very small quadrats and is usually used only in studies of single species. Visual estimation of cover has the great advantage of speed over more complicated surveying. Visual estimates are made more easily where you can look down on the vegetation. Cover may be hard to estimate with any degree of accuracy in tall vegetation such as scrub or forest, although it is possible if you can look up at the canopy and estimate the cover of individual trees. Visual estimation of cover has received a lot of criticism because it involves a subjective estimate, that is, cover is not measured directly. Some ecologists get very excited about this, going so far as to call visual estimation guessing. This is an extreme view, but it reflects the fact that different observers or one observer at different times may make different estimates. Some ecologists suggest that visual estimates of cover are not reproducible among observers, and indeed differences in the cover of a species of <20% may be due to variation among observers rather than real changes. Practically, you can reduce this variation by training observers on a common set of quadrats, having observers work in pairs, and checking on observers as they progress (Kercher *et al.* 2003). You can formally analyse differences among observers by having the observers survey the same set of test quadrats and using this information as a correction factor. If only you are doing the quadrats, it is worth getting one or two colleagues into the field to test your estimates on a few quadrats. Larger quadrats will reduce variation among observers since it will be easier to distinguish between cover values.

The precision of cover estimates has also been questioned, but on the whole they seem to compare well against objective measures of cover. Differences from objective measures seem to be inconsistent, with some authors reporting underestimation and others reporting overestimation of cover! If problems with visual estimation really worry you, a more objective method is to analyse digital photographs of the quadrat with image-analysis software (Olmstead *et al.* 2004). However, this is time-consuming and there can be problems with distinguishing photographed species.

Harvesting biomass has a large number of drawbacks and should be used only if you are certain that you need to measure the biomass of species. It is 'destructive sampling' and should be used only for vegetation suited to this sort of treatment (e.g. meadows), where the destruction is on such a small scale that it is unimportant, or where you and others do not mind the study area being destroyed. Large shrubs and trees are very difficult to harvest, transport and weigh. Very short vegetation, such as lawns, will be very difficult to harvest without massive error between samples caused by slight differences in cutting height. This method is really appropriate only for taller vegetation such as short-shrub, aquatic-macrophyte and meadow communities. Cutting height is extremely important and you should usually cut at or close to ground level. However, the ground surface is rarely level and you may easily cut too high in some places and too low (i.e. below the soil surface) in others, causing errors and contaminating the sample with soil. Corers for aquatic macrophytes cut below the substrate surface but still there will be variation in the depth of cutting. Grabs cut at very variable heights and also bring up a great deal of detritus and do not produce a reliable estimate of biomass at all. Sorting of species can be very time-consuming and difficult, especially if the plants fall apart, leaving you with fragments to identify. Fresh weight is not a good measure. It varies with the moisture content of the plants and the moisture loss from cut plants means that fresh weight will be strongly affected by the time since harvest. Dry weight is a much better measure.

Biases

It has been suggested that more conspicuous species, such as those in flower, those with a tall growth form, those forming clumps of individuals and those with broader leaves, may be given higher cover estimates than will dispersed, low-growing and fine-leaved species. This should not be a problem if you are careful. General biases for cover, biomass and density are given in the introduction. Frequency can be biased against species with a more clumped distribution. Harvesting of biomass can miss low-lying species if the cutting height is too high. If some species fall apart it is almost inevitable that you will lose and misidentify some plant parts. Fresh weight will bias the measure in favour of species with high water content.

Point quadrats

Point quadrats are used for estimating cover of grasses, herbs, mosses, etc. in short vegetation.

Method

A point quadrat is a thin rod with a sharpened tip, and should usually be made of metal for rigidity and strength. Good materials are thick gauge wire, welding rod, knitting needles and even bicycle spokes. The point quadrat is lowered vertically through the vegetation and various recording methods can be used to get various types of data. There are numerous measures, but many are confusing and difficult to interpret and I shall not discuss them. The most popular and acceptable measure is of the percentage cover of each species in the vegetation. To sample this you should identify the species of each living plant part the tip (and only the tip) of the point quadrat hits on the way down to the soil surface. The data recorded from that point quadrat are only the presence or absence of each species, i.e. whether or not the point quadrat hits a particular species; the number of hits on that species is unimportant.

The theory behind the use of point quadrats is fairly simple. Normal quadrats (see above) could be used to estimate cover as the percentage of quadrats covered by the species. For large quadrats, many will be only partially covered rather than either not covered or wholly covered, introducing an ambiguity into the measure. As quadrats are made smaller they become less likely to be only partially covered and this cover estimate becomes more accurate. Point quadrats are theoretically a quadrat of infinitesimally small area, i.e. a point. A plant can only be present or absent in an area of zero diameter and therefore there are no cases of partial cover. This gives a true value for cover. Of course, point quadrats do not give absolute points but have diameters and the diameter will affect the estimates of cover for particular species. You could get a zero-diameter point by using optical cross wires, like those used in rifle sights, but these are virtually impossible to use in the field. The practical solution is to use point quadrats of as fine a diameter as possible. Steel wire of diameter 1.5–2 mm survives quite well in field use and a sharpened tip will narrow the point quadrat further. It is very important to use point quadrats of the same diameter in all your sampling and, if you want to compare your work with another study, find out what point-quadrat diameter was used.

It is impossible to lower a point quadrat free-handed steadily and vertically through the vegetation while noting touches and identifying species. You should either stick the point quadrat into the soil and note touches along its length (it is important to have a point quadrat of very low diameter for this) or use a 'point frame'. This consists of a horizontal bar supported on one or two legs. The bar has holes along its length through which the point quadrat can be passed. The supporting legs are stuck into the ground so that the bar is level and readings are taken at the position of each hole in the bar. It is traditional, although not vital, for the frame to accommodate ten point quadrats and the presence/absence readings from the ten point quadrats are summed to give a score for the whole frame for each species. You should never treat each of the ten readings as independent measures. Whatever method is used you should be careful to avoid disturbing the vegetation when placing the point quadrat or taking readings; you could move plants onto or away from the point quadrat and cause errors in the sampling.

Advantages and disadvantages

There is a sounder theoretical basis for using point quadrats to assess percentage cover than there is for visual estimates in quadrats, although the two methods seem to give quite similar estimates (Brakenhielm & Liu 1995). Point quadrats are particularly useful in short vegetation, such as grasslands, especially when it is difficult to distinguish individual plants. The vegetation should never over-top the point quadrat and hence the point quadrats needed for some vegetation types such as hay meadows and tall-grass prairies will be too tall to be practical. This technique can be very slow and complex (especially in dense vegetation) and it involves crouching or lying on the ground for long periods. This seems to lead to variation in cover estimates among observers, which has been shown to be as much or more than variation in visual estimates of cover. Because a very small area is sampled, very many samples are needed when one wants to detect the rarest species. For this reason, the use of point quadrats is quite poor at detecting rare species compared with the expenditure of an equal amount of effort with normal quadrats (Brakenhielm & Liu 1995). For these reasons, point quadrats are not commonly used.

Biases

The biases involved in cover estimates are given in the introduction.

Transects

Transects can be used for a variety of survey purposes in any vegetation.

Method

Apart from the standard uses of transects (see Chapter 3), other transect-based methods can be used to survey vegetation. Transects are commonly used to survey changes in vegetation along an environmental gradient or through differing habitats. This can be done using belt transects or, for larger sample areas, gradsects. A second use is to estimate overall density or cover values of species in a single stand of vegetation by the line-intercept method. The length of the transect can be several centimetres or hundreds of kilometres, depending on the vegetation and the aim of the study.

The *line-intercept* method involves using the transect line as a surveying implement to measure density or cover. A variation is to lie a number of sticks 1–2 m long at locations throughout the study area (e.g. random locations or at intervals along a transect) and to use these as 'mini'-transects. For either approach, a simple measure is to count how many plants of a species touch the transect line to give a measure of plant density. For longer transects you could count only touches at certain interval points along the transect; for instance at 10-mm, 10-cm, or 10-m intervals. Alternatively, cover can be estimated by measuring the length of transect line occupied by each species and using this to calculate the percentage of the length of the transect 'covered' by a species. To increase the number of plants sampled per unit length, you can sample not just the line, but also the vegetation within a certain distance of the line, e.g. 5–10 cm.

Belt transects consist of quadrats of any size laid contiguously along the length of the transect. Cover, biomass, density or frequency can be estimated for each quadrat and the variation in the measure along the transect can be determined and correlated with the gradients in environmental factors.

Gradsects, or gradient-directed transects, are transects that are laid out to sample intentionally the full range of floristic variation over the study area. They are usually used to sample very large areas, sometimes being hundreds of kilometres long. To accomplish this, the transect is usually positioned to lie along a steep environmental gradient, for example one due to altitude, land use or geology. Gradsects can be placed using landscape features and maps to provide precise locations. However, Global Positioning System (GPS) receivers (hand-held devices that use satellite output to locate you precisely) are much more accurate and are now widely available. A wide range of census approaches can be used with gradsects. You can do species counts, do DAFOR, place quadrats at points along the gradsect, or do more intensive sampling at these points using local transects or placing multiple quadrats, point quadrats or line transects. Gradsects can be combined with larger-scale mapping through remote-sensing techniques (see below); for example, using remote-sensing to provide large-scale vegetation maps and gradsects to provide detailed information on changes in plant species (Sandmann & Lertzman 2003).

Advantages and disadvantages

For certain vegetation types it may be easier to use the line-transect method rather than normal or point quadrats. It can allow more productive sampling in sparse vegetation and can be more practical in tall vegetation. In either case sampling will be speeded up. It is directly comparable to visual estimates of cover in quadrats, having given similar cover values when compared for the same vegetation (Kercher *et al.* 2003). If the vegetation is at all dense then counting touches will take a very long time. If plants are tussocky, form definite clumps or are large and distinct then the length of transect occupied by a species can be measured reliably and simply. For example, this is a very useful and accurate method in deserts (Etchberger & Krausman 1997). Cover estimates will be very difficult in vegetation where plants are small and intermingled and hence different methods should be used. As with visual estimates of cover, there can be large variation among observers in estimates of cover. Similar methods should be used to reduce this variation (see above). Counting touches by individual plants is not only difficult in dense and intermingled vegetation where it is hard to distinguish individuals but also produces a measure with ambiguous meaning. It resembles a cover measure, being determined by the density and size of a species (bigger plants are more likely to touch the transect), but cannot be expressed as cover. I do not recommend this measure.

Although they produce very detailed data, belt transects can be very time-consuming and it is worthwhile to consider whether more sparsely spaced quadrats will fulfil your needs. Note the advantages and disadvantages in using quadrats discussed above.

Gradsects have a very precise use; to assess the range of vegetation types over a large area. They will yield more information than randomly placed transects and are commonly used to survey little-studied areas (Austin & Heyligers 1989). However, knowledge of the environmental gradients is needed, and you run the risk of being guided by environmental factors that are relatively unimportant in structuring the vegetation. This can lead to errors in determining the true determinants of the distribution of species and vegetation types (Sandmann & Lertzman 2003).

Biases

The count of touches or estimate of cover will often depend on the height of the line transect in the vegetation, different species having different vertical and horizontal structures. To comply best with the definition of cover as the vertical projection onto the substrate, the line should be placed on the substrate surface. Gradsects for an estimation of the range of floristic variation will be biased by which environmental factor is used to describe the gradient.

Mapping terrestrial vegetation

This technique is used for estimating cover of vegetation types over a large area.

Method

There are several ways to map and measure the cover of various vegetation types over a large area, i.e. from a few to thousands of km^2. The vegetation types used should depend on your needs and the scale of the survey.

'Vegetation type' can have a variety of definitions for the purposes of vegetation mapping and provides a way of categorising areas of similar species groupings and/or plant-growth form. Starting out with only a vague notion of what classification you will use will result in a waste of time and poor-quality information. You can devise your own categories, for instance based on the dominant species or species combinations or coarser criteria (e.g. high mangrove, mixed mangrove, salt marsh, savanna, pasture, etc.). Many standardised classification systems are available, the most famous being that of the Zurich–Montpellier school (Kent & Coker 1992). In Britain, the National Vegetation Classification provides the standard vegetation types (Rodwell 1998).

Ground mapping involves dividing up the survey area into manageable units, which act like 'mega'-quadrats. The size of these units should be determined by the total area to be surveyed and the detail you require. They can be measured out with reference to landmarks or as geographical entities (e.g. a hill side or the area between two streams), or you can use squares based on the grid of latitude and longitude. These can be located using a map and landmarks or triangulation, or by using a GPS. Another approach is to divide the study area by eye into obvious vegetation units, e.g. woodland, grassland, heath. Each of these units is surveyed as an entity.

Within mega-quadrats or vegetation units you can either estimate the cover of each vegetation type visually, or sketch their boundaries onto a map. The boundaries can also be positioned precisely using triangulation on landmarks or, more accurately and easily, a GPS. The areas on the map can be measured by hand or by using computer digitisers or image-analysis systems. The GPS readings can usually be read straight into Geographic Information System (GIS) software and combined with digital maps of landscape features to provide extremely accurate vegetation maps (Dominy & Duncan 2002).

Remote sensing by aerial photography or satellite images is highly technical (except for the most basic aerial photographs) and you generally need to consult experts to find the best methods. What is more, the technology is developing extremely rapidly, especially for satellite imaging. Introductory texts include Campbell (2002) and Lillesand *et al.* (2003) for remote sensing and Paine (2003) for aerial photography. Here, I shall give some background to allow you to decide whether you may find these techniques useful.

You could simply take a digital photograph from a plane or helicopter directly over the vegetation. There are other more technical and more accurate procedures requiring specialised equipment. Aerial photographs at a range of scales are commercially available in many countries either digitally, or as older 'hard-copy' photographs (Paine 2003). The photographs may be infrared, multispectral (i.e. colour) or monochrome and are divided into patches of different tone, texture and pattern that correspond in some way to the vegetation. These cover types must be related to vegetation types by field surveys of representative areas. The boundaries of each cover type can then be transferred to maps and their areas measured using tracing paper, by projecting the image onto a map, using computer digitisers (for instance, those associated with the GIS), using image-analysis systems or using specially designed transfer instruments. Digital photographs can be analysed and overlaid onto other maps directly on the computer.

Other forms of remote-sensing technology on aircraft and satellites include optical sensors, which record reflected solar radiation in visible and infrared spectra, and radar sensors, which transmit electromagnetic waves and measure characteristics of the backscattered radiation. Computer analysis relates the remotely sensed data to ground maps of vegetation types (groundtruthing), so that each vegetation type can be given a characteristic remotely sensed signal. The resolution (i.e. the spatial detail) of these images is determined by the pixel size used. The pixel is the area of land from which a single spectral image is taken and can now be very fine, e.g. 1 m (although 10 m is more commonly available). The vegetation types may be very coarse, e.g. forest, grassland etc.; but much greater detail may be achieved, e.g. differentiating conifer, deciduous and mixed woodland or wet heath, dry heath and bogs. Methods such as LIDAR can even detect individual trees.

Maps derived from ground mapping or remote sensing can be analysed (GIS programs are very useful for this) to provide data on, e.g., the number of patches of each vegetation type, the spatial pattern of vegetation types and the area, shape, boundary length, etc. of each patch. Vegetation mapping can be combined with more detailed censusing of selected vegetation blocks, e.g. using quadrats, transects, etc. This provides two scales of information, namely the distribution *and* composition of vegetation types in a landscape.

Advantages and disadvantages

Ground mapping is cheap in terms of equipment and can produce very accurate maps, especially if GPS devices and GIS programs are used. However, it is very time-consuming compared with the other techniques and for this reason cannot be carried out over very large areas. Until 2000, GPS accuracy was downgraded by satellite-signal errors introduced by the US government. These are no longer a problem and accuracy is now at a few metres. However, GPS satellite signals are poor in some regions of the world and can be attenuated under dense forest canopies. Climbing trees or using long antennae may help with the latter problem.

You can cover large areas by remote sensing (satellites can cover thousands of kilometres) and you can survey areas that are impenetrable on land. Once a system has been developed, analysis of remotely sensed images can be very rapid. However, remote sensing can be expensive. None of the remote-sensing techniques has the resolution of ground mapping and the cover types will not be as precise as can be achieved by a surveyor on the ground. For example, calcareous and acid grasslands can be identified easily by the observer on the ground, but may be indistinguishable in remotely sensed images. Remote sensing also has a higher degree of misidentification of vegetation. Aerial photographs tend to provide more detail than satellite images and so are better for mapping smaller areas precisely. However, satellite imagery is improving all the time (Mehner *et al.* 2004). In all mapping methods inaccuracies can arise in delimiting blocks of vegetation. In many landscapes, vegetation types do not have clear edges and large errors may arise, especially among observers, in attempting to locate boundaries for mapping. For example, on moving from woodland to grassland, one may pass through a narrow band of mixed trees and grass. It may be more useful to classify these boundaries as a separate vegetation type, although more sophisticated approaches can be used in remote sensing (Schmidtlein & Sassin 2004).

Biases

Vegetation types might not be distinguished by remote sensing due to similarities in appearance or in spectral image.

Mapping aquatic vegetation

This technique is used for estimating cover of aquatic macrophytes over a large area.

Methods

Aquatic macrophytes (plants identifiable to the naked eye, including all vascular plants and bryophytes and some algae) can be mapped using similar methods to those for terrestrial vegetation (e.g. ground mapping, aerial photographs) if they are visible from the water surface (i.e. they are emersed or are submersed to depth less than 4 m). If they are deeper, more technical equipment is needed, such as SCUBA and depth gauges or echo-sounding (Jager *et al*. 2004). Coastal, reef and open-sea vegetation should be censused with great care, using specialist equipment and with the help of experts (Miller *et al*. 2003). Particular, straightforward, ground (or water-surface) mapping methods for river and lakes have been developed and are described below. These methods can also be used to structure sampling for phytoplankton and benthic algae.

River macrophytes can be mapped by dividing the river (or the part of the river in which you are interested) up into sections, of any length in the range 5–500 m, depending on the total length to be surveyed. A common approach is to specify a stretch of 50–100 m that will represent the river. This is then divided up into smaller survey sections of, e.g., 5 m. The section length is measured along the riverbank either with a tape measure for short sections or, for longer sections, by pacing (having measured a pace length) or using a GPS device. If the river is narrow enough you can mark out the section by laying lines across the channel. Along each section you can score the abundance of each species within the river channel using DAFOR, percentage cover or a scoring system based on cover (see Table 4.2). In narrow or shallow rivers, you can sample the area using quadrats. The width of the river channel varies depending on the amount of water flowing, so it is usual to define the width of river channel for surveying as that which is submerged for >85% of the year (Scott *et al*. 2002). In practice this is done by surveying during periods when the river level is low – in summer and not after rain. This also makes it easier to see macrophytes (shallow water depth and low turbidity).

Surveying of narrow rivers can be done from the bank. Otherwise, shallow rivers can be waded, by describing a zig-zag from bank to bank. In deeper rivers you need to use a boat; if the macrophytes are visible from the surface then use the cover-estimation techniques described. If the river is too deep or turbid you should use a grab to sample at points.

Lakes can be mapped by running transects across the water body or laying out a grid (using the GPS) (Jager *et al.* 2004). Points are sampled using the shallow-, deep- or very-deep-water methods described above.

Advantages and disadvantages

Any surveying around open water should be done with great care, and the use of boats is more dangerous. Unless you are out in the (shallow) river channel itself estimates of cover may be inaccurate because you are not looking down on the plants, so it may be useful to use a scoring system to avoid the spurious accuracy of percentage cover (see the 'Quadrats' section). Identification at a distance may also be difficult, and you may need to retrieve samples using a grab from the bank. Problems with cover estimates are described in the 'Quadrats' section, and solutions such as training of observers to decrease errors among observers are recommended. Grab sampling at depth will be extremely inaccurate and more technical solutions should be considered. Because rivers are rarely exactly linear, measures of length may vary depending on where they are made. Technically, you should measure along the midline ('*Thalweg*') of the river, but, even where it is possible to walk in the river, you would probably stir up sediment and so make subsequent surveying difficult. In practice, aim to measure along the bank, staying as close to the river as is possible and safe.

Biases

Biases of cover estimates described in the 'Quadrats' section apply. If you are surveying from the bank, you are especially likely to miss small and small-leaved plants.

Seed traps

Seed traps are used to measure the density of the seed rain in terrestrial and aquatic communities.

Method

Seed traps are placed on the soil or water surface to estimate the density per unit time of seed arriving on that surface. The traps are sometimes placed vertically or at a slope but there is no benefit in doing this and it does not give a correct estimate of seed rain per unit area of the soil surface. Sticky traps comprise a sticky surface of known area applied to the ground, which traps all seeds that come into contact with it. The sticky substance should be non-drying and not toxic to the seeds if they are to be germinated or tested for viability. There are several permanently sticky petroleum-based substances that can be used, most commonly substances used to trap insect pests (available from horticultural suppliers). These are semi-liquid and can be smeared or aerosol-sprayed onto the trap surface. The surface used can be anything waterproof that can hold the sticky substance, for instance a pane of glass or polythene wrapped around wood or plastic. Sticky cards developed to trap flying insect pests can also be used without modification. To be

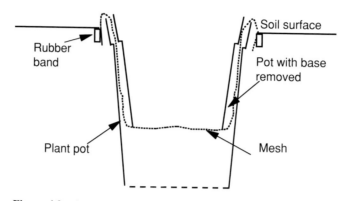

Figure 4.2   A seed trap made with plant pots.

manageable, and to trap sufficient seed, the trap should have a diameter in the range 10–30 cm. Unless the trap is heavy you should fix it in place, for instance with long pins. To avoid interference from vegetation, the traps can be fixed just above the canopy on some form of support. In water, this support can hold the trap above the water surface. An alternative is to place a circle of paper (e.g. filter paper) smeared with the sticky substance in a Petri dish and use this as the trap. The dish can be either pinned to the soil surface or nailed onto a length of doweling, which you push into the soil to fix the trap in place (again, the trap can be held above the canopy or water surface). You should make small holes in the dish to allow rain water to escape.

There are many alternatives to sticky traps. Plant pots modified with a mesh base (Figure 4.2) can be fixed onto or dug into the ground. Mesh suspended on a frame can be used for large seeds, e.g. in forests. Seed trays containing sterile compost can be laid out and seed numbers can be estimated by counting germinating plants. Seeds being transported in water can be trapped in nets held vertically in the water column. This can be done, in flowing water (including waves coming onto the shore), by attaching fine-mesh nets to supports that are either sunk into the sediment or fixed to natural or artificial structures. In non-flowing water it may be better to drag nets from a boat over designated sample areas (Stieglitz & Ridd 2001). Seed deposition in water-borne sediment (e.g. during floods) can be sampled by fixing pieces of artificial turf onto river banks or other areas reached by flood waters; these will trap the sediment.

As with quadrats, a number of replicate seed traps should be laid out in an appropriate design. Usually, the longer the traps are left in position, the more seeds will be trapped. So the measure you derive will be a rate. Traps should be removed or replaced after a few days or weeks. The period will depend on the rate of seed arrival. The seeds are removed from the traps, either by collecting the traps or by emptying them. On sticky traps this can be a complicated process, using forceps or a non-toxic solvent. All traps, but especially sediment traps, trap much more than seeds and so seeds need to be sorted. Because species have distinctive seeds, in principle the seeds can be identified as to species. However, there are no comprehensive seed-identification keys. You can make a collection of seeds from plants growing in your study area. However, related plants can have very similar seeds and some seeds may come from species not growing in the study area. So you must be sure that your identification is accurate. This method is most successful if you are interested only in a few particular species. Otherwise, you should germinate a sample of each seed type and grow on plants until they can be identified. Alternatively, you can avoid seed identification completely by sowing all the collected seeds into seed trays and then identify and count the resulting plants. This also tests for seed viability (see below). There are some seedling-identification guides that may help you (Muller 1978; Williams & Morrison 1987).

Some seeds trapped may be dead, so to get a true picture of the live seed rain you should test for viability using staining or by germination as described under the methods for estimating seedbanks. If sticky traps were used, clean off the sticky substance to allow water to penetrate. A simpler technique is also available for fresh seeds: non-viable seeds are usually partly or wholly empty and are therefore wrinkled or easily squashed. You should, however, test the accuracy of your judgement with the staining test. This method should not be used on older seeds because they may be intact yet dead.

Advantages and disadvantages

All methods for assessing the seed rain are labour intensive. The rain is often very patchy, so a good sample size is needed. Extra seeds and detritus may be blown or washed over the soil surface into the trap, causing an overestimate of the rain and clogging up the trap. This can be decreased by having a rim of about 1 cm on the trap, which protrudes above the substrate surface. Some traps may lose seeds over time by seeds being blown or washed out, or removed by animals. Therefore traps should be replaced or emptied frequently. Predation by large animals may be stopped by protecting traps with wire mesh, while insects may be excluded by a mini-fence around each trap or deterred by smearing touch insecticides onto the trap. Sticky traps are good at deterring seed predators, but can trap and kill huge numbers of insects or even small vertebrates, which is not desirable. Sticky traps also tend to become clogged up with insects and dirt, which reduces their seed-trapping ability, and clogging can be rapid in dusty habitats. If you wish to carry out long-term sampling of the seed rain it would be wise to use a sticky-trap design that allows you to remove and replace the sticky surface easily. The Petri-dish trap allows you to substitute a new paper circle for the old one very simply.

Germinating seeds involves the least work since it takes care of identification and viability testing, but has the problems of germination tests described under 'Sampling of seedbanks'. However, if you are also studying the seedbank then you can use these germination tests to get a straightforward comparison of the rain and the bank.

Seed rain is highly variable over time (e.g. season) and, if you want an estimate of the yearly seed rain, rather than that at a particular time, you must sample all year round (although in cool temperate zones you may be able to miss out the winter).

Biases

Very small seeds will almost certainly be missed if you sort and count the sample. Seed predators are selective, so, if these are not excluded, certain plant species will be under-represented. Germination tests have the biases described under the methods for estimating the seedbank.

Sampling of seedbanks

Sampling is used for estimating the density and distribution of seeds in the seedbank.

Methods

Soil samples of known area and depth are taken at sample points throughout the study area. Seed-banks exhibit great variation in seed density and species composition over small areas due to the patchiness of species distribution. You must therefore take many samples to achieve an adequate estimate. Dessaint *et al.* (1996) present a formal approach to designing seedbank sampling. You may want to sample seedbanks over a large area or even to map it. This can be done by positioning sampling points in relation to vegetation types or on a grid system and locating them using the GPS. There is also often great variation in the seedbank over time, reflecting temporal changes in seed production and germination, for instance due to the seasons. You should either take samples on at least two occasions to assess this variation or choose the sampling time carefully, for instance after the late summer peak in seed production and before autumn germination in cool temperate zones.

If it is large enough the sample can be dug out, using a quadrat to mark the area. Smaller samples can be taken using a sheet-metal cylinder that has one sharpened rim and is pushed into the soil to a certain depth to remove a soil core (Figure 4.3). The core should remain intact when it is removed from the soil. A removable cross-bar is useful to aid the pushing of the cylinder into the soil and to use to push the core out of the corer. Aquatic sediments can be sampled with cores in the same way, using SCUBA if necessary (McFarland & Rogers 1998). Alternatively, for deep water bodies, a dredge grab can be set to sample to a specific sediment depth. You should be able to estimate roughly the area sampled by the grab.

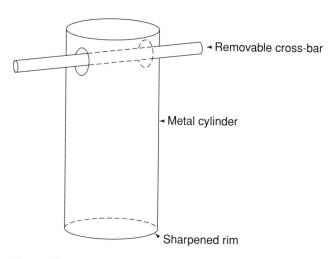

Figure 4.3   A soil corer.

There is no standard core diameter or depth but generally diameter is in the range 2–20 cm with a depth of 5–20 cm. The diameter should reflect the density of the seedbank and the size of seeds. Most studies in most habitats have used diameters of 2–5 cm. The smaller the cores you use the more samples you will have to take in order to sample a reasonable area, but this is usually desirable in order to sample the variation in the seedbank. The core depth should be decided after considering the aim of the study. Viable seeds are strongly concentrated in the top 2–3 cm of the soil and these are the seeds most likely to be recruited naturally into the community. The more deeply buried seeds will give a more complete picture of the seedbank and will also show the seeds available to be recruited following soil disturbance.

The soil cores are transported back to the laboratory in bags to avoid seed loss or contamination. Here the core can be separated into layers, if you wish, to find the vertical distribution of seeds in the soil. A common division is into layers of depths 0–2 cm, 2–5 cm and >5 cm. The top layer will contain litter and you could scrape this off and analyse the seeds in this separately. In some studies the seeds of the litter layer have been considered not part of the seedbank and discarded. There is no ecological basis for this idea.

The seedbank in the cores can be estimated by germination tests or by counting seeds. Methods involving flotation of seeds on a saline density gradient are inaccurate and difficult and I shall not discuss them.

In *germination tests* the sample is air-dried to kill any vegetation and the soil is spread over a seed tray. If you have large quantities of soil or sediment you can concentrate the seeds using the method of ter Heerdt *et al.* (1996). This involves breaking up the sample and forcing it through a coarse-mesh sieve and then through a fine-mesh sieve with a jet of water to remove both coarse and fine material. The material that passes through the coarse sieve but not through the fine is retained. The two mesh sizes are chosen to represent the extremes in seed size in the sample; ter Heerdt *et al.* (1996) used 4 mm and 0.212 mm. The sample can be spread onto a layer of seed compost or sand to decrease the proportion of seeds that remain buried. The seed trays are placed outside, in a glasshouse or in a growth room with simulated natural conditions (daylength, diurnal temperature variation, humidity, etc.) and the soil is kept moist by watering or misting. If the trays are outside you should protect them from herbivores and contamination by airborne seeds. Germination tests for samples from aquatic communities should usually be carried out in submersed seed trays, either in the original water body or under artificial conditions such as in water tanks, perhaps in standard culture solution (McFarland & Rogers 1998).

The trays should be monitored frequently (e.g. daily) and any seedlings that emerge are identified and removed. As the germination rate slows, monitoring can become more infrequent. The sample should be stirred occasionally to expose all the seeds. Seedlings are identified using the methods described in the 'Seed traps' section. It could be useful to maintain a reference collection of seedlings. Monitoring can stop when nothing has emerged for several weeks.

An extra procedure that may increase the germination rate of some plants of temperate regions is to chill the seeds. The soil samples are placed in a fridge at about 5 °C for 3–4 weeks and then germination is tested in seed trays as described above. The warming after chilling simulates the return of spring after the winter conditions of the fridge and this is often a trigger to break seed dormancy.

To make *counting of seeds* possible you need to sort the seeds from the soil using more complex sieving methods. The sample is passed through a graded series of sieves, starting with the largest mesh size. Seeds will be trapped in different sieves according to the seed size. The smallest mesh size should be fine enough to catch the smallest seeds but coarse enough to allow particles to pass through. A typical size is 0.2 mm. It is usually quicker to do this under a jet of water and to break up aggregates with your fingers. This method serves to concentrate the seeds and the material remaining in each sieve should be sorted to find these seeds. You should therefore use a range of sieve sizes to speed up this sorting, so that the majority of seeds will not be mixed in with too many stones and large particles.

The seeds are sorted from the remaining detritus, maybe using a binocular microscope for the smaller seeds, and are identified using the methods described in the 'Seed traps' section. It is usual to count how many of the seeds are still viable because the seedbank is defined only by the live seeds. To save time this can be done on a random subsample of the seeds to establish the proportion of the seeds of each species that are viable. Viability can be tested by germination tests in sterile soil or on dampened filter paper in a Petri dish or by staining with tetrazolium and indigocarmine. Tetrazolium (triphenyltetrazolium chloride) stains living tissue red and remains colourless if the seed is dead. Indigocarmine remains blue if the seed is dead but turns colourless in live tissue. These tests are not always successful, so it is best to divide up the seed tissue and to carry out both tests using separate tissue samples. One positive result from the two tests is enough to indicate that the seed is viable. These stains are mildly poisonous, so use them with care.

Tetrazolium

Fresh 1% 2,3,5-triphenyltetrazolium chloride is prepared in a phosphate buffer at pH 6–8. The seed is cut in half to expose its tissue and one half is incubated in the tetrazolium for 2 h in complete darkness. Excess stain is washed away with distilled water. Living seed tissue will be stained a red colour.

Indigocarmine

A 0.05% solution of indigocarmine is prepared in hot distilled water, filtered and allowed to cool. The seed tissue is submerged in the stain for 2 h in the dark and then the stain is washed off with distilled water. If the tissue is colourless, it is alive, but a blue colour indicates dead tissue.

Advantages and disadvantages

Seedbank sampling involves a lot of work no matter which method you use. Taking many samples and sampling on more than one occasion adds to the effort. For aquatic samples, grabs are much quicker and easier than taking cores, but there is much less accuracy in defining the area and depth of the sample.

Both methods for assessing viable seed numbers in the soil samples have drawbacks. It is virtually certain that not all the viable seeds will germinate in germination tests. While some species germinate readily, others exhibit seed dormancy, which can be broken only by specific environmental factors. Chilling may break dormancy, but other possibilities include after-ripening, scarification of the seed coat, high light intensity, widely fluctuating temperatures or even the severe heating of a fire. Seeds in certain Mediterranean habitats require smoke for germination and smoke-derived germination-stimulants are commercially available (Light & van Staden 2004). However, it is impossible to cover all the possibilities and most studies use just the standard conditions in the glasshouse or growth chamber. You must remember that this method therefore measures only the 'readily germinable fraction' of the seedbank. For these reasons it is also important to carry out germination tests under the same conditions for all samples.

By sieving the sample you will find most seeds, although the smallest seeds may be lost during the sieving and sorting processes. All stages in this technique are extremely labour intensive and for this reason it is rarely used. The germination test for viability has the drawbacks described above. The chemical tests avoid these problems and provide the best method for fully sampling the seedbank, although they take a long time even if you use only a subsample. However, they are impossible on small seeds (>2 mm) because you cannot see colour changes.

Biases

The depth of soil core will probably affect the proportions of various species in your sample. Seeds are generally older further down the soil horizon and thus they represent both the longer surviving species (species differ in seed longevity) and also the past composition of the plant community (which may be different from that at the present). The distribution of species in the soil can, however, be determined by separating the core into layers. The sample will also be biased towards those species that have produced seeds most recently. Sampling on several occasions will allow quantification of this bias.

The environmental conditions of the germination tests can strongly affect which species germinate and dormant seeds will not be noticed. The sieving method may underestimate the very-small-seeded species, such as the dust seeds of many of the Orchidaceae, for they might not be caught by the smallest mesh and may be too small to be seen during sorting and sieving. Because the chemical staining is a response to metabolic activity, dormant seeds with very low respiration rates might not stain.

Phytoplankton

This section concerns measurement of the density of phytoplankton (floating algae).

Method

There are many methods for taking and analysing samples of freshwater and marine phytoplankton, some of which involve specialised electronic equipment for *in situ* estimates. Analysis of samples is by cell counts. I shall describe only the simplest methods which use very little specialised equipment. You should refer to the technical literature and consider other procedures

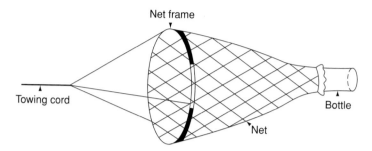

Figure 4.4   A plankton net.

(HMSO 1990; Porter *et al.* 1993; Hotzel & Croome 1999). Using the same techniques as for mapping macrophytes, the phytoplankton can be mapped in the water body.

The water body can be sampled using a jar or container on a line although a plankton net can be used to cover a larger area and to concentrate the plankton sample. A plankton net (Figure 4.4) consists of a fine-mesh net (e.g. 30 μm) ending with a bottle to hold the plankton. The net frame is usually square or circular, meaning that the sample area of the frame is easily calculated. The net should be small enough to be used easily and sufficiently small that the bottom of the water body is not disturbed (which is a problem especially in shallow streams). The net can be fixed either to a pole or to towing lines. For the latter the net can be towed either by you or by a boat. You can either hold the net against the current or sweep it through the water. The volume of water filtered depends on the area of the net frame, the speed of flow of the water, the speed you tow at and the time in the water. You should attempt to standardise this volume across all samples. Because of the fine mesh the net should be moved slowly (>1 knot) through the water. The phytoplankton will be stratified (i.e. each species exhibits a non-random vertical distribution in the water body) so the net should either be held at the same selected depth for all samples or be moved evenly through all depths. Alternatively, to sample a particular depth of water, a piece of flexible pipe (e.g. hose pipe) is lowered into the water with a line attached to the lower end. The top end is closed and the lower end is raised up until the pipe forms a U-shape. The whole pipe is taken out of the water and decanted.

Phytoplankton are not randomly distributed across a water body, due to variations in flow and nutrients. So samples should be taken at various locations; e.g. at various points across a river and near and far from the shore of a lake.

The algal cells will decay rapidly and should be preserved immediately after collection. Various fixatives are used, but the commonest is Lugol's iodine solution. A solution of 20 g potassium iodide in 200 ml distilled water is made and then saturated by adding 20 ml iodine. This is then acidified with 20 ml glacial acetic acid. The solution should be kept in a dark bottle to preserve the iodine. You should add 1–2 ml Lugol's iodine to every 100 ml of the sample (remember to allow for this extra volume when calculating cell densities).

Densities of species or cell types (a less precise classification) can be determined by counting through a microscope. If densities are low you can concentrate the cells by centrifuging or simply by sedimentation. The latter involves mixing up the sample to get an even distribution of cells and then removing a measured subsample to a measuring cylinder; narrower cylinders are better so that the final sample is more concentrated. At 20 °C cells will settle at a rate of two hours per 1 cm depth of water in the cylinder and a sedimentation time of 48 h is usually recommended for a standard 100-ml cylinder. After centrifuging or sedimentation you siphon off the top 90% of water, shake the remainder and take a further subsample for counting. Cells can be counted, but it may be better to count colonies or filaments of some species. The two simple methods are use of the haemacytometer and use of a counting chamber. The former is more usually used to count red blood cells. It holds a precise volume of liquid in which you can count the numbers of each species. Phytoplankton keys are widely available, but identification of algal cells is a skilled job. Several counts should be made for each sample and averaged to get a density. A Lund counting chamber comprises an ordinary microscope slide with two long, thin pieces of glass or another rigid material glued to the longer edges. This chamber is closed with a coverslip and a measured volume of the sample is placed in the chamber. Cells are then counted under a microscope. If you want a density of cells you should account for the concentration techniques used.

Advantages and disadvantages

These are the simplest and cheapest methods available. Netting phytoplankton will lose the smallest algae (nanoplankton); other methods can be used to avoid this. Pipe and net sampling requires two operators and a boat. Because algal cells vary widely in size ($<1->200$ μm) counting cells does not give a complete picture of the community. You can also estimate cell sizes (Hotzel & Croome 1999).

Biases

Poor control of the sampling depth of the sampling effort at each depth will bias the counts of the various species. The smallest species will not be sampled by netting.

Benthic algae

This section concerns measurement of the density or volume of benthic (bottom-dwelling) algae.

Methods

If the benthic algae are large, they can be censused in the same way as aquatic macrophytes, namely by quadrats, etc. Otherwise they should be sampled and the techniques described in the 'Phytoplankton' section can be used to count cells. Quadrats should be laid out and the algae in these sample areas removed. The removal technique depends on the substrate. More technical procedures than those given below are available, especially for deep water (HMSO 1984; Porter *et al.* 1993).

In mud, silt or sand the surface sediment should be drawn up using a large syringe. The sample is analysed as it is, by counting cells under the microscope. Stones and small rocks should be removed and then washed in freshwater to remove the algae. More tenacious algae can be removed by scrubbing with a brush or scraping with a scalpel. Lumps of algae should be broken up by shaking or with a spatula. Bedrock and other immovable objects must be scrubbed or scraped *in situ* and the dislodged algae should be collected with a large syringe.

Advantages and disadvantages

Microscopic enumeration of cells in sediment will be difficult since small algae will be hidden by the particles. It is very difficult to remove cells from substrates with larger particles (2–20 mm). Other methods should be used (HMSO 1984; Porter *et al.* 1993). Cells are certain to be lost in the scraping and brushing of rocks.

Biases

Certain species might not be counted. Smaller cells in sediments may be missed and the more tenacious species will be less likely to be dislodged from rocks.

Marking and mapping individuals

This techniqe is used to ascertain the fate of individual plants over a period of time in any vegetation.

Method

These methods will allow you to follow the fate or measure the performance (e.g. leaf, flower or seed production, change in basal area, clonal growth, etc.) of individuals in a population. Individual plants in the study area are marked and/or mapped so that they can be relocated and recognised at a later date. This approach is usually used in investigations on single species, such as demographic studies. You can study either some or all of the individuals in the study area. If you do the former you should either use quadrats to delimit sample areas or choose individuals at random or according to a defined protocol. Mapping and marking can be carried out at any scale; from hundreds of km^2 for a rare rainforest tree species to $100 \, cm^2$ for the tillers of a dominant pasture grass.

Mapping

The method you use should depend on the scale of the study and on the density of individuals. At large scales, you could simply note down the position of the plants on a map, locating them by coordinates (a GPS device could be used) or landmarks. This is best for scattered individuals within a large area. Sample areas must be relocated and it is best to use 'permanent quadrats' for this. A permanent quadrat is one that is fixed in place (e.g. using pins or tent pegs) or that can be replaced onto fixed markers, e.g. corner pegs. A good method is to use industrial survey markers, which are metal and can be fixed so as to be virtually immovable. For smaller quadrats, you can sink plastic or metal tubes (metal tubes can be relocated using a metal detector, but check beforehand that the metal detector can detect the tubes) into the ground and attach legs onto the quadrat corners, which can then be placed into these tubes. Plants can be mapped within the quadrat by fixing a scale onto the frame or dividing the quadrat into a grid (see under 'Quadrats') and determining coordinates in relation to the frame using rulers or measuring tape. Digital photographs can be taken and used to create computer maps. 'Permanent transects' are similar, and you mark the start and end of a transect using the same methods as for permanent quadrats. Plants are mapped in terms of their distance along the quadrat. Usually, you will also need to measure the perpendicular distance and direction of each plant from the transect line.

Marking

Individuals can be marked by a wide variety of methods; which is the best depends on the needs of the study, the plant size and structure and the materials available. If you want to distinguish every individual then they should be numbered. Numbers can be painted onto the bark of slow-growing trees, written onto posts, canes, plant labels or toothpicks (depending on the plant size) placed next to the plant, or put onto horticultural plant tags, plastic-coated wire rings or even bird rings, which you attach to the plant. You may wish to distinguish only groups of individuals, such as age cohorts. In this case you will not need to number the plants but can distinguish the groups by the colours of the markers. The posts and tags described above can be coloured and you can also apply differently coloured paints directly onto the plants. The paint should be non-toxic, fast-drying and hard-wearing, for instance acrylic paint. With care this can be used even on very small plants. Only small amounts of paint should be applied and only onto surfaces not important for light capture or gas exchange; the base of the plant is a good spot. The paint may need retouching during subsequent censuses.

On successive surveys each surviving individual is found again and you can evaluate measures of performance for every plant. If individuals are numbered then you can simply note the performance measures against each number on your datasheets. If you have mapped plants without marking them with individual numbers, you must either write the measures on your map, or identify the individuals by their coordinates or by their position on the map (determined by overlaying consecutive maps) while you are in the field. Therefore, unless you are looking simply at survival between censuses, which can be determined by comparing consecutive maps in the laboratory, you should mark the individuals in the field. If you are censusing and marking or mapping all individuals in the study or sample area you will locate new individuals at each census, which you can then mark or map. If you are using a sample of individuals then, if you want to locate new recruits (e.g. seedlings or young plants), you must resample at each census.

Advantages and disadvantages

Marking, locating and identifying individuals can be very time-consuming and detailed work in a dense population, especially where the plants are small. Despite this you must always take great care to disturb the plants as little as possible because you may alter their survival and growth by the process of measuring them.

Permanent markers will move over time due to soil movement (e.g. frost heave) and intentional or accidental interference from animals. Therefore, as time passes the position of a plant relative to the quadrat will change and mapped individuals may be lost or misidentified. This is a real problem only for dense populations. Certain types of markers may be lost through vandalism or interference from other animals. For instance, wire rings are lost very easily from grazed grass tillers. You should consider this problem and make the markers as permanent as possible. In most cases it is best both to map and to mark the individuals, either as a check that the mapping method has found the correct individual or as an aid to finding the marked plants.

If an individual is only mapped or if the marker is not fixed to the plant then, if the plant dies and a new plant grows up in the same place between censuses, you might mistake this new individual for the old one. New individuals may grow through wire rings to create the same problem. You must decide on the likelihood of this happening in relation to the vegetation type (low in forests, high in fertile grasslands) and the frequency of censuses.

In mapping by digital photography you must ensure that the camera is always perpendicular to the plot surface. If the angle varies between censuses, the relative positions of plants will appear to change. If the canopy height changes or certain individuals become smaller, the photograph will not detect those plants which have become hidden under the canopy.

Biases

The loss of markers or poor mapping may lead to underestimation of the survival of individuals. Misidentification of new individuals for old, dead ones may lead to overestimation of survival.

References

Austin, M. P. & Heyligers, P. C. (1989). Vegetation survey design for conservation: gradsect sampling of forests in north-eastern New South Wales. *Biological Conservation* **50**, 13–32.
Brakenhielm, S. & Liu, Q. H. (1995). Comparison of field methods in vegetation monitoring. *Water, Air and Soil Pollution* **79**, 75–87.
Campbell, J. B. (2002). *Introduction to Remote Sensing.* London, Taylor & Francis.
Dessaint, F., Barralis, G., Caixinhas, M. L. *et al.* (1996). Precision of soil seedbank sampling: how many soil cores? *Weed Research* **36**, 143–151.
Dominy, N. J. & Duncan, B. (2002). GPS and GIS methods in an African rain forest: applications to tropical ecology and conservation. *Conservation Ecology* **5**, article no. 6.
Etchberger, R. C. & Krausman, P. R. (1997). Evaluation of five methods for measuring desert vegetation. *Wildlife Society Bulletin* **25**, 604–609.

HMSO (1984). *Sampling of Non-planktonic Algae (Benthic Algae or Periphyton)*. London, Her Majesty's Staitionary Office.

(1990). *The Enumeration of Algae, Estimation of Cell Volume and Use in Bioassays*. London, Her Majesty's Stationary Office.

Hotzel, G. & Croome, R. (1999). A phytoplankton methods manual for Australian freshwaters. LWRRDC Occasional Paper 22/99.

Jager, P., Pall, K., & Dumfarth, E. (2004). A method of mapping macrophytes in large lakes with regard to the requirements of the Water Framework Directive. *Limnologica* **34**, 140–146.

Kent, M. & Coker, P. (1992). *Vegetation Description and Analysis*. London, Belhaven Press.

Kercher, S. M., Frieswyk, C. B., & Zedler, J. B. (2003). Effects of sampling teams and estimation methods on the assessment of plant cover. *Journal of Vegetation Science* **14**, 899–906.

Light, M. E. & van Staden, J. (2004). The potential of smoke in seed technology. *South African Journal of Botany* **70**, 97–101.

Lillesand, T. M., Kiefer, R. W. & Chipman, J. (2003). *Remote Sensing and Image Interpretation*. New York, John Wiley and Sons.

McFarland, D. G. & Rogers, S. J. (1998). The aquatic macrophyte seed bank in Lake Onalaska, Wisconsin. *Journal of Aquatic Plant Management* **36**, 33–39.

Mehner, H., Cutler, M., Fairbairn, D. & Thompson, G. (2004). Remote sensing of upland vegetation: the potential of high spatial resolution satellite sensors. *Global Ecology and Biogeography* **13**, 359–369.

Miller, M. W., Aronson, R. B. & Murdoch, T. J. T. (2003). Monitoring coral reef macroalgae: different pictures from different methods. *Bulletin of Marine Science* **72**, 199–206.

Morrison, D. A., Lebrocque, A. F. & Clarke, P. J. (1995). An assessment of some improved techniques for estimating the abundance (frequency) of sedentary organisms. *Vegetatio* **120**, 131–145.

Muller, F. M. (1978). *Seedlings of the North-Western European Lowland*. The Hague, Dr W. Junk.

Olmstead, M. A., Wample, R., Greene, S. & Tarara, J. (2004). Nondestructive measurement of vegetative cover using digital image analysis. *Hortscience* **39**, 55–59.

Paine, D. P. (2003). *Aerial Photography and Image Interpretation*. New York, John Wiley & Sons.

Porter, S. D., Cuffney, T. F., Gurtz, M. E. & Meador, M. R. (1993). Methods for collecting alga samples as part of the National Water Quality Assessment Program. Open File Report 93–409. Raleigh, NC, US Geological Survey.

Rodwell, J. S. (1998). *British Plant Communities: Volume 1. Woodland and Scrub*. Cambridge, Cambridge University Press.

Sandmann, H. & Lertzman, K. P. (2003). Combining high-resolution aerial photography with gradient-directed transects to guide field sampling and forest mapping in mountainous terrain. *Forest Science* **49**, 429–443.

Schmidtlein, S. & Sassin, J. (2004). Mapping of continuous floristic gradients in grasslands using hyper-spectral imagery. *Remote Sensing of Environment* **92**, 126–138.

Scott, W. A., Adamson, J. K., Rollinson, J. & Parr, T. W. (2002). Monitoring of aquatic macrophytes for detection of long-term change in river systems. *Environmental Monitoring and Assessment* **73**, 131–153.

Stieglitz, T. & Ridd, P. V. (2001). Trapping of mangrove propagules due to density-driven secondary circulation in the Normanby River estuary, NE Australia. *Marine Ecology-Progress Series* **211**, 131–142.

ter Heerdt, G. N. J., Verweij, G. L., Bekker, R. M. & Bakker, J. P. (1996). An improved method for seed-bank analysis: seedling emergence after removing the soil by sieving. *Functional Ecology* **10**, 144–151.

Williams, J. B. & Morrison, J. R. (1987). *ADAS Colour Atlas of Weed Seedlings*. London, Wolfe Publishing.

5 Invertebrates

Malcolm Ausden

Royal Society for the Protection of Birds, The Lodge, Sandy, Bedfordshire SG19 2DL, United Kingdom

Martin Drake

Consultant Entomologist, Orchid House, Burridge, Axminster, Devon EX13 7DF, United Kingdom

Introduction

As with other groups, the most important thing when censusing invertebrates is knowing why you are doing it and how you intend to analyse and use the results. The commonest reasons for censusing invertebrates are to

- evaluate the importance of a particular area for them and, in doing so, identify how the invertebrate fauna can be best conserved;
- monitor changes in the abundance and assemblages of species; and
- investigate the abundance of invertebrates as prey for other species, often birds.

Different aims require different approaches. The main requirement for evaluation surveys is that most or all habitats and micro-habitats thought likely to be important to invertebrates at a site are sampled. Evaluation surveys usually concentrate on searching specific habitats and micro-habitats considered important for species of high conservation value. They often use a range of complementary techniques to maximise the range of species recorded. Ideally, such surveys should be standardised as far as possible, although this can be difficult in practice, particularly if a large proportion of the surveyor's time is spent actively searching various habitats and micro-habitats at a site. At least the date, time spent surveying the site and weather conditions should be recorded. Several visits at various times throughout the active periods of the respective groups will usually be needed in order to obtain a reasonable impression of the importance of the site. It is often most efficient to survey one or a few key groups whose requirements will inform you most about the conservation value of the site for invertebrates and how it should be best managed. The results of evaluation surveys can be used to compare the relative conservation values of sites for particular groups by calculating for them indices based on the 'quality' of the species found (Fowles *et al.* 1999).

Monitoring changes in abundance of invertebrates over time presents particular challenges. Many insect species exhibit large year-to-year fluctuations in abundance, often in response to weather conditions, which can obscure long-term trends in abundance (Pollard 1988; Pollard

Ecological Census Techniques: A Handbook, ed. William J. Sutherland.
Published by Cambridge University Press. © Cambridge University Press 1996, 2006.

Yates 1993; Southwood *et al.* 2003). Furthermore, for many groups the only practical methods of censusing involve trapping techniques, whose catches might not necessarily provide a good measure of abundance. Activity of species is influenced by weather conditions and these will obviously vary between repeat periods of trapping. It can also be difficult to compare measures of invertebrate abundance for various types of vegetation using these techniques. One option is to monitor changes in *assemblages* of species, thus reducing the overall effect of short-term fluctuations in abundance of individual species. Sampling two or more sites over the same period of time can be used to determine the extent to which their assemblages are converging with, or diverging from, one another, for example to what extent the fauna at a habitat-creation site is converging with that of an existing area of the target habitat. The similarity of such assemblages can be assessed by calculating for each assemblage scores based on the ecological requirements of its component species. Alternatively, changes in invertebrate assemblages can be tracked using ordination techniques such as DCA/DECORANA (Hill & Gauch 1980) and CCA/CANOCO (ter Braak 1986, 1994).

Censusing invertebrates to provide a measure of the abundance of prey for other species is likely to require a different approach. At the outset it is important to consider how well the measure of invertebrate abundance obtained using the particular technique represents the abundance of prey to the particular predator. For example, will catches of invertebrates in pitfall traps, which are often dominated by large, nocturnal beetles, provide a reasonable measure of the abundance of prey to small diurnally feeding wading-bird chicks? Furthermore, the abundance of food in a given habitat might not necessarily be a good measure of its attractiveness to the given predator. For example, taller grass tends to contain more invertebrates than does shorter grass (see the review by Morris (2000)), but many bird species still prefer foraging in shorter vegetation, probably because their prey are more accessible and they have to spend less time being vigilant (Whittingham & Markland 2002; Butler & Gillings 2004).

In all cases of censusing invertebrates, other than the few conspicuous and relatively easy-to-identify groups such as butterflies and dragonflies, it is important to emphasise that the sampling is invariably the relatively quick and easy bit – it is the sorting and identifying of specimens that requires the time and expertise. Most invertebrate surveyors delay sample sorting and report writing until the end of the short and busy field season, and this should be borne in mind when commissioning surveys. An efficient way of dealing with large samples is to use non-experts to sort the trap contents into taxomonic groups and then send these away to relevant specialists for species identification.

When considering invertebrates as prey for other species, it will usually be necessary only to sort them into broad taxonomic groups. In some cases it might be sensible to sort potential invertebrate prey into size classes, thus recognising that larger invertebrates are likely to be more important as prey, and allowing one to ignore prey below a certain size threshold. This can considerably reduce the time spent sorting. The speed of sorting prey into size classes can be greatly increased by visually comparing them with specimens of known sizes (i.e. at the upper and lower end of each size class), rather than measuring each specimen individually.

When commissioning surveys by specialists, it is important to ensure that the contract includes the writing of a report highlighting important species found, their known requirements and recommendations on how the important invertebrate fauna can best be conserved. A report consisting solely of a list of Latin names is likely to be meaningless to most site managers and hence will not result in any conservation benefit.

There is a wide range of techniques available for censusing invertebrates (see Table 5.1). These fall into three main categories, which are discussed below.

Direct searching and collecting

These techniques include active searching and observation, sweep netting and beating. They provide indices of abundance, since they will not record all individuals from a given area and the susceptibility to capture will vary among species and with weather conditions. However, many of these techniques can be standardised to varying degrees to enable catches to be compared over time or between areas. This can be done by standardising the area searched and/or the level of sampling effort, together with the time of day and weather conditions under which the sampling is carried out.

Trapping

These techniques have the advantage that traps can be left at a site to catch specimens while the surveyor is occupied elsewhere. Trapping techniques provide only relative measures of abundance, since the catch is a product of the abundance of the particular species, its activity (which is usually heavily influenced by weather conditions) and the species' susceptibility to entering and remaining in the trap. Trapping effort can be standardised and replicated to allow statistical analysis. However, the efficiency of some types of traps (e.g. pitfall traps) varies greatly with vegetation structure, making it difficult to compare catches between different types of vegetation. Furthermore, even quite small modifications to trap design can alter their catching efficiency, for example, the age of malaise traps (Roberts 1975) and pitfall traps influences their efficiency. Used plastic pitfall traps get sufficiently scratched to provide a foothold for beetles and spiders to escape, so new traps catch more than do re-used ones (Luff 1975). Hence trap design needs to be standardised when comparing catches over time or between sampling areas.

Extraction from the substrate

These include digging, taking soil and benthic cores and litter samples and using extraction funnels. In most cases the majority or all of the invertebrates can be removed from the substrate. If the volume of the substrate is quantified, these methods can provide fairly absolute measures of abundance.

In the rest of this chapter we outline the most commonly used methods for censusing invertebrates. There are many additional, less commonly used techniques, particularly modifications of

Table 5.1. *A summary of methods suitable for various groups*

(a) Terrestrial & aerial

	Invertebrates on open ground and low vegetation	Invertebrates on low foliage of trees and scrubs	Invertebrates in the canopy of trees	Day-flying insects	Night-flying insects	Adult insects emerging from pupae in soil, dead wood, etc. and their parasitoids	Large invertebrates in soil	Small invertebrates in soil and litter	Page
Searching and direct observation	*	*		*					220
Pitfall traps	*								222
Sweep netting	*								225
Vacuum sampling	*								226
Beating		*							228
Fogging			*						228
Malaise traps				*					229
Window or interception traps				*					231
Water traps				*					232
Light traps					*				234
Other aerial attractant traps				*					236
Terrestrial emergence traps						*			237
Digging and taking soil cores							*		238
Litter samples and dessication funnels								*	239

(cont.)

Table 5.1. (cont.)

(b) Aquatic

	Invertebrates in the water column in still or slow-flowing water	Invertebrates amongst vegetation in still or slow-flowing water	Adult insects emerging from pupae in the water and benthos	Invertebrates in mud	Invertebrates in fast-flowing streams & rivers	Page
Direct searching	*	*		*	*	240
Pond netting	*	*				241
Cylinder samplers	*					242
Aquatic bait traps	*	*				243
Aquatic emergence traps			*			243
Digging, taking benthic cores and using grabs				*		244
Kick sampling					*	245

* Method usually applicable. The page number for each method is given.

trapping techniques described in this chapter. Readers are referred to Southwood (1988), Lott and Eyre (1996), McGavin (1997), New (1998) and texts on specific groups for further information on these techniques. Detailed information on the design and response of traps for flying insects (light traps, suction traps, Malaise and other interception traps etc.) is provided by Muirhead-Thompson (1991).

Storing, killing and preserving invertebrates

Live invertebrates should ideally be stored in glass or plastic specimen tubes or jam jars. If these are not available, clear plastic bags can be used. Aquatic invertebrates should be kept in containers part-filled with water taken from where they were found. Containers should be kept out of direct sunshine, since this can heat up the inside of the container and kill its occupants. Aquatic invertebrates, especially those from fast-flowing streams, are very sensitive to overheating, so containers should be kept sealed in a refrigerator. Dead and immobile invertebrates are more difficult to locate visually following, for example, wet-sieving (see p. 238).

Many invertebrates need to be immobilised or killed to allow identification. Furthermore, if a rare species is found, a voucher specimen should be preserved. Losses of invertebrates due to trapping and killing for identification need to be set against losses due to habitat destruction, legitimate use of chemicals in the wider countryside and any inappropriate management of conservation sites due to lack of knowledge of their invertebrate fauna. Only for butterflies, dragonflies and perhaps larger moths is there no justification for killing specimens for identification.

Insects and hard-bodied invertebrates can be killed and preserved by dropping them into 70% alcohol solution. Hot water can be used to kill large invertebrates that are slow to die in alcohol. Although most other invertebrate groups can be adequately preserved in alcohol, many are better 'fixed' beforehand, particularly if they are to be used in reference collections. Fixation is the process of stabilising protein constituents in body tissue to help maintain them in a similar condition to that when the animal was still alive. A wide range of chemicals is used for 'fixing', and these are described in texts on individual groups. Formalin (often referred to as formaldehyde solution) is an excellent fixative and is used as a preservative in some types of traps and for aquatic samples. Its use is not recommended, though, since it is a suspected carcinogen.

When using alcohol solution to store invertebrates, containers should be thoroughly sealed, since alcohol quickly evaporates. Cork stoppers are not suitable, because alcohol escapes through the pores in the cork. For storage of longer than a year it is advisable to add 5% glycerol to the alcohol solution to prevent specimens from becoming brittle or from completely drying out should all the alcohol evaporate.

Insects can also be killed by exposing them to ethyl acetate fumes. This is often done within a 'killing bottle'. This consists of a glass bottle (not plastic, since ethyl acetate reacts with some plastics), containing a layer of plaster of Paris onto which a few drops of ethyl acetate are dripped. Alternatively, ethyl acetate can simply be dripped onto a piece of crumpled tissue paper at the bottom of the bottle. Ethyl acetate causes irritation to skin and is harmful if swallowed or inhaled.

Lepidoptera should be preserved by pinning to prevent damage to the scales on their wings. Together with carding (gluing a specimen to a small piece of cardboard), this technique is often used to preserve insects required for display purposes. Details of these techniques are given in many entomology texts.

The best way to label specimens preserved in alcohol is in pencil on a piece of card placed with the specimen in the alcohol solution. Labels attached to the outsides of containers usually fall off eventually.

Searching and direct observation (terrestrial and aerial)

This method is used for various invertebrate groups.

Method

Searching the ground and vegetation by eye is an important method that is especially productive in fairly open habitats with sparse vegetation, e.g. coastal cliffs, dunes and quarries, in tussocks and at the bottom of tall vegetation where sweeping and other direct methods do not work. Direct observation is clearly the basis of recording day-flying butterflies and moths, dragonflies and damselflies, and many other large and conspicuous animals seen while engaged using other methods. Searching is superior to the use of traps and collecting devices for

- molluscs;
- sedentary species under bark and stones, and in dead wood, litter and moss, fungi and tussocks;
- crawling species in saturated habitats where pitfall traps may fill with water, e.g. fens, seepages and saltmarshes;
- phytophagous larvae and adults where association with host plants is useful, e.g. caterpillars, beetles, leaf miners and galls;
- single-species surveys where the exact niche of the species is known and it can be readily located without killing individuals; and
- conspicuous diurnal species.

Care should obviously also be taken to cause minimal disturbance to the habitat. In particular, excessive damage should not be caused to dead-wood during searching.

Tussocks of grass and other vegetation support large numbers of invertebrates, particularly during winter when many hibernate within them. These can be collected by slicing the tussock off at root level using a bread knife and then shaking and cutting it open over a white surface (e.g. a sheet or photographic developing tray). Smaller invertebrates are best located by taking tussocks indoors and sorting them under a strong lamp (see 'Litter samples and sieving'). Other techniques involve splashing water onto river banks and margins of other water bodies to expel beetles and providing artificial substrate to search at a later date, for example corrugated cardboard wrapped around tree trunks. Some groups such as molluscs are worth searching for under suitable conditions at night, when they are more active.

More active invertebrates, particularly insects, may require more active methods of search and capture by the surveyor; this is especially the case for the larger winged insects such as dragonflies,

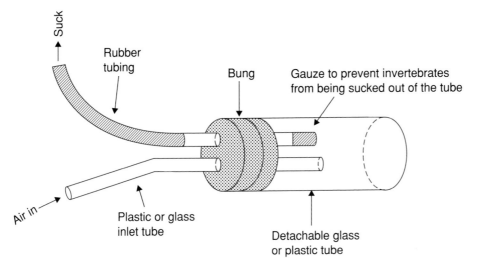

Figure 5.1 A pooter for picking up small invertebrates.

damselflies and moths. For flying insects it is often necessary to use a net. Wary fast-flying insects such as dragonflies are easiest to catch in the early morning or under cloudy conditions when they are less active, although they may be harder to locate under these conditions. Black or dark-green net mesh seems to be less detectable to many insects than white net mesh. A pooter or aspirator (Figure 5.1), pair of forceps or fine moistened paintbrush is useful for quickly and efficiently picking up small insects and arachnids. Pooters should never be used to suck invertebrates off dung or carrion because of the risk of inhaling bacteria or fungal spores.

In some cases it may be possible to standardise direct searching to some extent, in order to obtain relative population estimates, for example by counting numbers per unit effort or per unit of vegetation. The amount of foliage searched for less active invertebrates can be quantified by removing and weighing it. Quadrat boxes (a quadrat with high sides to help confine individuals while they are being counted) can be used to delimit areas of short vegetation searched.

Large, conspicuous, invertebrates can be counted along transects, for example, day-flying Lepidoptera, larval aggregations of Lepidoptera and dragonflies and damselflies (Pollard 1977, 1979; Pollard & Yates 1983; Thomas & Simcox 1982; Thomas 1983; Moore & Corbet 1990; Brooks 1993). When recording active invertebrates along transects it is important to set earliest and latest times of day for transects and minimum weather conditions under which the recording should take place. A standardised methodology has been developed for butterflies, which is used in national butterfly-monitoring schemes (Pollard 1977, 1979; Pollard & Yates 1993).

Counting exuviae (the cast-off skins of insects) is a useful method for providing an index of the productivity of insects such as dragonflies and damselflies that produce conspicuous exuviae in areas that can be easily and thoroughly searched. Exuviae should be collected at intervals as frequent as possible, and these collections should be made throughout the whole emergence period of the species. The precise timing of this will vary from year to year, mainly in response to weather conditions. In the cases of dragonflies and damselflies, the collecting of exuviae may be made easier by placing sticks in the water in easily accessible places, for use by emerging nymphs.

Advantages and disadvantages

Direct searching has the advantage, compared with trapping methods, of being selective. It also allows captured individuals to be released unharmed after their having been identified. For some groups, though, it will be less efficient in terms of numbers of individuals caught per time spent in the field than trapping methods.

Counting large invertebrates in the field is fast and does not require any equipment, but it does require prior identification knowledge of those invertebrates expected to be found. Removal of vegetation from the field is obviously destructive, and may require the transport of large amounts of vegetation. It can also be time-consuming. Counting conspicuous invertebrates or larval aggregations is quick and easy but suitable only in the few cases in which conspicuous invertebrate taxa occur at high enough densities for one to obtain meaningful results, but not high enough that counting becomes impractical.

Although searching for exuviae may be relatively quick, because this has to be carried out at regular intervals throughout the entire emergence period, the whole process can prove very time-consuming.

Biases

Direct searching is likely to locate the more visually obvious, active and large species, although the difficulty in catching large, active, flying insects may result in these being under-recorded. Small and cryptically coloured invertebrates are likely to be under-recorded.

Disturbance during searching may cause some more active insects to fly off, resulting in under-estimation of their numbers. Where foliage is removed, invertebrates may be missed or counted more than once.

Exuviae can be dislodged by rain or wind, leading to an underestimate of the productivity of the site.

Pitfall traps

These traps are used for active, surface-living invertebrates in low vegetation and bare ground, particularly larger beetles, spiders and ants.

Method

Pitfall traps consist of straight-sided containers sunk level with the surface of the ground into which invertebrates inadvertently fall (Figure 5.2). The rim must be flush with the soil – even a

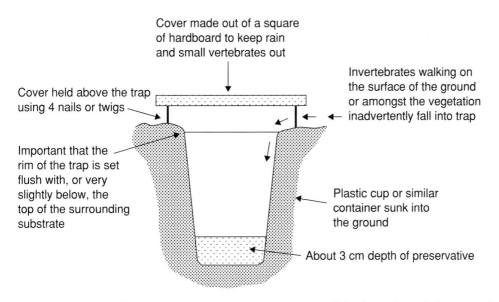

Cover made out of a square of hardboard to keep rain and small vertebrates out

Cover held above the trap using 4 nails or twigs

Invertebrates walking on the surface of the ground or amongst the vegetation inadvertently fall into trap

Important that the rim of the trap is set flush with, or very slightly below, the top of the surrounding substrate

Plastic cup or similar container sunk into the ground

About 3 cm depth of preservative

Figure 5.2 A pitfall trap for catching active, surface-living invertebrates in low vegetation and on bare ground.

small ridge or gap reduces the trap's effectiveness. Any size or type of container with smooth sides can be used, but tough polythene tubs about 8 cm in diameter and 10 cm deep with fitting lids have been found to work well and have the advantages that they are less easily crushed by stock, do not crumple when pushed into stony ground, and can be sealed quickly and replaced with a new tub without the bother of emptying the sample into a polythene bag. Traps with a larger circumference will catch more invertebrates, but this should be balanced against the need to set enough traps in different micro-habitats to ensure that the catches are representative of the site as a whole. Catches can be increased by erecting barriers that direct invertebrates to the traps.

Preservative solutions are usually put in the trap to arrest decay and prevent invertebrates from eating each other. The most reliable preservative is neat car antifreeze (ethylene glycol or propylene glycol). Ethylene glycol is toxic to humans by inhalation, ingestion and through skin absorption and should be handled with extreme care. Propylene glycol is considered safer. Where domestic animals could reach traps containing this preservative, the traps should be covered by wire mesh staked firmly in place. Ethylene glycol, if undiluted by rain water, will preserve most specimens for about four weeks, but spiders will decay in this time. It is recommended that the traps are emptied at fortnightly intervals in warm weather and checked to ensure that they have not become blocked up with leaves, have not overflowed with rain water, have not caught small vertebrates, which will act as carrion bait, and have not been upset by large vertebrates. Use of a cover can reduce these problems. Pitfall traps may be used without a preservative for short periods (a few days at most), but large animals will eat the small ones. Small vertebrates often fall in but this problem can be reduced using some sort of raised lid or grid over the top, leaving a small gap (about 10 mm).

There have been numerous studies on the efficacy of pitfall traps, whose authors have generally concluded that larger, uncovered traps catch more individuals and that the design and size of the trap and choice of preservative all influence catch rates and should be standardised if catches are being compared between sites or at the same site over time (Greenslade & Greenslade 1971; Luff 1975).

Pitfall traps can also be baited with raw meat, fish, cheese or fermenting fruit to attract beetles. Preservatives should not be used in baited traps, since these may mask the smell of the bait. Such traps therefore have to be checked daily.

If traps are to be emptied repeatedly, then a quick way to do this is to empty the trap contents through muslin. The trap liquid can then be re-used, although it may be necessary to top it up with preservative, especially, in the case of alcohol, if it has evaporated or become diluted with rain water. After traps have been emptied it is worth wiping their inside surfaces with a cloth, to keep them clean and smooth (particularly if slugs and snails have entered and left behind mucus). This will maintain the catching efficiency of the trap. Alternatively, if two tubs are put in together, the upper one can be replaced repeatedly without having to clean out material that inevitably falls into the hole as the container is withdrawn. Most studies use 5–10 traps per station. Traps are easiest to relocate if they are placed in a straight line at intervals of about five paces (or a similar distance measured with a tape) from a conspicuous marker (natural or a cane) in a known direction. If individual traps need to be set at random locations (see Chapter 2) then a hand-held global positioning system (GPS) is invaluable for relocating them. Even with the use of a GPS, traps can sometimes be surprisingly difficult to relocate amongst vegetation after having been left for more than a couple of weeks during the growing season.

Advantages and disadvantages

Pitfall trapping is probably the most commonly used trapping method for studying invertebrates and is a cheap, quick and easy method of catching very large numbers of specimens. Pitfall traps catch a range of animals that are rarely, if ever, caught using other standard methods, so they are particularly useful when used in combination with other methods. There is almost no substitute for this method.

As a method of comparing catches over time or between different areas, the use of pitfall traps has a number of disadvantages. In common with other active trapping techniques, catches reflect relative activity and vary with weather conditions. Catch rates also vary with the nature of the surrounding vegetation, because taller vegetation in the vicinity of the trap impedes invertebrate movement (Greenslade 1964). Hence it is possible to compare catch rates only between areas with vegetation of similar structure and from traps set over the same time period. Other animals may also interfere with pitfall traps.

Biases

Pitfall traps tend to catch proportionally more large (>3-mm-long) invertebrates. Some species of ground beetle, once caught, emit pheromones that attract other individuals to the trap (Luff

1986). In common with other active trapping techniques, catches reflect relative activity and susceptibility to trapping, rather than the relative abundance of the various species (Topping & Sunderland 1992). Decaying carcasses, especially of small mammals, act as bait, so carrion beetles can be over-represented.

Sweep netting

Sweep netting is used to catch invertebrates in low vegetation, particularly flies, bugs, spiders and small beetles.

Method

The method involves passing a sweep net through the vegetation using alternate backhand and forehand strokes. Nets come in two basic designs, a light-weight gauze, mainly for capturing insects on the wing and gentle sweeping through vegetation, and a heavy-duty manilla gauze and stout frame for sweeping through dense vegetation, mainly for beetles and bugs. The catch can be removed using a pooter or by tubing individual specimens. If all the contents are required then the contents of the net can be emptied into a killing bottle.

The results of sweep netting are not quantitative but, when used for sweeping through low vegetation with a fairly uniform structure, such as grassland and dwarf-shrub vegetation, catches can be standardised with reasonable precision by using a set number of back-and-forth sweeps (typically between 10 and 20) approximately 1 m in length taken every other pace while walking at a steady speed through the vegetation. The quantity of vegetation sampled can be more easily quantified by using a D-frame net swept relatively horizontally through the vegetation, making sure that the trap contents are not emptied out whilst doing so. The catch obtained will be influenced by the time of day and the speed, depth and angle at which the net is pulled through the vegetation. For example, many flies will avoid a slowly approaching net but be caught by a faster one.

When many samples are being taken, the efficiency of sampling can be increased by using detachable zip-off net bags, which can be quickly removed and stored for later sorting (Milne 1993).

Advantages and disadvantages

Sweep netting is a quick, low-cost, efficient and versatile way of collecting large numbers of invertebrates, making it well suited for surveying purposes. Sweep nets can be used on most vegetation, from low tree foliage to bare ground, although they are ineffective in some vegetation, e.g. tall reeds, very short turf and flattened vegetation and over-prickly plants. They are also ineffective in strong winds and on wet and trampled vegetation. Snails and seed heads, especially in dry grasslands and dunes, rattle about and severely damage the catch.

Biases

Sweep netting collects species that fly or sit on vegetation. Invertebrates that cling tightly to the vegetation (e.g. Lepidoptera larvae and sawfly larvae) are probably under-represented.

Vacuum sampling

This method is used for invertebrates in low vegetation, particularly flies, bugs, spiders and small beetles.

Method

Vacuum sampling involves the sucking up of invertebrates from vegetation through a wide, flexible tube into a collecting net inside it. Several models of sampler are available, ranging from the original and heavy D-vac sampler to modified suction machines designed for sucking up leaves in the garden and light, battery-operated vacuum cleaners sold for cleaning cars (Figure 5.3). These modified versions have suction power (and hence collecting efficiency) comparable to, and in many cases greater than, that of D-vac samplers (Wright & Stewart 1992; Stewart & Wright 1995). The wearing of ear protectors is advisable when using vacuum samplers.

Vacuum sampling provides a useful complement to sweep netting since it collects species living deep within the vegetation (e.g. in tussocks or plant litter) and on sparsely vegetated ground where sweep netting can be relatively ineffective. The great value of vacuum samplers is in their ability to provide relatively quantitative estimates of invertebrate numbers by using them to remove

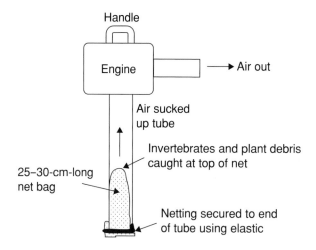

Figure 5.3   A vacuum sampler for sampling invertebrates in low vegetation, particularly flies, bugs, spiders and small beetles.

invertebrates from a defined area of vegetation. Two methods of operation can be used to achieve this. The collecting nozzle of the sampler can be pushed vertically downwards into the vegetation and held there for a standard length of time (e.g. 10–15 s) to suck up invertebrates from an area of vegetation the size of the sampler's nozzle (many samplers may also suck up invertebrates from outside this area, particularly as they are being lowered or raised). This can then be repeated a number of times. Some operators ruffle the vegetation using their hands to disturb animals as the sample is being taken. It is necessary to keep the sampler running between individual sucks to prevent collected specimens from escaping. Alternatively, a known area of vegetation can be defined and enclosed, and the collecting nozzle used to suck up the invertebrates from it for a standard length of time.

After the sample has been taken, the net bag containing the invertebrates should be sealed and placed in a killing bottle and its contents then removed and preserved, or alternatively the net can be emptied by inverting it into a large polythene bag. Since the net will inevitably also contain large amounts of plant debris sucked up with the invertebrates, the contents of the net bag will have to be sorted. Net bags can be sorted using two or three sieves down to 2 mm gauge, or by emptying the contents into a bowl of water and removing invertebrates using forceps. Alternatively, the live contents of the sampler can be emptied into a sweep net and a pooter can then be used to collect specimens as they crawl out of the debris.

Advantages and disadvantages

Vacuum sampling is ineffective in vegetation that has been flattened by wind, rain, or trampling. Like sweep netting, it cannot be used if the vegetation is very wet. Vacuum sampling collects fewer invertebrates per unit time spent in the field than sweep netting does, but it is easier to quantify the area sampled using vacuum sampling. Hence sweep netting is often the preferred method for surveying invertebrates and vacuum sampling the preferred method for proving quantitative measures of their abundance. The sampling efficiency of vacuum sampling will, though, vary to some extent with vegetation height and structure, so it is possible to make quantitative comparisons only of areas with vegetation of similar height and structure. The catch of suction samplers contains large amounts of debris, so samples can be time-consuming to sort.

Most petrol-driven samplers are heavy (about 10 kg), noisy and cumbersome, so they are not convenient to use far from vehicular access and on steep slopes. They also require refilling with a petrol-oil mix at frequent intervals. Being mechanical, they are also prone to breaking down. In particular, modified suction samplers are susceptible to getting their carburettor clogged with debris from the petrol tank. It is therefore important to strain the fuel through a funnel with a fine filter on it when refilling the petrol tank.

Biases

Vacuum sampling under-records large (>3-mm-long) invertebrates and probably those that cling tightly to the vegetation (e.g. Lepidoptera and sawfly larvae). It will also probably under-record species living low down in tall vegetation and species that can take evasive action when they sense the noisy sampler approaching.

Beating

Beating is used to catch invertebrates on the foliage of trees and bushes, particularly spiders, beetles bugs and Lepidoptera larvae and some dead-wood species.

Method

Beating involves sharply tapping branches with a stick or vigorously shaking them to dislodge invertebrates that are caught in a tray held beneath. A beating tray is a roughly 1-m² cloth-covered tray that is slightly sloping towards its centre and can be collapsed to ease transport. They can be bought or made, or alternatively an old umbrella or a sweep net can be used. Most animals are more easily seen on white beating trays, but grey or black is less dazzling in sunlight. Some insects initially remain motionless on the tray, so it is worth waiting a short while before discarding its contents. Most individuals are caught following the first sharp tap. The increase in number of individuals caught by further tapping needs to be balanced against loss of insects from the beating tray – many get up and crawl or fly away almost immediately after landing on the tray. Invertebrates can be collected from the tray using a pooter or by tubing individual specimens. Invertebrates on foliage can also be sampled by tapping branches and netting the insects as they fly away. For many groups, for example beetles and spiders, beating will collect a high proportion of the individuals present on the foliage. Beating pieces of dead branches can be used to recover dead-wood species and shaking swamp and fen vegetation over a white tray can be used to recover wetland snails and other species. Another variation on the technique is tapping branches and netting insects as they fly away.

Beating is difficult to use to compare catches over time or between areas, since it is virtually impossible to standardise the quantity of leaves and branches sampled.

Advantages and disadvantages

Beating is a quick and easy way to collect large numbers of invertebrates.

Biases

This method is biased towards species that are easily dislodged but do not fly away quickly when disturbed, such as beetles, bugs, spiders and Lepidoptera larvae.

Fogging

Fogging is used to catch invertebrates in the canopy of trees.

Method

This is the only really effective method for sampling the fauna of tree canopies. A knock-down non-persistent insecticide is sent into the canopy using a fogger or mist-blower raised up into it and the invertebrates that are killed or anaesthetised by it fall down and are collected on trays or sheets, or in funnels placed below the tree. The method can be standardised to provide indices of abundance. Since this is a relatively specialised method that will not be used often and has important health and safety considerations, the reader is referred to New (1998) and specific studies (Watt *et al.* 1997) for further details.

Advantages and disadvantages

The advantages of fogging are that it collects a large and representative sample of the resident fauna, including species that are difficult to sample using other methods; it is not dependent on insect activity, so samples contain a small proportion of non-resident species; and large samples can be collected. The disadvantages are the expense of the equipment; the need for a health-and-safety licence to use insecticides in this way; and that only a few samples can be collected in one day.

Biases

There is little evidence of any strong biases, other than that sessile invertebrates and those in epiphytes are poorly sampled (Majer & Recher 1988; Yanoviak *et al.* 2003).

Malaise traps

These traps are used to catch flying insects, particularly flies, bees and wasps, and smaller numbers of other flying insects such as bugs and beetles.

Method

Malaise traps are tent-shaped traps made out of fine netting with a black central screen suspended below a sloping ridge roof that leads to a collecting chamber at its upper end (Figure 5.4). Flying insects that hit the screen fly or walk upwards and along the roof to the chamber. Walls at either end of the screen reduce the numbers of insects escaping sideways. The screen is usually about 2 m × 2 m.

Because the standard design intercepts insects flying along one pathway, the screen should be set up at right-angles to obvious flight paths such as hedgerows and woodland rides, or at other points of high insect activity such as at junctions of two habitats. At the same time, the collecting chamber must be at the end that points towards the Sun (or the brightest part of the sky if in permanent shade) because the insects fly towards the brightest light. Traps work well when

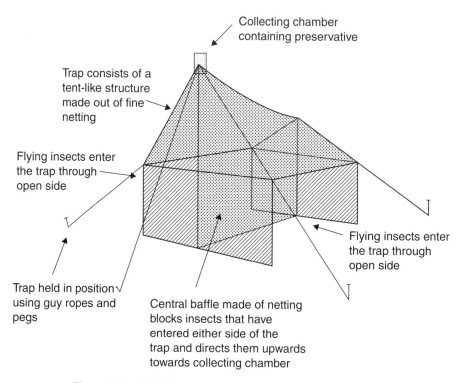

Collecting chamber
containing preservative

Trap consists of a
tent-like structure
made out of fine
netting

Flying insects enter
the trap through
open side

Flying insects enter
the trap through
open side

Trap held in position
using guy ropes and
pegs

Central baffle made of netting
blocks insects that have
entered either side of the
trap and directs them upwards
towards collecting chamber

Figure 5.4   A Malaise trap for catching flying insects.

hoisted into the canopy of woods, where they reveal a different fauna from that caught in traps at ground level (Hammond 1990). Smaller traps (1 m × 1 m) are convenient for this purpose. The central baffle can be treated as a window trap with a collecting trough placed below it. Since window/interception traps and Malaise traps collect different elements of the fauna, this simple addition is worthwhile.

Malaise traps can be left unserviced for long periods (monthly periods are commonly used) but catch so much material that the collecting chamber can fill completely if left too long during summer. Traps ought to be checked more frequently than monthly for damage.

If traps are serviced weekly, alcohol is a convenient preservative and it will evaporate only slowly because the collecting chamber is enclosed apart from the small entry hole. If traps are serviced at longer intervals, for example up to a month, a less volatile preservative (e.g. 70% ethylene glycol or preferably the less toxic propylene glycol) should be used.

As with other trapping methods, the details of the trap, such as the colour of the netting and even the age of the trap, influence catch rates (Townes 1962; Roberts 1975) and so they should be standardised when comparing catches between areas or over time.

Advantages and disadvantages

The advantages of Malaise traps are that they collect vast numbers of insects – one month's catch in a temperate environment will typically fill a litre container with many thousands of specimens. They also give a much better return for effort than any other single trap (and the longest species list for flying species), so short trapping periods of only a few days can give useful returns.

The time taken to sort and identify the catches is much longer than for other types of trap and needs to be taken into consideration in costing the survey. It is rarely practical, though, to replicate samples sufficiently to analyse statistically. Being conspicuous and flimsy, the traps should be sited away from the public and grazing animals, and not where they would be exposed to high winds. Birds trapped in the roof can tear the netting, allowing much of the potential catch to escape, and spiders often build webs across the entrance to the collecting chamber.

Biases

As with all trapping methods, the catches will be a product of the abundance, activity and susceptibility to capture of the particular species. Trap catches are biased towards strong-flying insects that are strongly attracted upwards towards light.

Window or interception traps

These traps are used to catch flying insects, mainly beetles.

Method

Flight-interception traps work by blocking flying insects with a screen of fine black netting, sheet of glass, or sheet of transparent plastic fixed vertically over a collecting trough or funnel-shaped construction, into which the insects fall (Figure 5.5). As with Malaise traps, the screen should be set up at right-angles to obvious insect flight paths. Designs that have been used successfully have windows 65-cm square made of clear acrylic supported on stakes, and windows 2.4 m × 1.1 m made of black terylene tied to convenient trees (Hammond 1990). The trough can be guttering or a plastic window box, which will need levelling but will be unwieldy to empty, or a series of aluminium baking trays, which are convenient to empty and do not require level ground. The preservatives used are as for water and pitfall traps and the traps can be checked at similar frequencies to these. Window traps have been used successfully in tree canopies. Window traps and Malaise traps may be combined into a single unit by placing a collecting trough below the central baffle of the Malaise trap. Muirhead-Thomson (1991) describes variations on these basic designs. As with other trapping techniques, the design of the trap will influence catch rates and so should be kept constant when comparing catches between sites or over time.

Advantages and disadvantages

Window traps are recommended in particular for studies of woodland beetles where direct collecting may damage dead-wood habitat. Malaise traps are better at catching smaller, more agile

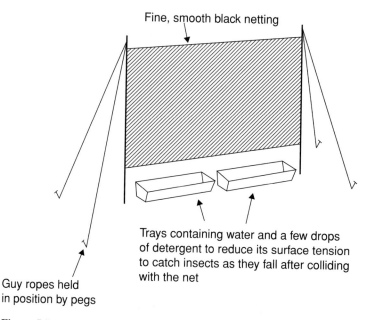

Fine, smooth black netting

Trays containing water and a few drops
of detergent to reduce its surface tension
to catch insects as they fall after colliding
with the net

Guy ropes held
in position by pegs

Figure 5.5 A flight-interception trap for catching flying insects.

flying insects, particularly flies and small Hymenoptera. The principal disadvantage of window/interception traps is their susceptibility to interference from passers-by.

Biases

Such traps are biased towards cumbersome flying insects, such as large beetles that fall downwards when hitting the vertical panel. As with all trapping methods, the catches will be a product of the abundance, activity and susceptibility to capture of the particular species.

Water traps

These are used to trap flying insects, mainly flies, bees and wasps, but also plant hoppers and hunting spiders if traps are set on the ground in vegetation that reaches their rim.

Method

Water traps are containers (bowls, trays, tubs) partly filled with water into which animals leap or fly and drown (Figure 5.6). A small amount of detergent (about 0.5 ml per litre of water) makes the animals become wet quickly and sink. Adding about 5% common salt reduces the swelling of some species.

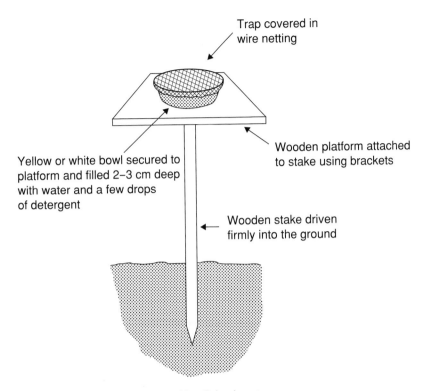

Trap covered in
wire netting

Wooden platform attached
to stake using brackets

Yellow or white bowl secured to
platform and filled 2–3 cm deep
with water and a few drops
of detergent

Wooden stake driven
firmly into the ground

Figure 5.6   A water trap for catching flying insects.

The depth of the container should be about 10 cm so that, when it is half-filled with water, there is enough liquid to counter evaporation in hot weather and enough free-board to prevent overflowing in wet weather. An overflow hole covered with coarse gauze near the rim will prevent the trap from overflowing and rain from splashing any floating specimens over the top of the bowl. The colour affects the size and composition of the catch (Kirk 1984). Yellow bowls are the best for catching both flies and Hymenoptera. White also attracts flies, but has a strong repellent effect on Hymenoptera. Alternatively, if 'neutral'-coloured bowls, such as brown, grey, or blue ones, are used, then these will have the least attractant (or repellent) effect on insects, and thus reduce the selectivity of the sampling (Disney *et al.* 1982; Disney 1986, 1987). Special day-glow yellow paint with a high ultraviolet (UV) reflectance ought to be more attractive, but it is not easy to get paint to stick to containers made of polythene. For sampling in woods it may be necessary to fix a wide-mesh gauze over traps to prevent leaves from falling in and affecting the attraction of the trap. A domed covering of coarse wire netting (of gauge 1–2 cm) should be used to prevent birds from bathing in the trap or eating the catch and also prevents leaves from falling in.

If traps are left for more than a day or two in warm weather, a preservative, such as 20%–30% antifreeze, is essential. Addition of antifreeze also prevents the liquid from completely drying out, but its colour presumably reduces the attractiveness of the trap. Good results have been obtained with its use, though. Traps with preservative can be emptied as infrequently as once a month.

Invertebrates can be removed from the traps by pouring the contents through a piece of muslin into a bowl. The muslin containing the specimens can then be removed and the water-trap liquid returned to the trap.

The species composition of water-trap catches varies with the height of the trap. Total trap catches are highest when the trap is just above the level of surrounding vegetation (Usher 1990). When a trap is set on the ground in vegetation that reaches the rim, crawling animals will be trapped, so this can give a more species-rich catch than would a raised trap.

Advantages and disadvantages

The principal disadvantages are that livestock and people often interfere with water traps, and birds may eat the catch. Some non-target species, notably butterflies, can be caught in embarrassingly large numbers.

Biases

As with all trapping methods, the catches will be a product of the abundance, activity and susceptibility to capture of the particular species.

Light traps

Light traps are used to catch moths, flies and other night-flying insects.

Method

Many night-flying insects, particularly moths, are attracted towards light, particularly at the UV end of the spectrum. Insects attracted to a light can be caught using a net or within a collecting container. Catches can be increased by positioning the light beside a white wall or next to or above a white sheet.

Alternatively, a light trap (most types are usually referred to as moth traps) can be used. Moth traps consist of a container with an opening and a light inside it (Figure 5.7). Insects attracted towards the light spiral around it and hit the baffles which cause them to fall down into the trap. There are three main moth-trap designs available commercially, namely the Heath, Robinson and Skinner types, and three main types of bulb, the 125-W mercury-vapour (MV) bulb, 15-W fluorescent (actinic) tube and the 125-W black bulb.

Heath traps are small, square collapsible traps usually fitted with a 15-W actinic tube as a light source. Catch rates of Heath traps fitted with actinic bulbs are between a third and half those of Robinson traps fitted with MV bulbs (Waring 1980), although some species appear to be more attracted to actinic than to MV bulbs. Their principal advantage over Robinson and Skinner traps

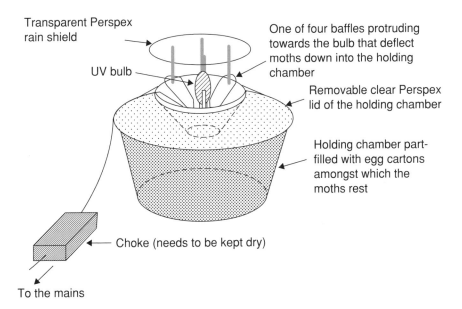

Transparent Perspex rain shield

One of four baffles protruding towards the bulb that deflect moths down into the holding chamber

UV bulb

Removable clear Perspex lid of the holding chamber

Holding chamber part-filled with egg cartons amongst which the moths rest

Choke (needs to be kept dry)

To the mains

Figure 5.7   A Robinson light trap for catching moths and other night-flying insects.

is that they can be run from a 12-V car or motorcycle battery (motorcycle batteries are far smaller and easier to carry) or the mains. They are also collapsible and more portable than Robinson traps. Actinic tubes do not run hot, and are less prone than MV bulbs to cracking if hit by rain.

Both Robinson and Skinner traps are usually fitted with a 125-W MV bulb or 125-W black bulb and run from the mains or a generator. Robinson traps consist of a circular container and retain a higher proportion of moths that enter and produce the highest overall catches of the three traps described. Addition of flaps to the Perspex sides of the square Skinner trap will increase the numbers of moths retained. The main advantage of the Skinner trap is that it is less than half the cost of a 'Robinson-type' trap and can be collapsed for portability and when not in use. The 125-W black bulbs differ from 125-W MV bulbs in that they contain a coating that absorbs most of the visible (to us) light and are therefore useful in built-up areas where bright 125-W MV bulbs might annoy the neighbours. The 125-W black bulbs are expensive though, and attract fewer moths than do 125-W MV bulbs. They also get hotter than 125-W MV bulbs and are therefore more prone to breaking if hit by rain. The MV bulbs will often still work if the outer envelope becomes cracked, but it is important to discard such damaged bulbs, since the outer glass envelope filters out UV radiation, which can seriously damage your eyes. Always seek the advice of a qualified electrician when dealing with the electrics of a moth trap and ensure that connections, sockets and plugs are fully waterproofed.

Catches will be maximised if the trap is positioned so that its light shines into as much habitat as possible. The response to light traps varies greatly among species. Some species tend to settle near the trap, rather than entering it, so it is worth searching vegetation in the vicinity of the trap to reveal additional species.

The New Jersey Trap is a light trap with a downdraft electric fan and is used to catch weakly flying insects, particularly mosquitoes, midges and sandflies (Mulhern 1985).

Advantages and disadvantages

Light trapping is the only practical method for recording moths, other than searching for larvae.

Biases

Light-trap catches are obviously biased towards species attracted to light and, if a trap is used, to those which also enter and remain in the trap. Catch rates reflect the activity of species, not just their abundance, and activity varies greatly with weather conditions and moonlight. The largest catches of most species are made on warm, humid, windless nights with little moonlight (Muirhead-Thompson 1991). If comparisons of catches on different nights are to be made, then these conditions must be taken into account. More detailed information on the use of light traps is provided by Muirhead-Thompson (1991) and Fry and Waring (2001).

Other aerial attractants and traps

These methods are used for flies, butterflies, moths, dung beetles and wasps.

Method

Various baits can be used to attract insects. These can be placed in traps that contain a funnel or narrow opening that guides the insects into a collecting chamber where they can be held alive or killed (Figure 5.8). Examples of baits include the following: rotting fruit for fruit flies and mainly tropical butterflies (over-ripe bananas are particularly good); dung for dung flies and dung beetles and some mainly tropical butterflies; carrion for carrion-feeding species; carbon dioxide (in the form of 'dry ice') wrapped in polythene for slow release for blood-sucking species; synthetically produced pheromones to attract males of particular moth species; and red-wine-soaked ropes and 'sugar' for moths. The basic recipe for 'sugar' consists of black treacle (or Barbados or molasses sugar) boiled up with over-ripe bananas or other rotting fruit with some alcohol (usually rum or beer), to which (optionally) a couple of drops of amyl acetate may be added just before use.

Advantages and disadvantages

Catches tend to be small and of a limited taxonomic range, but can be useful for targeted sampling of particular groups, such as dung beetles. Wasps and other larger Hymenoptera may enter traps, particularly those containing fruit, and kill and damage insects already caught inside. Ants also often enter the traps, especially in the tropics, and can kill and remove insects already caught. Sugaring attracts some species of moth that do not readily come to light.

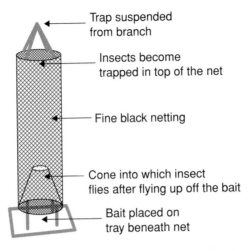

Trap suspended
from branch

Insects become
trapped in top of the net

Fine black netting

Cone into which insect
flies after flying up off the bait

Bait placed on
tray beneath net

Figure 5.8   A butterfly-cone trap for catching butterflies and other flying insects attracted to its bait.

Biases

Catches are obviously biased towards species attracted to the particular baits and, if a trap is used, those liable to enter and become caught in the trap. As with all active trapping methods, catch rates will also be influenced by weather conditions.

Terrestrial emergence traps

These traps catch mainly flies and beetles and their parasitoids.

Method

Emergence traps can be used to trap adult insects emerging from their pupae in soil and dead wood. The basic design of these traps consists of a dark container with a light area at its top towards which insects are funnelled into a collecting chamber. Owen emergence traps (Owen 1992) (Figure 5.9) and trunk-mounted interception traps (Kaila 1993) for erecting over dead wood and rot holes provide very good non-destructive ways of sampling dead-wood invertebrates, which are otherwise difficult to sample.

Emergence traps will provide absolute or relative estimates of numbers of emerging insects, depending on their efficiency. To standardise losses, traps should be emptied at regular intervals.

Advantages and disadvantages

Emergence traps are useful for identifying the particular habitats and micro-habitats used by the larval stages of individual species and for determining their emergence times. Traps are conspicuous and likely to be disturbed by passers-by, though.

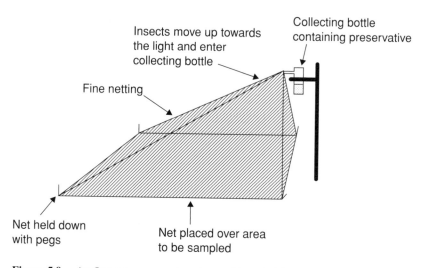

Figure 5.9   An Owen emergence trap for catching adult insects emerging from the ground or from other material, such as dead wood, placed within it.

Biases

There is little evidence of any serious biases.

Digging and taking soil cores

These methods are used for large soil invertebrates, particularly earthworms, fly larvae and beetle larvae.

Method

This method involves taking soil samples of a known volume. These can be dug using a spade (a quick way of measuring the surface area of samples is to dig around the outside of a wire quadrat), or by using a corer where the substrate is soft. A sample size of the order of 10 cm × 10 cm square is recommended for earthworms and leatherjackets (Tipulidae larvae).

The majority of earthworms and large invertebrates such as fly and beetle larvae can be removed while breaking the soil up by hand. To recover less conspicuous individuals the sample can then be 'wet-sieved' using a mesh size of 1 mm. The extra proportion removed by additional wet-sieving will depend on the nature of the soil and the sizes and visibility of the invertebrates being removed. In many cases it may be most efficient to sort the entire sample by hand but to wet-sieve only the root layer, which can be more difficult to sort thoroughly by hand and usually contains a large proportion of the individuals within the sample, particularly small, inconspicuous surface-living earthworm species. When sampling earthworms and leatherjackets it is advisable to record their biomass (fresh weight), rather than density. The proportion of very small earthworms located will depend on the diligence of the observer, and this can result in considerable observer bias. However, although small individuals can in some cases comprise a high proportion of the total numerical density of individuals, the majority of the biomass is usually constituted by large, conspicuous individuals. Hence measures of biomass are less affected by observer bias.

Advantages and disadvantages

Sorting dug samples is the only reliable means for estimating earthworm and tipulid abundance. Chemical extraction, for example using mustard solution, is often used to expel earthworms from the soil to provide measures of their abundance (Gunn 1992). However, it is not possible to calculate earthworm densities or biomass per known volume of soil using chemical extraction, since it is impossible to define the depth to which the chemical infiltrates the soil and also because species of earthworm vary in their response to the irritant (Chan & Munroe, 2001). Furthermore, the depth of infiltration will vary with soil type, structure and moisture, making it impossible or at least very difficult to compare earthworm densities in different soils. Digging and carrying large numbers of soil samples is time-consuming and tiring, though.

Biases

The proportion of worms extracted using this method varies with soil type. Raw (1960) found that hand sorting using flotation recovered 52% of the total number of worms (84% of the total weight) in good-quality soils. Small and dark-coloured worms can be under-recorded (Raw 1960; Nordström & Rundgren 1972), as can large earthworms that retreat to below the depth of the sample removed (Chan & Munroe, 2001).

Litter samples and desiccation funnels

These methods are used for small soil- and litter-inhabiting invertebrates, particularly spiders, mites, beetles, molluscs and pseudoscorpions.

Method

Many ground-dwelling species, particularly spiders, beetles and snails, live or hide in leaf litter, flood debris, old bird's nests, etc. Handfuls of such litter can be collected and searched on a large white polythene or cloth sheet or tray, preferably in the sun, which makes the animals more active. Alternatively the sample can be put into polythene bags, brought back to the laboratory with the animals alive and the animals sorted by eye, or extracted using desiccation funnels.

Sorting by eye is easiest if the total amount of material is separated into manageable fractions of similar-sized particles by sieving. For beetles, bugs and spiders, a single, small garden sieve with a mesh size of about 10 mm is adequate. The material is sieved onto a white tray, where the beetles run about and can be collected using forceps or a pooter. For snails, a stack of sieves with mesh sizes of about 10, 3–4 and 1 mm is needed because the snails are not conspicuous among particles of widely varying sizes. Damp litter is easier to sieve after partly drying it on newspaper. Some invertebrates, such as beetles and pseudoscorpions, become motionless when disturbed, so it is best to wait a short time before discarding sorted material. It is also useful to sieve under a strong light or in bright sunshine. Not only does this make it easier to see, but also, since some groups are negatively phototactic and are also stimulated by heat, they become more active under a strong light and hence are easier to find.

Desiccation funnels work by creating warm, dry and light conditions at the top of the funnel, which encourages cool-, shade- and moisture-loving invertebrates to move down the funnel away from the light source, until they eventually fall out of the bottom of the funnel into a collecting bottle. If live specimens are required then a large piece of filter paper should be placed in the collecting container. Funnels are usually left in operation for a week or so and, if live specimens are being collected, they should be checked daily. There is a wide variety of designs of this basic apparatus, including the 'Berlese funnel', which, instead of having a light source above it, consists of a metal funnel encased on its outside by a jacket through which hot water is passed, and the 'Baermann funnel', which is used mainly for collecting nematodes and groups such as rotifers. These and other modifications to desiccation funnels are described in more detail in Southwood (1988) and New (1998).

Saturated moss and other wet vegetation from ponds can be wrung out gently and shaken or sieved over a polythene sheet. This is useful for beetles but most other animals will be damaged by this treatment.

Three ways of standardising the sample size can be used:

- collecting a fixed volume of material (the usual sample size is 2–5 litres of litter, preferably the larger volume unless many replicates are taken);
- collecting all the litter within a quadrat; and
- sieving for a set period of time (e.g. 15 min) or until no new specimens are recorded after, for example, 5 min.

Advantages and disadvantages

Sorting through litter samples and using extraction funnels is the only practical way to recover small invertebrates.

Biases

The numbers of individuals of various species recovered will be influenced by the diligence of the person sieving and sorting. The catch from desiccation funnels will be affected by the size of the funnel, larger funnels tending to extract relatively more large invertebrates. This may to some extent be due to a greater proportion of smaller invertebrates becoming desiccated within the funnel before they reach the collecting tube.

Searching and direct observation (aquatic)

This method can be used for a wide range of aquatic invertebrates.

Method

Direct searching can be used to find a wide range of aquatic species. The most productive places to search are on and under stones (for mayfly, lacewing, alderfly, stonefly and caddis-fly larvae and for leeches), amongst aquatic vegetation (particularly in the axils of leaves of emergent plants and in the sheathing leaf bases of tall monocotyledons) and amongst sticks and roots of marginal vegetation.

Invertebrates living amongst aquatic vegetation can be sampled by placing the vegetation in a shallow, white tray (e.g. a photographic developing tray) and removing invertebrates displaced from it. For greater effect, the tray can be painted with a large chess-board pattern of black and white, since some invertebrates are more easily seen on a black background than on a white one. The greatest variety of invertebrates can be found by leaving the weed overnight in a water-filled, covered bucket. The subsequent oxygen depletion and slight fouling of the water encourages previously hidden invertebrates to come to the water's surface, where they are more easily visible.

Advantages and disadvantages

The advantages and disadvantages of this method are similar to those for searching and direct observation of terrestrial and aerial invertebrates (p. 220).

Biases

The biases also are similar to those for searching and direct observation of terrestrial and aerial invertebrates (p. 220).

Pond netting

Pond netting is used to catch aquatic invertebrates in still or slow-flowing water.

Method

Pond nets can be used as a quick method of catching large numbers of aquatic invertebrates. Most aquatic species spend the majority of their time amongst vegetation or on the bottom of the water body in shallow water. Hence netting should be aimed at these areas. Flour sieves and tea strainers can be useful for deftly catching water beetles and bugs.

After taking the net out of the water, it should be allowed to drain and the net contents emptied onto a white tray (e.g. a photographic developing tray), taking care to wash all the specimens off (some nymphs cling very tightly). Specimens can be picked up using a small pipette and placed in temporary storage containers. It is important to wash the net carefully before moving on to the next site to prevent invertebrates inadvertently being recorded from the wrong site.

Catches may be compared between sites or over time by standardising the effort used for each sample. This can be done by taking a fixed number of sweeps, netting for a set period (typically up to 3 min, during which time the net may need emptying as it becomes clogged with vegetation), or continuing to sample and sort the catch for a set period (typically 30–45 min).

Advantages and disadvantages

Pond netting is easy to use in any water that is close to a bank or shallow enough to wade into. It is particularly useful for carrying out quick surveys of sites. For example, Furse *et al.* (1981) found that sampling of ponds using a hand-held net for just 3 min collected 62% of families and 50% of species that could be found by 18 min of netting.

Sampling in water deeper than about 1 m is difficult without thigh and chest waders, and this activity can be dangerous because of the potential for the waders to fill with water and make it difficult to move. Sampling with nets from small boats can also be dangerous, because the strong motions needed to get good net samples will rock the boat.

Biases

In open water catches will vary with the speed of the net movement. Slow movement will fail to catch invertebrates that are able to avoid the approaching net. Fast movement will push forwards the water in front of the net and again fail to catch a significant proportion of invertebrates in its path. The invertebrate fauna will vary with depth, and the depth at which the net is pulled through the water can be difficult to keep constant. Unless it is specifically dragged close to the bottom of the water, standard netting will underestimate numbers of invertebrates resting on the substrate (e.g. resting prawns, shrimps and corixids).

Cylinder samplers

Cylinder samplers are used to collect small nektonic crustaceans and other zooplankton.

Method

Cylinders can be used to delimit and remove columns of water of known diameter and depth, containing small crustaceans and other zooplankton. The cylinder should be lowered vertically or horizontally (in the case, for example, of sampling invertebrates in gaps between aquatic plants), quickly enough that invertebrates are unable to flee from the approaching cylinder, but not fast enough to cause unnecessary turbulence in the water. A bung should then be placed into the bottom of the cylinder (or both ends if it is being used horizontally), then the cylinder should be removed and its contents emptied into temporary storage containers. Small invertebrates are best sorted by subsampling the water collected.

Advantages and disadvantages

Cylinder samplers are practical only for use in water shallow enough to wade into or areas very close to the water's edge. Zooplankton often have a very patchy distribution and hence it is usually necessary to take a large number of samples to obtain a reasonable population estimate.

Biases

Some zooplankton may move away and avoid capture because of disturbance caused during sampling. Others may actively avoid the approach of the cylinder. The speed at which the cylinder is moved through the water will influence the latter.

Aquatic bait traps

These traps are used to catch scavenging planarians, leeches and crustaceans.

Method

Traps baited with meat can be used to catch a variety of scavenging invertebrates. The bait (a slit earthworm, fresh liver, meat, tinned sardines, etc.) should be placed in a jam jar with one or more openings made in its lid. The diameter of the opening(s) can be used to determine what size of and hence which groups of invertebrates are caught (e.g. small holes will allow triclads in but exclude leeches and crayfish). If newts or salamanders are present, then traps should be positioned with an air-trap at one end and checked daily.

Advantages and disadvantages

This method is easy, but difficult to standardise for use in comparing numbers of invertebrates between sites or at the same site over time for the reasons given below.

Biases

There will be a bias towards species attracted by the particular bait and also towards species likely to enter and become caught within the trap. Numbers caught will be a reflection of both the activity of invertebrates and the distance over which they can detect the bait and also, to some extent, of the availability of alternative food sources.

Aquatic emergence traps

These traps are used to catch flies, mayflies and caddis flies.

Method

Emergence traps can be used to trap adult insects emerging from the water. The design is similar to that of terrestrial emergence traps (p. 237).

Aquatic emergence traps will provide absolute or relative estimates of numbers of emerging insects, depending on their efficiency. To standardise losses, traps should be emptied at regular intervals. They also need regular cleaning, otherwise they catch disproportionately more species living on the underwater part of the trap (weed and surface dwellers) than those living in the benthos.

Advantages and disadvantages

Aquatic emergence traps are useful for identifying the particular areas used by the larval stages of aquatic species and their emergence times. Aquatic emergence traps can be damaged by wave action, and are usually difficult to use at sites with widely fluctuating water level. They are particularly conspicuous to passers-by and consequently very prone to disturbance.

Biases

Larvae about to emerge may be attracted to the trap (owing to the favourable, sheltered conditions for emergence that it might provide), resulting in an overestimate of numbers of emerging insects, or may be repulsed (many larvae and pupae are attracted towards light and will therefore move away from any shadow cast by a trap), resulting in an underestimate. Numbers of insects caught will also depend on the frequency of collection. Even under very favourable conditions the proportion of insects lost from the traps will increase the longer the traps are left. Loss occurs mainly through decay and sinking, and the activity of predators, and will be increased by wave action.

Digging, taking benthic cores and using grabs

These methods are used for benthic invertebrates in still or slow-moving shallow water, sand and mud.

Method

On estuaries and sandy or muddy shores, large invertebrates, such as various polychaete worms living at low densities, can be surveyed and monitored by digging substrate samples in the same way as described for digging soil samples. Invertebrates can then be extracted by wet-sieving. Large polychaete worms rapidly retreat deep into the substrate when sensing disturbance, so a quick, levering action should be used when digging for these.

Smaller invertebrates and those occurring at higher densities are best sampled by taking smaller substrate cores. The corer should be sunk into the substrate to the required depth. If the substrate is firm enough, the core can then simply be lifted up. Otherwise, a thin piece of metal or wood should be slid across the bottom of the corer to prevent material from falling out, and the corer then lifted up.

Benthic invertebrates in deep water can be sampled from a boat using a corer similar to that described above, but longer and with a bung that fits tightly into its top end. The corer should be pushed into the benthos, and then the bung pushed into its top end, creating a vacuum, which prevents the corer's contents from falling out. It should then be gently rocked from side to side to free it, and then lifted up. The corer should be pushed into the benthos to a depth greater than that of the required sample (e.g. to 10–15 cm for a 5-cm-deep sample). The benthos may then be carefully removed from the corer, taking care not to damage the delicate invertebrates within it. The lower, unwanted portion can be discarded. Particular care should be taken to retain the water from within the corer since this can contain invertebrates displaced from the benthos, for example, 60% of chironomids in a study by Euliss *et al.* (1992). There is also a wide range of grabs, dredges and air-lift samplers available for taking samples of benthos from deep water (Elliott & Tullett 1978; Elliott & Drake, 1981a, 1981b).

Benthic invertebrates are best extracted by 'wet-sieving', i.e. spraying a jet of water through the benthos during sieving, using a short length of pipe attached to a tap. A sieve of 0.5-mm mesh size is suggested to recover invertebrates down to a length of 1 mm (these will pass through a sieve of 1-mm mesh, but not one of 0.5-mm mesh). The sievings should then be washed into a white tray (a photographic developing tray is ideal) and invertebrates removed using forceps. It is worth checking the sieve after its contents have been emptied, since small oligochaete and polychaete worms will entangle themselves amongst the mesh of the sieve.

Advantages and disadvantages

Digging, taking cores and using grabs are the only quantitative methods for sampling invertebrates in the benthos. Ekman grabs work well in mud, although other, larger grabs often give poor results (Elliott & Drake 1981a). Sieving samples can be very time-consuming and will vary with the nature of the substrate.

Biases

Some large, active invertebrates may detect disturbance caused by the removal of samples, retreat lower in the substrate, and hence not be sampled.

Kick sampling

Kick sampling is used for invertebrates in riffles in fast-flowing water, particularly mayfly, caddis-fly and stonefly larvae.

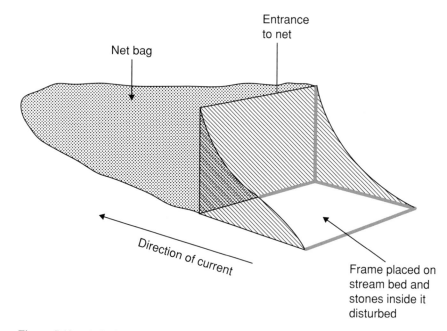

Figure 5.10   A Surber sampler for sampling invertebrates in running water.

Method

Most invertebrates in fast-flowing streams and rivers are found amongst stones and gravel on the stream bed. Kick sampling involves dislodging invertebrates in the stream bed by kicking and disturbing the substrate and catching the dislodged invertebrates in a net held a short distance downstream. The technique is easily standardised and widely used to obtain 'scores' of macro-invertebrates for use in water-quality assessment. A widely used method is to kick-sample all the micro-habitats along a stretch of river for a total of 3 min using a 0.9-mm-mesh pond net (Armitage *et al.* 1983).

A quantitative method of kick sampling involves the use of a Surber sampler (Figure 5.10). This is a quadrat with two sides, an open front that faces upstream and a collecting net on the downstream side. The bottom is disturbed by hand and dislodged invertebrates are swept into the net.

Advantages and disadvantages

Kick sampling is a quick method to obtain relative estimates of invertebrate abundance.

Biases

Kick sampling will tend to under-record invertebrate species firmly attached to stones and heavy species that are unlikely to be carried by the water and caught in the net, such as stone-cased caddis-fly larvae.

References

Armitage, P. D., Moss, D., Wright, J. F. & Furze, M. T. (1983). The performance of a new biological water quality score system based on macro-invertebrates over a wide range of unpolluted running-water sites. *Water Research* **17**, 333–347.

ter Braak, C. J. F. (1986). Canonical correspondance analysis: a new eigenvector technique for multivariate direct gradient analysis. *Ecology* **67**, 1167–1179.

(1994). Canonical community ordination. Part 1: basic theory and linear models. *Ecoscience* **1**, 127–140.

Brooks, S. J. (1993). Review of a method to monitor adult dragonfly populations. *Journal of the British Dragonfly Society* **9**, 1–4.

Butler, S. J. & Gillings, S. (2004). Quantifying the effects of habitat structure on prey detectibility and accessibility to farmland birds. *Ibis* **146** (suppl. 2), 123–130.

Chan, K. Y. & Munroe, K. (2001). Evaluating mustard extracts for earthworm sampling. *Pedobiologia* **45**, 272–278.

Disney, R. H. L. (1986). Assessments using invertebrates: posing the problem. In *Wildlife Conservation Evaluation*, ed. M. B. Ussher, London, Chapman & Hall, pp. 271–293.

(1987). Rapid surveys of arthropods and the ranking of sites in terms of conservation value. In *Biological Surveys of Estuaries and Coasts*, eds. J. M. Baker & W. J. Wolff. Cambridge, Cambridge University Press, pp. 73–75.

Disney, R. H. L., Erzinclioglu, Y. Z., Henshaw, D. J. de C. *et al.* (1982). Collecting methods and the adequacy of attempted fauna surveys, with reference to the Diptera. *Field Studies* **5**, 607–621.

Elliott, J. M. & Drake, C. M. (1981a). A comparative study of seven grabs used for sampling benthic macroinvertebrates in rivers. *Freshwater Biology* **11**, 99–120.

(1981b). A comparative study of four dredges used for sampling benthic macroinvertebrates in rivers. *Freshwater Biology* **11**, 245–261.

Elliott, J. M. & Tullett, P. A. (1978). *A Bibliography of Samplers for Benthic Invertebrates*. Occasional Publication. No. 4. Ambleside, *Freshwater Biological Association*. (Supplement, 1983, Publication No. 20).

Euliss, N. H., Swanson, G. A. & Mackay, J. (1992). Multiple tube sampler for benthic and pelagic invertebrates in shallow wetlands. *Journal of Wildlife Management* **56**, 186–191.

Fowles, A. P., Alexander, K. N. A. & Key, R. S. (1999). The Saproxylic Quality Index: evaluating wooded habitats for the conservation of dead-wood Coleoptera. *Coleopterist* **8**, 121–141.

Fry, R. & Waring, P. (2001). *A Guide to Moth Traps and Their Use*. Amateur Entomologist 24 (2nd edn). London, Amateur Entomologists' Society.

Furse, M. T., Wright, J. F., Armitage, P. D. & Moss, D. (1981). An appraisal of pond net samples for biological monitoring of lotic macroinvertebrates. *Water Research* **15**, 679–689.

Greenslade, P. & Greenslade, P. J. M. (1971). The use of baits and preservatives in pitfall traps. *Journal of the Australian Entomological Society* **10**, 253–260.

Greenslade, P. J. M. (1964). Pitfall trapping as a method for studying populations of Carabidae. *Journal of Animal Ecology* **33**, 301–310.

Gunn, A. (1992). The use of mustard to estimate earthworm populations. *Pedobiologia* **36**, 65–67.

Hammond, P. M. (1990). Insect abundance and diversity in the Dumoga-Bone National Park, North Sulawesi, with special reference to the beetle fauna of lowland rainforest in the Toraut region. In *Insects and the Rainforests of South-East Asia*, ed. W. J. Knight and J. D. Holloway. London, Royal Entomological Society, pp. 197–254.

Hill, M. O. & Gauch, H. G. (1980). Detrended correspondence analysis: an improved ordination technique. *Vegetatio* **42**, 47–58.

Kaila, L. (1993). A new method for collecting quantitative samples of insects associated with decaying wood or wood fungi. *Entomologica Fennica* **29**, 21–23.

Kirk, W. D. J. (1984). Ecologically selective colour traps. *Ecological Entomology* **9**, 35–41.

Lott, D. A. & Eyre, M. D. (1996). Invertebrate sampling methods. In *Environmental Monitoring, Surveillance and Conservation Using Invertebrates*, ed. M. D. Eyre. Newcastle upon Tyne, EMS Publications.

Luff, M. L. (1975). Some factors influencing the efficiency of pitfall traps. *Oecologia* **19**, 345–357.

(1986). Aggregation of some Carabidae in pitfall traps. In *Carabid Beetles: Their Adaptation and Dynamics*, ed. P. J. den Boer, M. L. Luff, D. Mossakowski & F. Weber. Stuttgart, Gustav Fischer.

Majer, J. D. & Recher, H. F. (1988). Invertebrate communities in Western Australian eucalypts: a comparison of branch clipping and chemical knock-down procedures. *Australian Journal of Ecology* **13**, 269–278.

McGavin, G. C. (1997). *Expedition Field Techniques. Insects and Other Terrestrial Arthropods.* London, Expedition Advisory Centre, Royal Geographical Society.

Milne, W. M. (1993). Detachable bags for multiple sweep net samples. *Antenna* **17**, 14–15.

Moore, N. W. & Corbet, P. S. (1990). Guidelines for monitoring dragonfly populations. *Journal of the British Dragonfly Society* **6**(2), 21–23.

Morris M. G. (2000). The effects of structure and its dynamics on the ecology and conservation of arthropods in British grasslands. *Biological Conservation* **95**, 129–142.

Muirhead-Thomson, R. C. (1991). *Trap Responses of Flying Insects.* London, Academic Press.

Mulhern, T. D. (1985). New Jersey mechanical trap for mosquito surveys. *Journal of the American Mosquito Control Association* **1**, 411–418.

New, T. R. (1998). *Invertebrate Surveys for Conservation.* Oxford, Oxford University Press.

Nordström, S. & Rundgren, S. (1972). Methods of sampling lumbricids. *Oikos* **3**, 344–352.

Owen, J. A. (1992). Experience with an emergence trap for insects breeding in dead wood. *British Journal of Entomology and Natural History* **5**, 17–20.

Pollard, E. (1977). A method for assessing changes in the abundance of butterflies. *Biological Conservation* **12**, 115–134.

Pollard, E. (1979). A national scheme for monitoring the abundance of butterflies: the first three years. *British Entomological and Natural History Society, Proceedings and Transactions* **12**, 77–90.

Pollard, E. (1988). Temperature, rainfall and butterfly numbers. *Journal of Applied Ecology* **25**, 819–828.

Pollard, E. & Yates, T. (1993). *Monitoring Butterflies for Ecology and Conservation.* London, Chapman and Hall.

Raw, F. (1960). Earthworm population studies: a comparison of sampling methods. *Nature* **187**, 257.

Roberts, R. H. (1975). Influence of trap screen age on collections of tabanids in Malaise traps. *Mosquito News* **35**, 538–539.

Southwood, T. R. E. (1988). *Ecological Methods, with Particular Reference to the Study of Insect Populations*, 2nd edn. London, Methuen.

Southwood, T. R. E., Henderson, P. A. & Woiwod, I. P. (2003). Stability and change over 67 years – the community of Heteroptera as caught in a light trap at Rothamstead, UK. *European Journal of Entomology* **100**, 557–561.

Stewart, A. J. A. & Wright, A. F. (1995). A new inexpensive suction apparatus for sampling arthropods in grassland. *Ecological Entomology* **20**, 98–102.

Thomas, J. A. (1983). A quick method for estimating butterfly numbers during surveys. *Biological Conservation* **27**, 195–211.

Thomas, J. A. & Simcox, D. J. (1982). A quick method for estimating larval populations of *Melitaea cinxia* during surveys. *Biological Conservation* **22**, 315–322.

Topping, C. J. & Sunderland, K. D. (1992). Limitations to the use of pitfall traps in ecological studies exemplified by a study of spiders in a field of winter-wheat. *Journal of Applied Ecology* **29**, 485–491.

Townes, H. (1962). Design for a Malaise trap. *Proceedings of the Entomological Society of Washington* **64**, 253–262.

Usher, M. B. (1990). Assessment of conservation values: the use of water traps to assess the arthropod communities of heather moorland. *Biological Conservation* **53**, 191–198.

Waring, P. (1980). A comparison of the Heath and Robinson M. V. moth traps. *Entomologists' Record and Journal of Variation* **92**, 283–289.

Watt, A. D., Stork, N. E., McBeath, C. & Lawson, G. (1997). Impact of forest management on insect abundance and damage in a lowland tropical forest in southern Cameroon. *Journal of Applied Ecology* **34**, 985–998.

Whittingham, M. J. & Markland, H. M. (2002). The influence of substrate on the functional response of an avian granivore and its implications for farmland bird conservation. *Oecologia* **130**, 637–644.

Wright, A. F. & Stewart, A. J. A. (1992). A study of the efficacy of a new inexpensive type of suction apparatus in quantitive sampling of grassland invertebrate populations. *British Ecological Society Bulletin* **23**, 116–120.

Yanoviak, S. P., Nadkarni, N. M. & Gering, J. C. (2003). Arthropods in epiphytes: a diversity component that is not effectively sampled by canopy fogging. *Biodiversity and Conservation* **12**, 731–741.

6 Fish

Isabelle M. Côté

Department of Biological Sciences, Simon Fraser University, Burnaby, B.C.,
Canada V5A 1S6

Martin R. Perrow

ECON Ecological Consultancy, School of Biological Sciences, University of East Anglia,
Norwich NR4 7TJ, UK

Introduction

Fish are the most abundant, widespread and diverse group of vertebrates, comprising about 22 000 species with a dazzling variety of form, size and habits. To exploit this profusion of potential food, humans have devised a wide array of gears to capture fish. Ecologists rely heavily on modified forms of these methods to census fish populations, since fish are often difficult to observe in their natural habitat.

Sampling fish requires a high level of resources (e.g. time, labour, cost of equipment), and this increases with the size of the habitat (e.g. a pond versus the sea). Many commercial-scale techniques (e.g. deep-sea seines and trawls) are beyond the scope of ecological sampling, but they can be scaled down to suit smaller habitats. To census fish in the largest aquatic systems, we recommend using data from commercial catches, where available, or visiting markets where fish are landed. The use of local knowledge and technology, particularly where resources are limited, is always recommended.

Methods for capturing fish fall into two categories: *passive* methods, which rely on the fish swimming into a net or a trap; and *active* methods, in which fish are pursued. The fact that fish are cold-blooded influences the choice of method and timing of sampling. For example, in temperate zones, active methods may be more successful in winter when fish are less mobile, whereas passive techniques may work best in summer when fish are more active. The choice of method will also be guided by gear selectivity. Most methods are selective, and the limitations of many, even those in widespread use, are unresolved. There are few truly quantitative methods, so statistics such as catch per unit effort (CPUE) are commonly used to generate indices of abundance (Cowx 1991). Selectivity stems from the physical features of the gear, e.g. mesh size, but it is also influenced by the ecology and behaviour of the target species. Variability in swimming speed, seasonality as a result of migration, diurnality and patchiness of distribution all influence how catchable a species is. It is therefore crucial to have at least some knowledge of the natural history of fish in the habitat to be sampled.

Ecological Census Techniques: A Handbook, ed. William J. Sutherland.
Published by Cambridge University Press. © Cambridge University Press 1996, 2006.

Finally, the choice of method and the manner in which it is used (i.e. technique) will depend on the habitat to be sampled. Factors such as depth, clarity, the presence of vegetation and speed of current will need to be considered. A hydrographical survey prior to sampling may therefore be necessary. Note that a single habitat may be surveyed using more than one method (McClanahan & Mangi 2004), hence recognition of method-specific biases is important.

This chapter provides only a general outline of the methods and techniques that we consider most likely to be used in the course of ecological censusing of fish. We strongly recommend Murphy and Willis (1996) for more detailed discussion of these and other methods. Poisons (e.g. rotenone) and explosives have been used to census fish; however, we consider them too destructive, and they are not included here. Most methods are aimed at censusing juvenile and adult fish, but many can be modified to sample larval fish as well. Several methods also exist for counting fish eggs. A few are covered in this chapter, but Smith and Richardson (1977) provide a good review of standard techniques for surveying eggs and larvae.

Bankside counts

Bankside counts are used for conspicuous fish in pools and slow-moving, shallow freshwater streams and small rivers.

Method

It is sometimes possible to census fish without catching them and without getting wet. Bankside counts are a good technique in shallow, slow-moving and clear waters with little vegetation, such as streams and even lake shores.

The stretch of water to be surveyed is usually divided into contiguous but non-overlapping sections. The sections should be small enough that all fish can be counted from a single vantage point. The use of landmarks on the shore helps in delimiting the sections. Observers, wearing polarising sunglasses, to reduce glare, and dull-coloured clothes, should move slowly to the vantage point and conceal themselves behind riparian vegetation. Artificial structures, such as docks and bridges, may be used when available. Once in position, the observer should wait motionless for at least 5 min, to minimise the effects of disturbance, before counting. Counts are best made on sunny days or at least bright overcast days. Intense sunshine can create problems with glare and shadows can betray the presence of the observer, while rain, wind and surface ripples make observations nearly impossible. After the count, the observer should slowly move away from the bank before proceeding to the next observation point. Fish density can be calculated by measuring the area surveyed.

Advantages and disadvantages

Shore-based visual counts are cheap, fast and easy to carry out. They are particularly appropriate when the water is too shallow to be sampled easily by other techniques and are especially useful for censusing particular age-classes of fish (e.g. young-of-the-year) that seek shallow-water habitats.

Table 6.1. *A summary of the census methods suitable for fish in various habitats*

	Shallow	Deep	Still	Slow flow	Fast flow	Freshwater/ Brackish	Saline	Open	Vegetated	Coral reef	Page
Bankside counts	*		*	*		*		*			251
Underwater observations	+	*	*	*	?	*	*	*	?	*	253
Electric fishing	*		*	*	+	*		*	*		254
Seine netting	*		*	*		*	*	*			257
Trawling		*	*	*		*	*	*			260
Lift-and-throw nets	*	+	*	+		*	*	*	+	+	263
Push nets	*		*	*		*	*	*			263
Hook and line	*	*	*	*	*	*	*	*		*	265
Gill netting	*	*	*	+		*	*	*	+	+	261
Traps	*	*	*	*		*	*	*	+	*	269
Hydroacoustics		*	*	*		*	*	*	+		271

* Method usually applicable, + method often applicable, ? method sometimes applicable. The page number for each method is given.

Table 6.2. *A summary of the census methods suitable for fish eggs of various groups*

	Benthic eggs in nest	Pelagic eggs	Page
Visual estimate	*		273
Volumetric estimate	*		273
Plankton nets		*	274
Emergence traps	*		275

* Method usually applicable, + method often applicable, ? method sometimes applicable. The page number for each method is given.

Direct observation (see also 'Underwater observations', p. 253) allows parameters such as fish size and sex and many features of the environment (e.g. water depth, speed of current, substratum type, vegetation cover) to be recorded. It also allows detailed behavioural observations to be made. Stress to the fish is minimal.

One potential problem is that fish are very aware of human presence, and any disturbance will reduce the accuracy of visual estimates. Fish can take a long time to leave cover and resume their activities following disturbance. Visual counts from banks typically have high inter-observer variability (Hankin & Reeves 1988), although having a few, well-trained people counting fish can reduce this problem.

Biases

Big and brightly coloured fish are easier to see than smaller, duller-coloured fish. This may introduce a sex bias since males are often larger and/or more brightly coloured than females. Age-related behaviour may also introduce a bias if some age-classes seek cover more than others. Fish may be overlooked in deeper water, turbulent areas, turbid waters, areas of high surface glare or dense habitat.

Underwater observations

Underwater observations are used for fish in clear, calm, shallow marine (especially coral reefs) water or freshwater.

Method

There are two ways of gathering underwater observations of fish: snorkelling and SCUBA diving. The choice depends mainly on the clarity of the water and the depth at which observations must be made. Observations made deeper than 1–1.5 m in turbid freshwater lakes or 3–4 m in clear tropical waters will generally require SCUBA.

With both techniques, fixed transects, roving transects or point counts may be used. For fixed transects (see Chapter 3), the observer swims along a fixed measuring tape or chain marked at 1–5-m intervals, laid on the substratum along a depth contour. After waiting 5–15 min for disturbance created by setting the transect line to subside, the numbers and sizes of fish of each species occurring within a given distance of the transect line are recorded. In clear tropical waters, this distance can be up to 5 m on either side of and above the transect, whereas in temperate waters, the lower visibility may reduce the transect width and height to 1–2 m. Narrow transects are preferable for small, cryptic species. It is best to record fish that are somewhat ahead of the observer (the distance will depend on the visibility), since some species will leave the transect corridor if the observer comes too close. Swimming speed is important when running transects because the faster you swim, the less accurate the survey becomes. Swimming speed should therefore be slow and constant.

Data are most easily recorded by writing with a soft lead pencil in a waterproof notebook or on a Perspex sheet from which the glaze has been removed by abrasion (sandpaper or steel wool works well). The writing may later be rubbed off with an eraser.

Roving transects (Jones & Thompson 1978) are timed swims that begin at a random location in the habitat. They typically last 50–60 min. The census period is divided into 10-min intervals and the name of each species is recorded in the interval in which it is first seen. Numbers of individuals are not recorded. Instead, species can be given an abundance score depending on the interval in which they were recorded (i.e. species recorded in early intervals are given higher scores than those seen in later intervals). They are identical to timed species counts (Chapter 3).

Point-count sampling (see Chapter 3) is undertaken from a given location underwater and is generally done using SCUBA. The observer stretches out a measuring tape along the substratum to mark the sampling diameter (usually 10–15 m, or less in poor visibility), and then takes position at the middle mark. While rotating, the observer then records for 5 min the names of all species within an imaginary cylinder stretching from the bottom, at the ends of the tape, to the surface. Numbers and sizes should be recorded only for those species swimming through the cylinder. The numbers and sizes of other species can be recorded after the initial 5-min period.

Fixed transects and point counts give good measures of abundance and density. Random transects give good measures of species diversity. A-priori practice at estimating the length of either PVC sticks or fish models of various sizes underwater usually results in accurate size evaluation during surveys. Remember that things look about 30% larger underwater than they really are.

Advantages and disadvantages

Underwater observations are a favoured way of censusing fish because of the ecological insights gained by sharing the fish's environment. Underwater observations are not destructive and cause minimal disturbance. It is relatively easy to train people to dive or snorkel, but the reliability of underwater censuses depends on accurate species identification. This can be challenging, particularly in the tropics, where species diversity can be great. While underwater, several other facets of the environment and fish behaviour can also be recorded.

Snorkelling equipment is cheap compared with SCUBA gear. Although there are inherent risks to both techniques (e.g. drowning or being eaten by a shark), there are more safety concerns with SCUBA. SCUBA requires certification, the presence of a diving partner, nearby tank-refilling facilities and a decompression chamber within reach. While there is no time limit to snorkelling or diving near the surface, there are strict rules about the time that can safely be spent deeper than 10 m. The deeper the dive, the shorter the permitted bottom time, which will limit the duration of surveys. Getting wet invariably means getting cold, even in tropical waters, although the use of wet or dry suits may help to alleviate this problem.

Biases

Conspicuous species, ages and sexes are more likely to be recorded. Sedentary species may be surveyed with greater accuracy than more mobile species (Samoilys & Carlos 2000). The timing of a census is important; on coral reefs, for example, the composition of fish communities changes markedly over the course of the day or night. Some species may be attracted to the observer. In heavily fished areas, especially where spearfishing takes place, large fish may avoid the observer.

Electric fishing

Electric fishing can be used for all fish in shallow, relatively clear, fresh or brackish waters, particularly vegetated areas.

Method

Electric fishing (or electrofishing in common parlance) involves passing an electric current through water via electrodes (anode and cathode), which stuns nearby fish, leading to their disorientation and easy capture. Power is supplied by an electrical generator (or batteries in the case of backpack units) and is converted to the required current and wave form by an electric-fishing unit or box. The circuit is completed by on/off switches on the anodes. Several types of current may be used, each producing slightly different effects. The most commonly used is DC, because it attracts fish to the anode and causes fewer harmful effects to the fish than AC. However, because DC requires large generators and boxes, pulsed DC is often used instead.

During electrofishing, anodes are often hand held, while the cathode (a multi-strand copper wire or even a metallic plate) trails behind the boat or operator. The charge is usually kept on during fishing. The *key* to electrofishing is always to be close enough to the target fish to induce a response and to 'explore' all available habitat with the anodes. In addition, the operator should always work in an upstream direction since disturbed sediments then flow away from the sampling area and stunned fish drift towards the operator. In narrow (relative to the span of the electrodes) streams and rivers, fish are captured efficiently (Kennedy & Strange 1981) and absolute measures of abundance may be generated.

In small streams, the operator typically wades in the water wearing rubber waders, with the battery-powered gear on his or her back, holding the anode in one hand and a dip net in the other. If greater power is required, generators and boxes may be left on the bank, with long leads connecting them to the anode. Stop-nets can be stretched across the stream both to delineate the area to be sampled and to prevent fish from swimming out of the area. Fishing operators then systematically and continuously search the stream with the anodes moving upstream from the downstream stop-net. Stunned fish are removed with dip nets and placed in containers of water to recover. This usually takes a few seconds, but may take several minutes for large fish. On warm days or when many fish are encountered, the water in the containers should be aerated and refreshed regularly. The whole fishing procedure is then repeated a fixed number of times (usually three) at the same site. This type of 'depletion fishing' allows an absolute estimation of the population size and/or density of fish (see Chapter 3).

In deeper streams or canals, electrofishing can be undertaken from a non-metallic boat carrying both staff and gear. The operator is usually at the bow holding the anode ahead of the boat, while a crew steers with oars or engines. The boat may also be pulled on ropes by operators on the banks. If the river is much wider than the boat, a tight zig-zag pattern of search can be adopted, which allows a more thorough sampling of all habitats. Alternatively, the river may be divided into manageable sections with stop-nets, more than one team may operate in parallel, or multiple anode systems may be mounted on booms (Cowx *et al.* 1990).

In very large rivers and lakes, the use of transects or point sampling (Chapter 3) is preferred. Both may be random or systematic. During transects, the power is usually kept on as the boat cruises along a predetermined route, and stunned fish are collected as they encounter the electric field. The area sampled may be calculated from the distance travelled and the effective width of the anodes (see below). In point sampling, the power is supplied only at predetermined points as the boat moves across the water body. The effective area may be calculated by recording, with a volt meter, the distance from the anode at which the voltage gradient drops to <0.12 V, which is thought to be the value at which forced swimming to the anode is induced (Copp & Peñáz 1988).

Point sampling is particularly effective in vegetated areas and for larval and small fish, although it will also work for larger individuals and species (Perrow *et al.* 1996). Good estimates of density and biomass may be obtained by sampling a large number of points, e.g. over 50 (Garner 1997).

Frightening fish from the area to be sampled is a potential problem that can partly be overcome by using over-sized anodes to increase the effective sampling zone or by altering the method of propulsion (e.g. oars versus engine). Depending on the circumstances, it may pay to move either quietly and slowly or extremely rapidly so that fish have little time to react.

In large systems, relative measures of abundance, such as the CPUE, are more appropriate than absolute estimates. The CPUE is also valuable in monitoring population change over time if the time of fishing and area covered are standardised.

Several environmental factors, such as the conductivity of the water body, influence the efficiency of electrofishing, and there are various ways to adjust the gear to suit particular environmental conditions (Box 6.1). The capture of stunned fish is greatly affected by the clarity of the water body. Fish cannot be sampled in turbid water and, even if clarity is not limiting, experience suggests that the maximum depth for efficient capture of fish is about 2 m. Vision (and concentration) is greatly improved by the use of polarising sunglasses and large hats to reduce glare. Water temperature affects the swimming speed of fish and hence the likelihood of their escape. Each group of fish appears to have an optimal temperature range for capture. Salmonids, for example, are best caught at 5–10 °C, and temperate cyprinids at 10–20 °C (Zalewski & Cowx 1990).

Box 6.1. The effect of the conductivity of water on electric fishing, and how to adjust gear to maximise stunning efficiency

- As conductivity decreases, there is a corresponding decrease in electrode current and a reduction in the stunning radius.
  Solution: Increase voltage output and/or increase anode size.
- At very low conductivity (<50 microsiemens), the fish is more conductive than the water when close to the anode, resulting in potential injury.
  Solution: Use a higher-frequency (600-Hz) system.
- At very high conductivity (>1000 microsiemens), the current density is greater, with an increased potential gradient along the length of the fish, giving a good, low-threshold response and thus requiring lower voltage. However, wastage of current between the electrodes is high.
  Solution: Increase current supply to the electrodes (i.e. bigger generator and units) or, alternatively, reduce energy requirements by decreasing the size of the electrodes (at the cost of a decrease in stunning efficiency).

Advantages and disadvantages

Mixing water and electricity is inherently dangerous, and people have died while carrying out electric fishing. This method therefore requires a high level of training and rigorous safety standards. In many countries, only licensed operators may electrofish. Basic safety procedures include wearing rubber gloves and boots at all times and avoiding immersion of any unprotected body parts. Equipment should have automatic cut-out switches on the anodes, immersion switches in backpack units and emergency stop buttons, all of which are designed to break the circuit immediately if an operator falls into the water. For safety reasons a relatively high number of personnel is required (a minimum of three in the UK). The cost of good-quality equipment is high and regular maintenance of gear is not only critical to ensure high-quality performance, but may be required by law. Such maintenance may be difficult in remote areas.

Electric fishing may injure or kill fish and other aquatic wildlife (including mammals, birds and amphibians) if they are exposed to current for too long. Injury to fish occurs most often when large numbers are encountered simultaneously. Operators should always be prepared to turn off the gear and collect all stunned fish before resuming fishing.

Electric fishing is a flexible method that may be used in a wide range of techniques in virtually any freshwater and brackish-water bodies. Point sampling is particularly adaptable and may be used in a range of habitat types (e.g. littoral margins and open water) within a water body, allowing the measurement of a range of parameters, from species diversity to actual abundance. It may even be used at night when many species are more effectively captured. Electric fishing generally disturbs the habitat less than other active methods do and does not subject fish to abrasion (as in netting and trawling). Its main advantage over most other methods is that it draws fish from cover, making it the method of choice for sampling in vegetated waters.

Finally, to be performed to its full potential, electric fishing requires high motivation, concentration, visual acuity and good hand–eye coordination, physical endurance and the ability to work in a team.

Biases

Large fish respond better to electrofishing than do small fish, and long, thin fish are stunned more efficiently than short, stocky fish of equivalent weight (Zalewski 1983). Benthic species and those preferring dense cover might not rise to the surface, making their retrieval difficult. Territorial species may attempt to maintain their position when approached and are thus relatively easy to capture. In contrast, shoaling species tend to flee, which makes them difficult to catch, but, if they are encountered, many individuals may be captured simultaneously. Large fish and brightly coloured fish are easier to capture than small, cryptic fish. Operators may vary considerably in their ability to stun and collect fish.

Seine netting

Seine netting is used to catch pelagic and/or demersal fish in open, still or slowly flowing water.

Method

A seine net is a wall of net fitted with floats at the top (the float line) and a weighted line (the lead line) on the bottom, generally with a bulging section (bunt or bag) at the back of the net to hold the catch (Figure 6.1(a)).

The first step in seine netting is to encircle a known area of water. To do this, the net is generally fixed at one end, which can be the shore, a boat or a buoy, and the rest of the net is laid out from a boat, or by walking when in very shallow water, in an arc or semi-circular fashion and returning to the fixed point (Figure 6.1(b)). In shallow water, seine nets can also be dragged through the water by two people, one at each end, towards a fixed point (a boat or the shore, Figure 6.1(b)). A major potential problem is that the fish may be frightened away from the area while the net is being set. One way to alleviate this is to set the first part of the net as *quietly* as possible and then to close the gap to the fixed station as *quickly* as possible.

The second step, hauling, is usually done by at least two people, each pulling on one end of the net (the number of people will depend on the net length and drag.) If the seine spans the entire water column (i.e. in shallow water), only the float line is pulled at first, to ensure that the lead line remains in contact with the bottom. In the closing stages of the haul, the lead line is pulled ahead of the float line and out from underneath the net, trapping the fish in the bag. If the seine spans only the top portion of the water column (i.e. in deeper water), both ends of the net must be hauled swiftly and simultaneously until only the bag remains in the water. During hauling, swimmers or boats may be used to frighten the fish into the bag. This is especially useful where fish try to jump over the float line. If the net is hauled too slowly, fish may escape from the mouth of the net. If it is hauled too quickly, the lead line may lift off the bottom or the float line may sink, again allowing fish to escape. This occurs particularly where fine sediments or macrophytes increase water resistance.

If a snag is encountered during hauling, it is best to stop and send out a snorkeller, diver or boat to investigate. Pulling from a different angle, particularly vertically, may be effective and thereafter hauling may resume. However, if this is unsuccessful, it is best to abandon the haul and pull the net back into the boat, since damaging a net can be inconvenient and costly.

The final step, removing the catch from the net, is best achieved in small stages: removing fish with dip nets, pulling the net in a little closer, removing fish with dip nets, etc. Where large amounts of sediment are encountered, sluicing water through the net or rocking the net back and forth in the water may be effective, although great care must be taken not to damage the fish. Once most of the fish have been removed, the net may be pulled onto the bank or boat. The net should then be cleaned and stacked neatly like an accordion (with floats and leads separated) in preparation for the next haul.

Seines come in a huge range of lengths and depths. When sampling in shallow systems, the net depth should be 1.5 times the depth of water (Buckley 1987). This tends to prevent the net from lifting off the bottom during hauling. In deeper water, a purse seine (Figure 6.1(c)) may be more appropriate to catch fish in middle and surface waters. Purse seines have rings attached to the lead line, through which a drawstring (purse line) passes. Once the net has been set, the purse

(a)

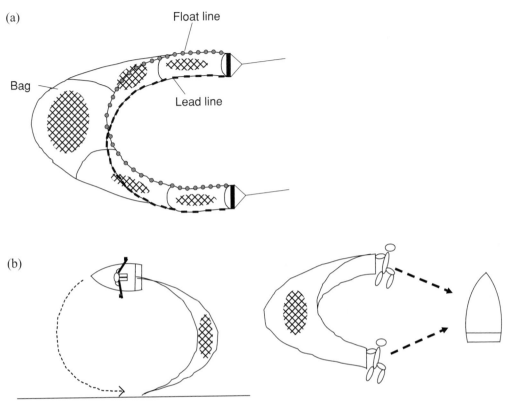

(b)

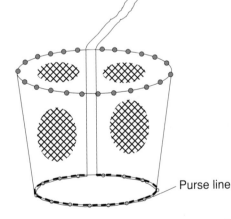

(c)

Figure 6.1 (a) A seine net. (b) Various methods of deploying a seine net: from a boat using a fixed point on the shore (left); and in shallow water with two people dragging the net towards a boat or shore (right). (c) A purse seine.

line is pulled and the bottom of the net is gathered together and pulled upwards, thus trapping the fish.

The size of the net should be related to the size of the water body. In large systems, use the largest net that can be managed, but remember that the larger the net, the greater the effort required to haul it. In ponds and lakes, two people may be able to pull a 50-m-long net, but a 200-m-long one may require between six and eight people. Large nets may be hauled mechanically, but hand pulling is usually preferred since this provides a better 'feel' for how the net is performing, the nature of the bottom and whether any obstructions have been encountered.

Make as many hauls as possible, to give a reasonable number of samples. Several small hauls are better than one large one.

Advantages and disadvantages

Seine netting can be adapted to suit many situations and, as long as the net is small, it is efficient in terms of cost and labour. With large nets, the method quickly becomes expensive and labour-intensive. Furthermore, nets should be designed for specific water bodies (known depth of water, length required, specific mesh sizes to avoid sediments, etc.); hence a truly general-purpose net does not exist. One way of making a seine more versatile is to have detachable sections, perhaps with different mesh sizes, so that a net of the required length can then be 'made up' by tying particular sections together.

Seine nets can be used only in waters that are free of natural (trees, rocks, dense macrophytes, etc.) and artificial (wrecks, litter and other human rubbish) obstructions. A hydrographical survey before sampling is advisable. Seining can cause great disturbance to the habitat sampled when macrophytes and sediment are swept into the net. Moreover, nets can damage and abrade fish, especially where sediment and gravels are retained in the net. Fine sediments can quickly deoxygenate the water and fish can suffocate while waiting to be removed in the final stages of the haul. Repairing damaged seine nets (and they will get damaged!) is a skilled operation, although small repairs can be done quickly with nylon rope or cable ties.

Biases

This is a relatively unselective technique with few biases, although fish in the littoral margins are not sampled adequately. There may be a tendency to miss fast-moving species during setting of the net and species that seek refuge in the substratum during hauling.

Trawling

Trawling is used to catch slow-moving demersal and bottom-living species in large rivers and lakes and in the sea.

Method

Trawling involves towing a cone-shaped net along the bottom or through the water column at a specific depth. For mid-water or surface trawls, the mouth of the net is fitted with floats at the top (the headrope) and leads at the bottom (the groundrope) to keep it open (Figure 6.2(a)). The simplest bottom trawl, the 'beam trawl' (Figure 6.2(b)), has a horizontal bar (beam) on the headrope, and sometimes a rectangular frame hanging from it helping to keep the mouth of the net open, and rows of chain (tickler chains) set in front of the groundrope to disturb fish buried in the sediments. Another common bottom trawl is the 'otter trawl', in which the mouth of the towed net is held open by water pressure against boards attached at an angle to the towing lines or warps (Figure 6.2(c)). Bottom trawls are sometimes mounted on sled runners, which allow them to glide more smoothly over the substratum.

For both bottom and mid-water trawls, the net is deployed either from the stern or from the side of a boat. It is essential that the deck area be clear and that the net be folded properly prior to fishing. In particular, make sure that the warps are neatly coiled in storage containers (e.g. plastic dustbins). While cruising slowly, the cod end, marked with a buoy, should be introduced first. The net should then be fed into the water by hand by at least two people, one on each side of the net, keeping the net fairly tight. When the net is out, the otter boards (in the case of otter trawls) should be lowered into the water and kept in the correct position by keeping warps tense. The warps can then be fed out evenly under tension by two persons, one on each warp. Marks at intervals along the warps help to determine how much line to feed so that the net ends up at the required depth. The warps are then tied to the corners of the stern. The angle of the warps should indicate that the otter boards are keeping the net open.

The net must be towed at a speed faster than the fish can swim. This may vary from around 1.5 knots for slow-moving bottom fish to 5 knots or more for faster-swimming species. The length of the warps for bottom trawls should be approximately three times the depth of the water (Pereyra 1963). For mid-water trawls, the warp length will obviously depend on the depth of sampling. Towing a mid-water trawl at a specific depth is a trade-off between towing speed and the sinking speed of the net. A quick way to determine the trawl depth is to measure the angle ϕ of the warps relative to vertical. The trawl depth is equal to the length of the warps multiplied by $\sin(90 - \phi)$.

Once the trawl has been completed, the engine is cut or put on tick over and the warps are retrieved by hand, with one person on each warp, stacking them away as they come in. The otter boards should also be stacked away as soon as they come aboard. Once the net approaches the boat, care should be taken to ensure that it does not over-run the boat. This may mean moving the boat forwards or turning the boat in an upstream direction, taking care not to foul the net on the propeller. If the catch is large, the net should be emptied by hand nets before retrieving the cod end. If the catch is small, the cod end can be lifted directly onto the deck. If much detritus, e.g. leaf litter, has been picked up, the cod end and the body of the net are best towed to the bank and the catch sorted there.

To estimate the volume of water (for mid-water trawls) or area of bottom (for bottom trawls) sampled, the distance covered by the net must be known. This can be determined either by towing between two predetermined markers on the bank, or by towing at a given speed for a set period

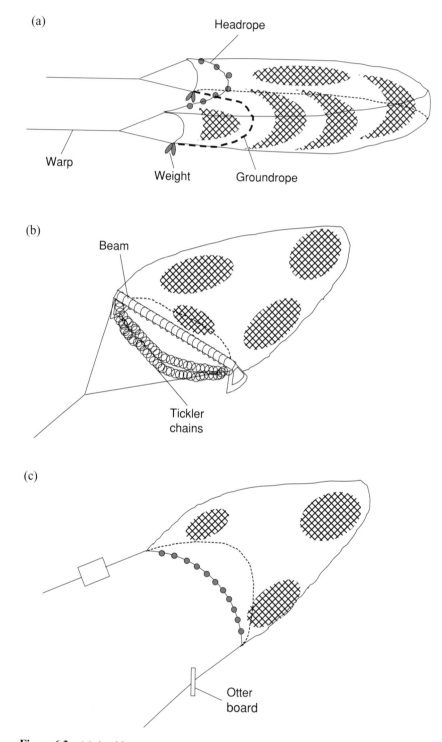

(a)

Headrope

Warp

Weight   Groundrope

(b)

Beam

Tickler
chains

(c)

Otter
board

Figure 6.2   (a) A mid-water trawl net. (b) A beam trawl (for bottom trawling). (c) An otter trawl (for bottom trawling).

of time. Trawling is at best a semi-quantitative method for estimating numbers and biomass; otherwise, CPUE statistics are most appropriate.

Advantages and disadvantages

Trawling lends itself well to censusing in large water bodies where large areas can be covered in a short time. It is valuable in slowly flowing systems, where flow would hinder the use of seine nets, and in tidal rivers that are too large or too saline for electric fishing.

Trawling is limited to waters that are relatively free of obstructions. It is time-consuming, requires expensive equipment and uses considerable fuel (particularly in the case of otter trawls). The chains on bottom trawls are often heavy and can cause considerable environmental damage, which can be reduced by using electrified beam trawls. Fish tend to get damaged in the net, through crowding and abrasion.

Biases

Slow-moving fish and those with poor escape responses are more likely to be caught, particularly in mid-water trawls. Bottom-living species are sampled most reliably.

Lift, throw and push netting

Such nets are used to catch small fish in shallow (or surface), still or slowly flowing water.

Methods

Lift nets fall into two basic types: hand-held scoop nets (Figure 6.3(a)), which are simply inserted below the water surface and brought up sharply; and buoyant nets, which are allowed to lie on the bottom for a set period before coming up abruptly to the surface. As their name implies, buoyant nets are naturally positively buoyant. This buoyancy is temporarily counteracted by attaching the net to a heavy frame or weight by means of dissolving tape or mint sweets with a hollow centre (e.g. 'Polo mints' in the UK). Once the attachment has dissolved, the net pops up to the surface, catching fish on the way.

Cast nets are circles of netting, with weights around the perimeter. They usually have a central line that is retained in the hand for hauling the net after casting. Casting the net from the bank or especially from a boat requires great manual dexterity and practice to achieve distance and the correct shape of the net in the air to maximise the area sampled. Fish are captured by tangling in the mesh as the net collapses when it is hauled. One particularly efficient design is fitted with a purse line (as in purse seines), which is used to draw the net together to prevent the escape of the fish during hauling (Figure 6.3(b)).

Push nets are similar to trawl nets in having a pocket-shaped net attached to a triangular, rectangular, or D-shaped frame, which keeps the mouth of the net open (Figure 6.3(c)). The frame

(a)

(b)

Purse lines

(c)

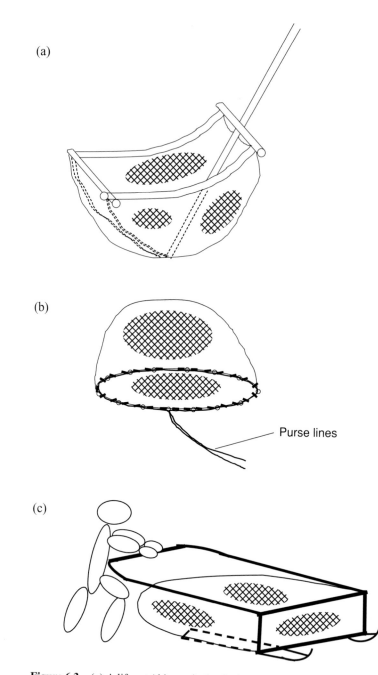

Figure 6.3   (a) A lift net (this particular design was used to fish elvers from the lower river Severn, UK). (b) A cast net with purse line. (c) A push net.

is attached to long handles and can be pushed by wading or from a boat in shallow water. The frame is often fitted with rollers or skis to facilitate pushing. Push-netting sampling effort can easily be standardised by always taking the same number of paces while pushing.

These techniques are typically used for small, shoaling fish in shallow waters, or near the surface of deeper water. They can generate absolute measures of abundance or biomass through discrete point sampling (Chapter 3).

Advantages and disadvantages

Lift, cast and push netting methods are simple to perform and easily repeatable after acquiring some practice. The equipment is fairly cheap, particularly if made with local materials. All these methods are used extensively in developing countries, and tapping into local knowledge will be a great advantage.

With a large number of samples, the methods may become labour-intensive. Lift and cast netting may be usable in highly vegetated habitats, which are difficult to sample by other methods. Push netting is limited to sparsely vegetated, hard-bottomed habitats. A major disadvantage of lift and cast netting is that little is known of their efficiency and selectivity.

Biases

Since the area sampled by lift, cast and push nets is small, only small fish are likely to be captured effectively. Fast-swimming fish are more likely to escape.

Hook and lining

This method is used to catch large, predatory fish at low density, especially in fast-flowing rivers, deep lakes and the sea.

Method

This technique relies on catching fish with a baited hook attached to a line. For predatory fish, the bait may be live invertebrates, such as worms, or pieces of meat, fish, etc. Artificial lures, which imitate worms, flies, fish, frogs and mammals, may also be used. Bread, fruit, seeds, etc. may be used for other groups.

The line may be short and simply attached to the hand, or longer and strung to a pole. Rods with reels are also often used. Using poles and rods means that the fish do not have to be approached so closely and are less likely to be disturbed. They also give more control of the hook. The line with baited hook is then 'cast', from the shore or from a boat. The line can be periodically pulled to mimic the movement of live prey. Many hooks may be used on a line to increase the chance of catching fish.

When a fish bites at the hook, the line becomes tense. A float attached to the line can give a clear indication that a fish is biting since it will sink when a fish pulls on the hook. Rapid upward action then ensures that the hook becomes embedded in the fish's mouth. As soon as the fish nears the surface of the water, it should be scooped out with a hand net. The hook should be removed carefully from the fish. Fish caught can be kept in water-filled containers or keep-nets until the sampling is done.

Hook and line can generate CPUE statistics for targeted groups or species. With mark–recapture methodology it may also be possible to generate more quantitative information on population parameters (Chapter 3).

As a general rule, always exploit local knowledge when using hook-and-line sampling. This will involve talking to local fishers for advice on rigs and baits. This will also help ascertain whether the method is going to be of use for the species or group of interest and what the best techniques are. Anglers participating in tournaments may sometimes provide the labour force for a census.

Advantages and disadvantages

Hook and line is cheap and can be a good method to census large, predatory species, species occurring at low density, or species that live in habitats that are difficult to sample by other methods. It is unsuitable for monitoring entire fish communities as a result of its high selectivity (see 'Biases' below). If many poles and hooks are used, it may quickly become labour-intensive. The chief disadvantage of hook-and-line fishing is that fish are inevitably damaged and subjected to considerable stress. The use of barbless hooks, which can be removed from the mouth more easily, lessens injury to the fish. Hook and lining may also present a significant risk to other groups, especially birds (e.g. albatrosses in marine systems), that are attracted to the baited hooks.

Biases

Hook and lining is highly selective, targeting only focal species. The type of bait used will often determine which species is caught. The success of angling (like other passive methods) depends heavily on environmental conditions such as water temperature, pressure and fish behaviour (particularly timing of foraging). It depends also on the fisher's experience and natural ability.

Gill netting

This method is used to catch mobile fish species in freshwater or seawater, under fairly calm conditions.

Method

Gill netting is a passive method that relies on fish trying to swim through diamond-shaped apertures in a net set vertically in the water column. In theory, the apertures are large enough to allow the

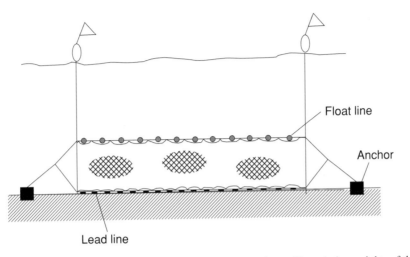

Figure 6.4 A gill net set to sample the lower water column. The relative weights of the float line and lead line can be altered to sample off the bottom at any required depth.

fish's head through, but not the rest of the body, so the fish becomes trapped since it is unable to back out of the net because of its flaring gill covers. A trammel net is a special type of gill net, which is made of a small-meshed panel hanging loosely between two large-meshed panels. When a fish hits the net from either side, it passes through the large mesh and carries the small-mesh net through the large mesh on the other side, forming a pocket in which the fish is trapped. In practice, both with gill and with trammel nets, fish may become wedged across the centre of the body or entangled by fins or spines.

Gill nets usually have a lead line along the bottom edge and a float line along the top (Figure 6.4). By varying the weight of the lead line, gill nets can be made to sink to the bottom, or to stay at the surface of the water, allowing the capture of species at various depths. In shallow water, gill nets can be anchored to the bottom to sample the entire water column.

To deploy a gill net, one end of the net should be attached to a fixed object (such as a buoy or bank). As the boat moves slowly, the net is fed out by hand or simply allowed to peel out of its storage container. If the feeding out is set too fast, there is little chance of correcting any net entanglement. Once the net is nearly stretched out, the second end is also secured to a fixed point. Do not set the net so tightly across the anchor points that the fish will simply bounce off. In shallow waters over a hard bottom, gill nets may be set by wading, although further checking and retrieval should ideally be done from a boat. The position of gill nets should be marked with buoys at both ends.

To keep the fish alive, the net needs to be checked regularly while it is set, by leaning over the side of the boat and lifting sections of net to look for gilled fish. When the net has caught enough fish, or after the required amount of time has elapsed, the net is retrieved starting at the downcurrent or downwind end first, storing each section as it comes aboard in a container or on a gill-net hook. Fish should be removed from the net as they come out of the water. A small hook can be used to lift the mesh over the opercula of the fish and then slide it over the head. Although speed is important to minimise stress to the fish, great care must be taken not to damage fish while freeing them from the net. The net should then be cleared of debris.

For census purposes, gill-net catches are usually expressed as the CPUE, where effort is calculated as net length × time for which the net is set. Although one may expect the catch to increase with net length and set time, beware that these effects are not necessarily linear (Minns & Hurley 1988).

Advantages and disadvantages

Gill netting is a low-cost method of censusing fish. Gill nets are relatively cheap and long-lasting, although regular maintenance is necessary. Removing gilled fish and rubbish (detritus, plants, twigs) from the net is time-consuming, particularly if the net is long.

Gill nets are the most selective of nets because the mesh size determines exactly the body diameter of fish that will be caught (rather than just the minimum size caught as for other nets). This can be an advantage if one wants to survey a particular segment of a fish population, or a disadvantage, if a more general survey is required. One way to decrease selectivity is to use nets with variable mesh sizes. These nets have a dozen or more mesh sizes arranged in blocks across the net. Another is to use several nets of various mesh sizes. Trammel nets are less selective.

Although there are numerous published studies of selectivity measurements for many species and net sizes, selectivity for the particular net used and the particular species sought may need to be established. This can be very time-consuming. Hamley (1975) gives a thorough discussion of gill-net selectivity.

A major disadvantage of gill netting is that the fish caught in gill nets often die, especially if the net was set for too long or too many fish were caught. If fish are just caught by their gills or perhaps tangled on spines or fins, robust species such as percids may be removed without undue stress and mortality, but sensitive species that lose scales easily (e.g. many cyprinids) are unlikely to survive. Trammel nets tend to damage fish less than gill nets do. In addition, gill nets can entangle untargeted animals, including turtles, birds and mammals. Losing a gill net may be disastrous since it will continue to capture fish and other animals until it finally sinks to the bottom.

Gill nets are most effective in lakes and rivers with little current when the target species is highly mobile. Low visibility increases capture efficiency. In areas of strong water current, gill nets are less effective because the lead line may be lifted off the bottom or the float line dragged down so that the net is not maintained in a vertical position. Wind and waves also make the deployment of a gill net difficult. Gill nets may be set in gaps or channels in vegetated areas, although be prepared for the time-consuming process of removing plant matter from the net when it is lifted.

Biases

Gill nets are highly selective with respect to fish size and also bias the catch in favour of mobile, active, fast-swimming fish. Bottom-dwelling fish are not generally caught. Multifilament gill nets catch slightly larger fish than do monofilament or monotwine nets (Hylén & Jakobsen 1979), and thinner twine may catch smaller fish than thicker twine (Yokota *et al.* 2001).

Trapping

Trapping will catch most species under most conditions.

Method

Traps for censusing fish fall into three general categories: pot gear, fyke nets (Figure 6.5) and trapping barriers (weirs). However diverse these may be in design and building materials, virtually all traps operate on the 'funnel' or 'maze' principle, with fish passing easily through an entrance hole, but being confused by the blind endings within the trap and being unable to find their way out. Fyke nets exploit this principle further by having a succession of funnels, which concentrates the fish into the final section of the net. They may also have one or two wings of netting (leaders) attached to the first hoop, which lead the fish to the entrance of the trap (Figure 6.5(b)). Pot traps and fyke nets are usually set temporarily, whereas weirs are more permanent structures that are often used along sea coasts or rivers to catch migratory species such as salmonids.

Pot traps can be set easily by throwing them from boats or from the bank, or by wading in shallow water. Depending on the build material, they may need to be weighted down with rocks or bricks. The position of each trap should be marked with a float and, where the water is deep, the trap should be on a rope to allow it to be retrieved easily. Alternatively, pot traps can be set in midstream and supported on frames to exploit the action of the flow bringing fish to the traps. This is especially useful for migratory species.

Fyke nets require more care in being deployed. When set from a boat, the net should be placed on the bow with the leader(s) on top. The end of the leader is staked or anchored into position and, as the boat moves backwards, the leader is let out until it is fully extended. The hoop net is then put overboard, taking care to keep the float line of the leader upright. The net is then stretched out and its end staked in firmly (Figure 6.5(b)). Ideally, one of the leaders should extend perpendicular and near to shore to prevent fish from swimming around the trap. Fyke nets may be set singly, in pairs with leaders end to end, or in lines (or 'gangs') with the leader of one net near the end of the previous net (Figure 6.5(b)).

Traps are often baited with pieces of fish or meat for predatory species and bread, rice or fruit for omnivorous/herbivorous species. Traps should be checked at least daily to prevent predation or cannibalism by some species (e.g. eels) and to reduce general stress and mortality. Even for a one-off survey, traps really need to be in position at least from dawn to dusk, preferably much longer, to account for short-term variations in environmental factors such as weather and water temperature.

With all traps, CPUE statistics can be used with confidence, as long as trap design and size are standardised. Moreover, with individual traps, mark–recapture techniques may be used to generate data on absolute population size or biomass (Chapter 3). Brandt (1984) gives an excellent review of fish trapping.

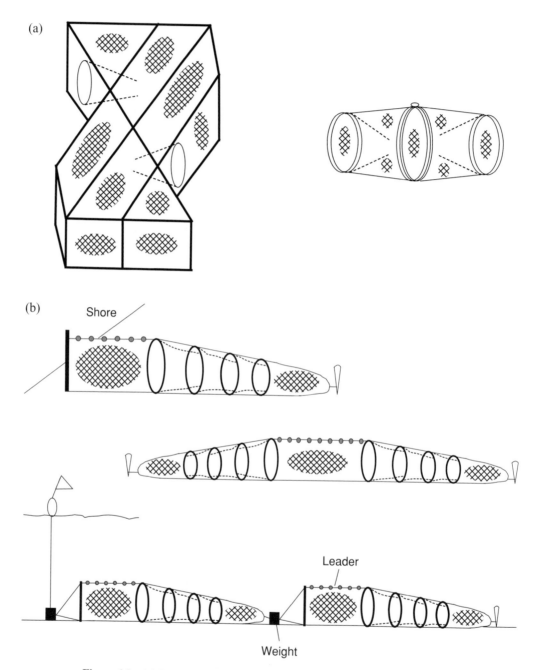

Figure 6.5   (a) Pot traps: an artisanal design with two funnels used widely in the Caribbean (left, view from the top) and a commercially available 'minnow' trap (right), detachable in the middle for easy retrieval of fish. (b) Fyke nets, set singly against the shore (top), in a pair (middle) or in a gang (bottom), used especially for eels.

Advantages and disadvantages

Trapping is one of the most versatile methods for censusing fish. It can be used in a wide variety of habitats, from strongly flowing rivers to stillwater lakes and wetlands dominated by vegetation, from coral reefs to featureless estuaries, to catch a wide variety of species. It is particularly effective for species that occur at low density or are active at night. Since traps sample fish passively, the ratio of effort to return can be good. Traps are generally cheap to buy and even cheaper to make. Because trapping has been used for centuries in most areas of the world, adopting local trapping technology usually saves time and money.

Trapping may become labour-intensive when many traps are used, the catches of fish are high, or regular maintenance of traps is required. In highly vegetated marshy area, substantial damage to the traps may be caused by mustelids, if present. Traps can also be highly selective (see 'Biases' below), which can be either an advantage, if a particular species is targeted, or a disadvantage, if a more general census of fish populations or community composition is required. Since trapping success varies greatly with trap construction, sampling location and set time, all these factors must be standardised to yield meaningful CPUE statistics.

Biases

Trap selectivity depends on the size of the trap entrance and size of mesh. Various species, depending on their mobility and activity patterns, territoriality and inquisitiveness, have different capture efficiencies. Even within a species, trappability is influenced by season, sex, age, habitat availability, etc. Little is known of these biases except for a few taxa, e.g. perch *Perca fluviatilis* in Windermere (Bagenal 1972) and snappers and emperors on the Great Barrier Reef (Newman & Williams 1995).

Pot gear is most effective for catching bottom-dwelling species that are seeking food or shelter. Fyke nets also target cover-seeking species, but usually more mobile species than those targeted by pot traps. Weirs catch mostly migratory species that follow shorelines.

Hydroacoustics

Hydroacoustics can be used to detect most large fish in deep, fresh and marine waters.

Method

Hydroacoustics relies on a sonar system. A transmitter produces a pulse of electricity, which is converted by an underwater speaker (the transducer) into an audible sound (usually a 'ping' or 'beep'). This sound travels as a beam through the water and, on hitting an obstacle (the bottom, a fish or a submarine!), bounces back towards the boat as an echo. This echo is picked up by the transducer and reconverted to electrical signals. These are then modified and amplified by a receiver and displayed on a paper recorder, oscilloscope, or cathode-ray tube.

The strength of the echo (or target strength) is related to the size of the fish, and the length of time taken for the echo to return is related to depth. Hydroacoustic surveys can therefore yield information not only on the number and sizes of fish, but also on their distribution. The numbers of fish present can be estimated by counting the number of returning echoes, if echoes are predominantly from individual fish, or by echo integration, if echoes are from schooling fish. Echo integration is mathematically complex – formulae for its calculation are given by Ehrenberg (1973) – and is greatly affected by errors in estimation of target strengths.

Echo-sounding equipment is usually installed on the hull of a boat and used vertically to census fish occurring below the boat. It may also be rigged to sample the water column horizontally, which is more useful in rivers because fish generally orientate themselves against the flow. Surveying can be undertaken from moving boats or from fixed points (e.g. a buoy or the bank). The area sampled can be calculated from the dimensions of the beam, although, when in a moving boat, the duration-in-beam of the target must be calculated from the boat speed. Echo-sounding allows one to cover the entire area of the water body to be surveyed or to undertake transects (p. 145) or point sampling (p. 148).

Advantages and disadvantages

Although the initial cost of the equipment may be very high, the operational costs of hydroacoustic surveys in terms of ship time and labour are relatively low, giving cost-efficient information on fish density and biomass. Absolute population size can be estimated, even in traditionally difficult habitats such as large, fast-flowing lowland rivers, large and deep lakes and the sea. The technique is non-destructive and fish can be censused *in situ*. Results are obtained quickly and variance between replicate censuses is generally low.

A serious disadvantage is the inability to distinguish easily between species (see Horne (2000) for a review). Any discrimination relies on detailed knowledge of species distribution, size classes and composition of stocks as determined by other techniques. The equipment needs to be calibrated in the system to be sampled to correct for environmental noise in the signals. Bobek (1990) illustrates the differences among various sounding methods and the influence of diurnal activity patterns and fish orientation on hydroacoustic data.

Hydroacoustic equipment is complex and requires a high level of operator training. To convert echo number reliably into population estimates, target strengths need to be known for all fish likely to be encountered. Evaluating the target strengths of particular species can be difficult and time-consuming (Stepnowski & Moszynski 2000).

Biases

Fish with swimbladders have target strengths some 10 decibels higher than those of fish of the same size without swimbladders. This may bias estimated lengths and biomass when species-specific target strengths are unknown. Fish swimming at the surface or sitting on the bottom are not sampled with vertical echo-sounding. This can be partly alleviated by horizontal sounding. Some fish may also avoid moving boats and swim at a faster speed than can be sampled.

Visual estimates of eggs

This method enumerates benthic eggs spawned in a mass or a single layer.

Methods

Egg number can be assessed visually by a variety of means for species spawning eggs in nests. For species that lay eggs in a single layer (e.g. blennies, gobies, damselfishes), the area of the nest covered by eggs can be measured and then converted to egg number if egg diameter is known. For species that lay eggs in a mass (e.g. centrarchids), a scoring system can be devised in which increasing numbers on a scale of, say, 1 to 5 represent increasing numbers of eggs. All nests can then be scored in this way and, if the scale is an absolute rather than a relative one, egg scores may be compared among areas or years. The scores can then be validated by collecting and counting eggs in several (3–5) entire nests of each egg score. Entire nests may be collected with a suction gun (in North America, the devices used for basting turkey work well), large pipettes or scoops, and the contents transferred underwater into labelled plastic bags. It may also be necessary to collect vegetation, rocks and debris from the nest since eggs may have adhered to them. The eggs should then be counted the same day in a laboratory.

Advantages and disadvantages

Visual assessment is done quickly and becomes more accurate with experience. Consistency of scoring, both within and among observers, can be a problem. Inter-observer correlations should be examined if more than one individual is collecting the data. Validation of the technique is time-consuming.

Biases

The size of the egg mass relative to the size of the nest may bias visual assessments: small masses in large nests may appear smaller than masses of the same size in smaller nests.

Volumetric estimates of eggs

This method is used for benthic eggs spawned in a mass.

Method

Entire nests may be collected as described under 'Visual estimates' (see above) and the volume of the egg mass determined by putting the eggs into a graduated cylinder. Large egg masses may exceed the capacity of the cylinder and may need to be divided. The volume of the egg mass can then be converted into number of eggs by measuring the volume of a small, known number of eggs (50–200, depending on egg size) in the same way.

Advantages and disadvantages

This method is relatively rapid, but in waters deeper than 1–2 m will require the use of SCUBA to collect the eggs.

Biases

The presence of debris to which the eggs can adhere may artificially inflate the estimate.

Plankton nets for catching eggs

This method is used for pelagic eggs.

Method

Nets used for plankton may also be used for catching floating fish eggs. These nets consist of a funnel of fine-mesh netting held open at the mouth by a circular, metallic ring. The nets are towed, pushed, pulled, buoyed up, or held stationary in flowing water (see p. 208). As water is filtered through the mesh, organisms larger than the mesh size concentrate at the end of the net (the 'cod end'). Nets vary tremendously in mouth diameter, but Bowles *et al.* (1978) suggested that nets with 1-m-diameter mouth sampled pelagic eggs better than any other type of gear.

Monofilament netting with relatively square apertures is generally preferred. The mesh sizes used most commonly to sample fish eggs range between 0.3 and 0.8 mm (up to 1 mm in inland waters). The cod end can consist of a removable bag, jar, or bucket with screened windows. The latter is preferred for collecting fish eggs since it allows more efficient concentration of the filtrate.

To be able to convert the number of eggs caught into a quantitative measurement, the volume of water filtered must be known. This can be achieved most accurately by attaching a flow meter in the mouth of the net. The rate of flow is then multiplied by the area of the net mouth and the length of time of the tow.

Eggs can be damaged easily, hence extreme care must be taken when transferring the filtrate to a storage container. The eggs can be preserved in a solution of 3%–5% formalin (1%–2% formaldehyde) in water. Alcohol is not recommended since it causes shrinkage of the eggs, and egg diameter and shape may be important parameters in species identification. The number of eggs caught can then be counted by eye, or with an electronic particle counter.

Advantages and disadvantages

Nets allow a large amount of water to be sampled in a short time. They are also either inexpensive or only moderately expensive. Clogging can be a major problem, leading to passive avoidance by eggs and fish larvae of the net. Performance is also highly dependent on hydrographic characteristics, such as turbulence. Egg density may be used to estimate spawning stock biomass if the sex ratio of the fish population, the proportion of females spawning and the relationship between fecundity and fish size are known (Parker 1980).

Biases

Egg densities may appear lower if some eggs were extruded through the mesh under pressure.

Emergence traps for eggs

This method is used for benthic eggs.

Method

Emergence traps of the type used to catch aquatic insects (see p. 237) may also be used to catch fish larvae hatching from eggs deposited on the bottom. The apparatus is anchored to the bottom over the area of the nest after spawning has taken place. This method is most appropriate for sampling eggs of species, such as salmonids, that scatter or bury their benthic eggs and do not provide parental care lasting until the eggs hatch. Young salmonids may swim down in the gravel and some may be missed unless the edges of the trap are buried deeply.

Advantages and disadvantages

If nests are discrete, each will require its own trap. This has the advantage of providing an estimate of egg number for each nest, but can become expensive since many traps will be required to achieve a good sample size of nests. There is a risk of traps becoming dislodged by strong currents if left over nests for a long period.

Biases

Traps might not cover the whole area of a nest if the nest is large, thus leading to an underestimate of the number of eggs spawned. This method actually measures the number of eggs hatching rather than the number spawned.

References

Bagenal, T. B. (1972). The variability in the number of perch, *Perca fluviatilis* L., caught in traps. *Freshwater Biology* **2**, 27–36.

Bobek, M. (1990). Applied hydroacoustics in cyprinid research. In *Fisheries in the Year 2000*, ed. K. T. O'Grady, A. J. B. Butterworth, P. B. Spillett & J. L. J. Domaniewski. Nottingham, Institute of Fisheries Management.

Bowles, R. R., Merriner, J. V. & Grant, G. C. (1978). *Factors Associated with Accuracy in Sampling Fish Eggs and Larvae*. Ann Arbor, Michigan, US Fish and Wildlife Service.

Brandt, A. von, (1984). *Fish Catching Methods of the World*. Farnham, Surrey, Fishing News Books.

Buckley, B. (1987). *Seine Netting. Advisory Booklet from the Specialist Section – Management*. Nottingham, Publications of the Institute of Fisheries Management.

Copp, G. H. & Peñáz, M. (1988). Ecology of fish spawning and nursery zones in the flood plain, using a new sampling approach. *Hydrobiologia* **169**, 209–224.

Cowx, I. G. (ed) (1991). *Catch Effort Sampling Strategies – Their Application in Freshwater Fisheries Management*. Oxford, Fishing News Books, Blackwell Scientific Publications.

Cowx, I. G., Wheatley, G. A. & Hickley, P. (1990). Developments of boom electric fishing equipment for use in large rivers and canals in the United Kingdom. *Aquaculture and Fisheries Management* **19**, 205–212.

Ehrenberg, J. E. (1973). *Estimation of the Intensity of a Filtered Poisson Process and its Application to Acoustic Assessment of Marine Organisms*. Seattle, Washington, University of Washington Sea Grant Publication WSG 73-2.

Garner, P. (1997). Sample sizes for length and density estimation of 0+ fish when using point sampling by electrofishing. *Journal of Fish Biology* **50**, 95–106.

Hamley, J. M. (1975). Review of gill net selectivity. *Journal of the Fisheries Research Board of Canada* **32**, 1943–1969.

Hankin, D. G. & Reeves, G. H. (1988). Estimating total fish abundance and total habitat area in small streams based on visual estimation methods. *Canadian Journal of Fisheries and Aquatic Sciences* **45**, 834–844.

Horne, J. K. (2000). Acoustic approaches to remote species identification: a review. *Fisheries Oceanography* **9**, 356–371.

Hylén, A. & Jakobsen, T. (1979). A fishing experiment with multifilament, monofilament and monotwine gill nets in Lofoten during the spawning season of Arcto-Norwegian cod in 1974. *Fiskeridirektoratets Skrifter, Serie Havundersøkelser* **16**, 531–550.

Jones, R. S. & Thompson, M. J. (1978). Comparison of Florida reef fish assemblages using a rapid visual technique. *Bulletin of Marine Science* **28**, 159–172.

Kennedy, G. J. A. & Strange, C. D. (1981). Efficiency of electric fishing for salmonids in relation to river width. *Fisheries Management* **12**, 55–60.

McClanahan, T. R. & Mangi, S. C. (2004). Gear-based management of a tropical artisanal fishery based on species selectivity and capture size. *Fisheries Management and Ecology* **11**, 51–60.

Minns, C. K. & Hurley D. A. (1988). Effects of net length and set time on fish catches in gill nets. *North American Journal of Fisheries Management* **8**, 216–223.

Murphy, B. R. & Willis, D. W. (eds.) (1996). *Fisheries Techniques*, 2nd edn. Bethesda, Maryland, American Fisheries Society.

Newman, S. J. & Williams, D. M. (1995). Mesh size selection and diel variability in catch of fish traps on the central Great Barrier Reef, Australia – a preliminary investigation. *Fisheries Research* **23**, 237–253.

Parker, K. (1980). A direct method for estimating northern anchovy, *Engraulis mordax*, spawning biomass. *Fisheries Bulletin US* **78**, 5541–5544.

Pereyra, W. T. (1963). Scope ratio–depth relationships for beam trawl, shrimp trawl, and otter trawl. *Commercial Fisheries Review* **25**, 7–10.

Perrow, M. R., Jowitt, A. J. D. & Zambrano González, L. (1996). Sampling fish communities in shallow lowland lakes: point-sample electrofishing versus electrofishing within stop-nets. *Fisheries Management & Ecology* **3**, 303–313.

Samoilys, M. A. & Carlos, G. (2000). Determining methods for underwater visual census for estimating the abundance of coral reef fishes. *Environmental Biology of Fishes* **57**, 289–304.

Smith, P. E. & Richardson, S. L. (1977). *Standard Techniques for Pelagic Fish Egg and Larva Surveys*. Fisheries Technical Paper 175. Rome, Food and Agriculture Organization of the United Nations.

Stepnowski, A. & Moszynski, M. (2000). Inverse problem solution techniques as applied to indirect in situ estimation of fish target strength. *Journal of the Acoustical Society of America* **107**, 2554–2562.

Yokota, K. Fujimori, Y., Shimode, D. & Tokai, T. (2001). Effect of thin twine on fill net size-selectivity analyzed with the direct estimation method. *Fisheries Science* **67**, 851–856.

Zalewski, M. (1983). The influence of fish community structure on the efficiency of electrofishing. *Fish Management* **14**, 177–186.

Zalewski, M. & Cowx, I. G. (1990). Factors affecting the efficiency of electric fishing. In *Fishing with Electricity – Applications in Freshwater Fisheries Management*, ed. I. G. Cowx & P. Lamarque. Oxford, Fishing News Books, Blackwell Scientific Publications.

7 Amphibians

Tim Halliday

Department of Biological Sciences, The Open University, Milton Keynes MK7 6AA, UK

Introduction

The habits and life histories of amphibians are such as to pose a number of major problems for anyone seeking to estimate their abundance accurately (Table 7.1). Most are highly secretive in their habits and may spend the greater part of their lives underground or otherwise inaccessible to biologists. The limbless caecilians, for example, live entirely beneath the ground surface and little is known about most aspects of their biology. When amphibians do venture out they typically do so only at night. They have low food requirements and so can afford to emerge only when conditions are optimal, typically when the weather is warm and wet. Their activities are highly seasonal; most temperate amphibians hibernate over winter and many, notably desert species, aestivate during hot, dry periods.

Amphibians are typically most evident, and thus most easily censused, when they breed, but breeding activity is characteristically seasonal and may be very unpredictable. In some temperate amphibians breeding is 'explosive', with annual breeding activity being completed in one or two days. In such species, effective censusing can be achieved by intensive fieldwork over a limited period, provided that the censuser is alert to the climatic conditions that stimulate breeding. In tropical species, however, breeding may occur over an extended period of the year, sometimes sporadically, so censusing work has to be maintained over many weeks or months. In some desert species, breeding does not occur for one or more years if favourable wet conditions do not occur.

Breeding, in many species, offers good opportunities for censusing amphibians because they aggregate in large numbers at a limited number of breeding sites, such as ponds. This applies to those species that lay their eggs in or close to water and have an aquatic larval stage. There are, however, many amphibians, such as terrestrial-breeding frogs and salamanders, in which eggs are not laid in water, so there are no focal points for breeding activity at which censusing efforts may be directed. For pond-breeding species, it cannot be assumed that the amphibians in and around a particular pond represent a demographically independent population. If there are other ponds close by, they may be a sub-population of a larger population, which should be regarded as the appropriate demographic unit for censusing (Petranka *et al.* 2004).

For some species, eggs or larvae are more readily detected, observed and counted than adults and it is tempting to focus censusing efforts on them. This can be highly misleading, however. In

Ecological Census Techniques: A Handbook, ed. William J. Sutherland.
Published by Cambridge University Press. © Cambridge University Press 1996, 2006.

Table 7.1. *A summary of methods suitable for various groups*

Method	Aquatic salamanders and newts	Terrestrial salamanders	Ground- and water-breeding frogs and toads	Tree-breeding frogs	Caecilians	Page
Recognising individuals	+	+	+			280
Drift fencing	+	?	+			285
Scan searching	+	*	+	+		286
Netting	+		+			287
Trapping	+	+	+			288
Transect and patch sampling	*	*	*	+		289
Removal studies	+	*	+	?		290
Call surveys			+	+		290

* Method usually applicable, + method often applicable, ? method sometimes applicable. The page number for each method is given.

many amphibian species, some individuals, especially females, do not attempt to breed in a given year. Many amphibians have very high fecundity, but there can be enormous mortality among eggs and larvae. Amphibians are typically long-lived and breed several times during the course of their lives and, even in a stable and viable population, there may be little or no survival of eggs and larvae in a majority of years. Under these circumstances, counts of eggs and larvae can give a highly misleading measure of a population, especially if they are made only in a single year (Pechmann *et al.* 1991). An analysis of the population dynamics of three anuran species by Biek *et al.* (2002) suggests that post-metamorphic survival is a particularly important determinant of population size and that this stage of the amphibian life cycle requires more attention than it generally receives.

The need to census amphibians has never been more urgent than it is now. Among herpetologists, a growing awareness that amphibians are declining and becoming extinct in many parts of the world, in many instances in areas of apparently pristine and protected habitat, led to the formation in 1991 of the Declining Amphibian Populations Task Force (DAPTF), set up under the auspices of the IUCN/SSC. Since 1991, it has become apparent that the situation is even worse than expected. The IUCN Global Amphibian Assessment (GAA) has revealed that 32% of the world's 5743 amphibian species are threatened with extinction (Stuart *et al.* 2004).

The phenomenon of amphibian decline has prompted many studies that seek to assess amphibian diversity at the species level, and to track changes in the number of species over time. Some studies are concerned with determining the number of populations of a species in a given area; others seek to determine the number of species. While such studies are beyond the scope of this book, they use many of the methods that are used to census populations and encounter many of the same problems. One aspect of the work of the DAPTF has been the production of a guide to techniques for monitoring amphibian populations (Heyer *et al.* 1994). While that book is largely concerned with methods for assessing amphibian diversity in terms of numbers of species in various habitats, it contains a great deal of information about techniques for censusing and monitoring individual populations.

Recognising individuals

To derive an accurate estimate of a population's size requires, in most censusing techniques, being able to recognise individuals, as in mark–recapture studies (see Chapter 3). Until recently the most widely used method for recognising individual amphibians was toe-clipping, but less invasive methods, such as exploiting the fact that most amphibians have highly variable skin markings, are becoming more widely used.

Toe-clipping

One or more fingers or toes are removed from an individual, with scissors or toe-nail clippers, using a simple code. It is imperative that, when marking anurans (frogs and toads), the thumb of males is not removed, since these bear nuptial pads that are important in mating. Removal of a single digit can be used in mark – recapture studies, simply to identify individuals that have been caught. Tissue from toe-clips can also be used to extract DNA for genetic studies. Toe-clipping is much more effective as a long-term marking technique for anurans because removed digits do not regenerate; in urodeles (newts and salamanders) digits will re-grow, usually within a year.

The advantages of toe-clipping are that it is simple, inexpensive and provides tissue that can be very useful for life-history and genetic studies. Though toe-clipping is widely used, there is increasing concern that it is harmful to amphibians; there is evidence that it decreases survival (Clarke 1972) and that it causes tissue damage (Golay & Durrer 1994). In view of accumulated evidence that toe-clipping reduces survival in some species, an effect that increases as more toes are removed (McCarthy & Parris 2004), it is recommended that no more than one toe ever be removed and that toe-clipping is not used for more delicate species such as treefrogs; some would ban toe-clipping altogether (May 2004). In many countries it requires a licence and anyone doing it should be aware that it causes public disquiet, and that many journals regard it as a serious ethical issue.

Recognising skin patterns

Several herpetologists have used the natural variation present in the skin patterns of amphibians to recognise up to several hundred individuals. Each individual is photographed or photocopied and entered in a register so that it can be recognised when next caught (Figure 7.1). If a population is large, it is helpful to remove one digit from each registered animal, so that registered and new individuals can be differentiated. Digital photography has made visual identification much easier and cheaper and it is now widely used (for examples see Table 7.2).

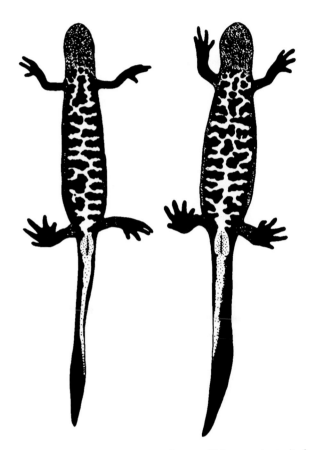

Figure 7.1 Two female crested newts (*Triturus cristatus*), showing the belly pattern used in individual recognition. (Source: the author, from John Baker.)

Other methods

A variety of methods has been used to mark and identify amphibians individually; these are summarised in Table 7.2. It should be noted that each method has been used on a limited number of species, in some cases only one; that a method that is effective for one species might not work with another; and that any method needs to be tested and validated for a particular species before it is used on a large scale. Methods vary in terms of cost, both financial and in terms of the time required. A comparison of these costs for two methods, visual recognition and PIT tags, is provided by Arntzen *et al.* (2004).

Detection probability

Being generally secretive creatures that are only occasionally active, amphibians are typically much more difficult to detect at some times, and in some habitats, than others. It is very important

Table 7.2. *Methods used for marking and recognising individual amphibians*

Method	Procedure	Advantages	Disadvantages	References
Toe-clipping	One digit cut from hands or feet	Inexpensive	Digits of urodeles re-grow; may cause tissue damage or reduce survival	Martof (1953)
Elastic wastebands	Fitted to individuals according to their size; can be colour-coded or numbered	Inexpensive; large numbers of individuals can be recognised at a distance	Suitable only for short-term studies (a few days or weeks)	Emlen (1968), Davies & Halliday (1979)
Knee-tagging	Small tag tied to knee with stretchable thread	Inexpensive; large numbers of individuals can be marked	Suitable only for anurans	Elmberg (1989)
Fluorescent colour marks	Coloured fluorescent dust applied to skin with compressed air	Large numbers of individuals can be recognised at a distance and in the dark	Relatively expensive; marks are temporary, lasting 1–2 years	Nishikawa & Service (1988), Harvey (2003)
Skin transplants	Small piece of ventral skin exchanged for piece of dorsal skin	Permanent; inexpensive	Time-consuming; requires expertise; relatively few animals can be marked; suitable only for species with contrasting dorsal and ventral colours	Rafinski (1977)
Skin staining	Dye sprayed onto skin with dental (Panjet) injector	Inexpensive	Can cause injury; relatively few animals can be marked; marks are short-lived	Wisniewski *et al.* (1980), Gittins *et al.* (1980)
Visual implant elastomer (VIE)	Coloured implants inserted beneath the skin	Inexpensive; large numbers of individuals can be recognized	May require a licence	Bailey (2004)
Acrylic-polymer marking	Acrylic polymer injected into skin of body or tail	Inexpensive; long-lasting; good for larvae and small adults	May increase predation risk	Johnson & Wallace (2002)

Table 7.2. (*cont.*)

Method	Procedure	Advantages	Disadvantages	References
Tattooing	Dye injected beneath skin with electric tattooer	Inexpensive; large number of individuals can be marked; marks last for at least 3 years	Not suitable for dark-skinned species	Joly & Miaud (1989)
Branding	Red-hot or frozen metal applied to skin	Inexpensive	Can cause severe wounds; marks are short-lived in some species	Daugherty (1976), Nace & Manders (1982)
PIT tags	A passive integrated transponder is inserted beneath the animal's skin and a unique radio signal is read with a portable scanner	Very large numbers of individuals can be individually recognised	Very expensive; suitable only for larger species; usually requires a licence	Camper & Dixon (1988)
Natural variation in skin patterns	Individuals are photographed or photocopied to provide a register of known individuals	Non-invasive	Labour-intensive	Hagstrom (1973), Kurashina et al. (2003), Bailey (2004)

Note: most of these methods have been used successfully for only one or a few species (see references) and it should not be assumed that any method is suitable for any given species.

to address this issue before setting up a census or monitoring programme, because many methods currently used to count amphibians fail to take variation in the detection probability of their target species into account. Schmidt (2003, 2004) suggests that most of the data used to infer changes in amphibian abundance, locally and worldwide, are unreliable because they do not allow for variation in detection probability and consequently underestimate the true population size.

Variation in detection probability may arise for biological reasons that are intrinsic to particular species (amphibians are commonly more active on warm, wet nights, for example) or for procedural reasons intrinsic to the people carrying out the study. For example, Hairston and Wiley (1993) suggested that much of the variation that they observed in the apparent abundance of woodland salamanders over several years was due to variation in the motivation of the students who carried out the work. A more recent study of terrestrial salamanders has revealed variation in detection probability among species, locations and years (Dodd & Dorazio 2004).

Variation in detection probability is an issue not only for census studies of single populations, but also for those studies that score the number of species in an area and those that count the

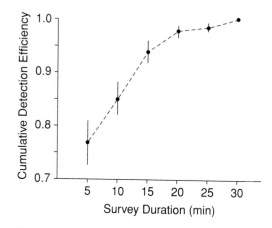

Figure 7.2 Detection efficiency and call-survey duration, showing the number of frog species detected during call surveys as a function of their duration. (Source: Pierce & Gutzwiller (2004)).

number of populations. In a long-term study looking at number of species, Skelly *et al.* (2003) found that re-surveying an area one year after their initial survey indicated a decline in number of species by 45%. A further re-survey after two years indicated a decline of 28% and repeated surveys over five years suggested a decline of 3%. This effect was most likely to have been due to the fact that several species were not detectable in the study area every year. In a study focusing on censusing the number of populations in an area, Pellet and Schmidt (2005) found that call surveys (see below) reliably detected most populations of the European treefrog (*Hyla arborea*), but were poor at detecting the natterjack toad (*Bufo calamita*). This example shows that detection probability is also a property of the census method that is used.

The first step in overcoming the problem of detection probability is to be aware of it and to analyse carefully how it might apply to the species you wish to census and the method you wish to use. The more you know about the natural history and behaviour of your species, the more you are likely to be aware of when and where it is most easily detected. Some census methods are less subject to error due to variation in detection probability than others; mark–recapture is least affected (Schmidt *et al.* 2002; Schmidt 2004). Researchers are strongly recommended to use more than one census method, since this quickly reveals the limitations of each (see below).

Variation in detection probability also has important implications for how long a census study should be conducted to provide reliable data. An assessment of two monitoring programmes carried out in Canada revealed that, because detection probability varies among species, it requires 24 visits to a given site to detect 90% of the species present; 12 visits detected only 80% (de Solla *et al.* 2005). By assessing the reliability of call surveys (see below), Pierce and Gutzwiller (2004) found that 15-min surveys detected 94% of species; 30-min surveys were no better, but 5-min surveys were worse (Figure 7.2).

Statistical methods for estimating population sizes that allow for variation in detection probability are provided by MacKenzie *et al.* (2002) and Royle (2004).

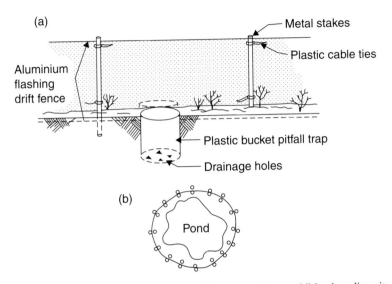

Figure 7.3 A drift fence with a pitfall fence around an amphibian breeding site. (a) Positions of pitfall traps in relation to the fence. (b) A continuous fence around a breeding pond. (Source: Heyer *et al*. (1994), p. 127.)

Drift fencing

Drift fencing is used for amphibians that migrate periodically to small water bodies to mate and/or lay eggs.

Method

A fence, supported by posts, is built so as to encircle a breeding site (Figure 7.3). The fence should be at least 35–40 cm high and should be dug into the ground to a depth of at least 20 cm. The fence can be made of a variety of materials, including aluminium sheet, plastic sheet, or hardware cloth. Many amphibians are very adept at climbing and it is desirable that the fence has an overhang at the top, at least on the outside, preferably on both sides. The fence should be positioned as near the water as possible, but must be above the expected high-water mark. Pitfall traps are dug into the ground, inside and outside the fence, at a spacing of 3–5 m along the fence. These may be plastic buckets or metal cans and must be strong enough not to collapse under the weight of the soil around them. To prevent them filling with water, they should have small holes made in the bottom. The pitfalls should also contain cover objects, such as broken flower pots, under which amphibians can hide.

It is vital that pitfall traps are checked frequently: at least once a day, preferably twice. Pitfalls trap predators of amphibians, such as shrews, snakes, beetles and spiders. Placing sticks in pitfall traps reduces mortality among small mammals and does not reduce their effectiveness for catching amphibians (Perkins & Hunter 2002). In dry weather, amphibians can die from desiccation in pitfalls; in wet weather they may drown in pitfalls that fill with water. Particularly in temperate habitats, most amphibians move towards breeding sites immediately after dark, and an early evening surge of animals is usually followed by a steady trickle throughout the night. The best times to check pitfalls are just after the early evening surge, about 2–3 h after dusk, and as early as possible after daybreak.

It is desirable to back up a drift-fence study with trapping or collecting of animals in the breeding site itself. If all animals caught at the fence are identified in some way, then the ratio of identified to unidentified animals caught inside the fence provides a measure of the effectiveness with which the fence samples the population. However, at some amphibian breeding sites, some individuals may live very close to the water's edge during the terrestrial phase and so may never cross the fence at all (Weddeling *et al.* 2004).

Advantages and disadvantages

Drift fences can catch a very high proportion of a population, so, combined with some form of mark–recapture analysis, they can provide a very accurate estimate of population size. In addition, they yield data on a variety of aspects of the behaviour and life history of amphibians, such as the direction(s) in which they migrate to and from breeding sites, sex, age and body-size differences in migration times and in time spent at the breeding site, and, if they are kept in place for several years, individual variation in the frequency of breeding. A well-built drift fence will catch all individuals in a population, except of those species that are able to jump or climb over it. The ability of amphibians to find ways across drift fences should not be underestimated and it cannot be assumed that even the best-constructed fence catches every animal.

Drift fences are quite expensive to build; they are labour-intensive, both in building them and in maintaining and checking them. They are easily destroyed by large animals and small children and so are not suitable on farmland or near urban areas unless they are protected by conventional fencing. If not checked frequently, they can cause high mortality among amphibians, through predation, desiccation or drowning.

More detailed accounts of the design and construction of drift fences, and of the analysis of the data that they yield, are provided by Gibbons & Semlitsch (1981) and Heyer *et al.* (1994).

Biases

Drift fences provide less reliable population estimates of species that are good climbers or jumpers. Very small individuals, e.g. metamorphs, often hide close to a fence and are thus likely to be overlooked. It is possible that, in some species, animals encountering a drift fence will turn away and go elsewhere.

Scan searching

This method is used for amphibians that aggregate at breeding sites or are relatively abundant in a given area.

Method

Fieldworkers walk around or through a breeding site or some other prescribed area, systematically searching for animals. The number of animals detected per unit of person-hours provides an approximate estimate of numbers. This is not a suitable method for estimating population size accurately unless it is combined with a mark–recapture study. For many species, scan searches are best conducted at dusk or at night, using torches, when the animals are most active (Cooke & Arnold 2003). A more accurate variant of scan searching is double-observer estimation (DOE) in which each observation by a primary observer is confirmed and recorded by a second observer (Jung *et al.* 2002b). For those anurans in which males call, numbers of males may be estimated by counting the number of calling males seen or heard. It should be remembered, however, that, in many species, not all males call within a given time period, and that many frogs stop calling in the presence of humans. In species in which females lay eggs in discrete batches or clumps, female numbers can be estimated by counting egg masses, though it must first be established whether females of the species under study lay one or more clutches in a breeding episode. Long-term monitoring of natterjack-toad (*Bufo calamita*) populations in the UK has been done by counting spawn strings (Buckley & Beebee 2004).

Advantages and disadvantages

Scan searching is simple and cheap but is time-consuming. It is a good method for counting larvae, which are easily damaged by netting or trapping.

Biases

Scan searching is subject to two major sources of bias. First, it is a common observation that many amphibians are less wary of humans when they are aggregated in large numbers, so this method can provide an underestimate of numbers when small or dispersed populations are being sampled. Second, males are typically more active than females at breeding sites, may thus be more readily detected, and may thus appear to be more abundant. Furthermore, the two sexes may exhibit very different temporal patterns of arrival at a breeding site, so repeated sampling over many days or nights is essential.

Netting

Netting is used to catch amphibians that aggregate to breed in water bodies.

Method

A water body is sampled by means of hand nets or seine nets with the aim of catching as many animals as possible. This is not a suitable method for estimating population size accurately unless combined with a mark–recapture study. Every effort should be made to net all parts of the water body equally thoroughly, but care must be taken not to damage aquatic vegetation, which may provide cover and spawn sites for amphibians.

Advantages and disadvantages

Netting is simple and cheap but, for many species, is not very effective because they elude nets quite easily. It can also be very destructive both of the animals and of their habitat. It should not be carried out on a large scale when externally gilled larvae are present since they are easily killed by netting.

Biases

Differences in the behaviour and distribution of males and females while at a breeding site may lead to very different numbers of the two sexes being caught, leading to a spurious estimate of the sex ratio.

Trapping

Trapping is used to catch amphibians that aggregate at breeding sites or are relatively abundant in a given area.

Method

Traps are set out in a water body or over an area of ground and are checked frequently. Since amphibians eat only live food, it is generally impractical to bait traps; trapping relies instead on the fact that many amphibians move extensively around their habitat when the weather is suitable. Trapping is not a suitable method for estimating population size accurately unless combined with a mark–recapture study. In water bodies, a variety of traps suitable for fishes is widely used. Since many species need to go to the water surface to breathe, it is essential that traps are either only partially submerged or that they provide some kind of access to the surface, especially in warm weather. Small amphibians and larvae can be sampled by simple traps made out of plastic drink bottles (Griffiths 1985). Buech and Egeland (2002a) compare the efficacies of a variety of funnel-trap designs. Terrestrial amphibians can be sampled by deploying pitfall traps similar to those used at drift fences (see above) and by setting out a number of artificial cover objects such as pieces of wood or paving slabs. Wooden boards are good for sampling terrestrial salamanders, but frequent sampling leads to reduced density and they should not be checked more often than once a week (Marsh & Goicochea 2003).

The positioning of traps can be crucial and they are much more successful if they are positioned so as to intercept amphibians during regular daily movements. For example, many aquatic salamanders and newts move from deep water during the day to shallower pond margins at night. Funnel traps are thus most effective if they are set so that their entrances point towards deep water at dusk, towards the bank at dawn.

Advantages and disadvantages

Trapping is simple and cheap but can be labour-intensive. Traps typically provide a low return, in terms of the number of animals caught, unless a population is particularly dense. Mortality can occur in traps, because of lack of oxygen, over-heating or predation, so it is essential that traps are checked regularly and that animals do not remain long in traps on warm days.

Biases

Differences in the behaviour and distribution of males and females while at a breeding site may lead to very different numbers of the two sexes being caught, leading to a spurious estimate of the sex ratio.

Transect and patch sampling

These methods are used for terrestrial amphibians.

Method

Transect sampling is most effective if combined with other methods described above, such as drift fencing or trapping. A line is marked out over an area of ground and fieldworkers walk along it at regular intervals, systematically searching for animals. Cover objects within a specified distance of the line (e.g. 2 m) are turned over and checked, as are any traps set along the transect. The number of animals detected per unit of person-hours provides an approximate estimate of numbers. This is not a suitable method for estimating population size accurately unless combined with a mark–recapture study. A particular strength of transect sampling is that it can be used to relate amphibian abundance to habitat variables, such as altitude and vegetation.

Patch sampling is essentially the same method, but involves systematically traversing and searching one or more marked-out patches rather than a linear transect. The patches may take the form of quadrats, a series of small squares laid out at randomly selected sites within a habitat. When quadrats are placed randomly and in sufficient number, quadrat samples enable statistical inferences about the abundance of amphibians and their distribution in relation to habitat variables to be made. In aquatic habitats, quadrats take the form of netted boxes, a technique called enclosure sampling (Mullins *et al.* 2004).

Advantages and disadvantages

These are cheap and simple methods, but are labour-intensive (Rödel & Ernst 2004). They may yield a low return in terms of numbers of animals found, except where a species is particularly abundant. Quadrat sampling is especially suitable for species living on the forest floor (Heyer *et al.* 1994). In the context of using transects to determine the number of species in tropical habitats, at least 20 walks along a transect are required in order to detect all species present (Veith *et al.* 2004).

Biases

Transects or sample patches might not cross habitats of species with very specific habitat requirements, but researchers should avoid areas that 'look good' for amphibians. Individuals of very active species may escape before they can be sampled, and their numbers may therefore be underestimated.

Removal studies

This method is used for amphibians that aggregate at breeding sites or are relatively abundant in a given area.

Method

A water body or a marked-out area of ground is systematically searched, netted or trapped and all animals found are removed and held elsewhere, to be returned subsequently. This process is repeated at predetermined intervals. The ratio of numbers caught in successive captures is used to estimate the population size (Heyer *et al.* 1994).

Advantages and disadvantages

This is a cheap and simple method, but is labour-intensive. If a pond or area of ground is to be sampled thoroughly, it can be very destructive of the animals' habitat, and ponds should not be subjected to such treatment when eggs or larvae are present. Animals removed from a site must be kept under suitable conditions until they are returned. Even if no mortality occurs, the successful reproduction of animals temporarily removed from the field can be compromised; some amphibians undergo a marked stress reaction to capture and quickly go out of breeding condition. An advantage of removal studies is that, while animals are held in captivity, much useful data can be collected, such as age, sex and breeding condition.

Biases

This method assumes that animals are equally catchable at all times; because the behaviour of amphibians varies considerably with changes in the weather, this assumption is rarely true.

Call surveys

In recent years there has been a proliferation in the use of call surveys (acoustic sampling) to monitor the abundance and diversity of amphibians. For example, call surveys are a major component of the Amphibian Research and Monitoring Initiative (ARMI) in North America (Nelson & Graves 2004). Typically, a group of volunteers is trained to recognise the species-specific calls of frogs, before going out at night along roads or transects (Rödel & Ernst 2004), or along a river in a boat (Jung *et al.* 2002a). Call surveys are inexpensive but are applicable only to those frog species that call. This method is very susceptible to the detection-probability problem. Some frogs call continuously and so are easily detected; others call sporadically and hence may be missed. Calling behaviour is strongly influenced by environmental factors such as temperature and humidity, so frogs are more easily detected on some nights than they are on others. A major advantage of call surveys is that they require volunteers and are thus a very good way to increase local awareness of amphibians and of declines in amphibian populations.

When used to determine the number of species in an area, call surveys are clearly better for some species than they are for others (de Solla *et al.* 2005). An evaluation of call surveys in Hungary concluded that they detect most species, but not all (Anthony 2002). The authors of a study in France found that call surveys were good for detecting populations of the European treefrog (*Hyla arborea*), but poor for the natterjack toad (*Bufo calamita*), and failed to detect the rarest species (Pellet & Schmidt 2005). Similar results were found in Rhode Island, USA (Crouch & Paton 2002). Whether or not sporadically calling or rare frogs will be detected by a call survey depends largely on the duration of the survey. Pierce and Gutzwiller (2004) found that 15-min surveys detected more species than 5-min surveys, but that extending the duration to 30 min yielded no improvement.

Using call surveys to determine population size is problematic (Shirose *et al.* 1997). Typically, only male frogs call, so call surveys cannot census females. In some species, calling may occur on nights when no mating occurs and so can be a poor measure of reproductive activity. In many species, not all males at a chorus call but some may be silent 'satellites'; moreover, the frequency of satellites is often greater in larger choruses. To determine whether calling activity is a reliable indicator of population size, it is essential to validate it by simultaneously taking a census using another method. Nelson and Graves (2004) used mark–recapture to validate call-survey data for the green frog (*Rana clamitans*) and found call data to be reliable. In contrast, counting egg masses revealed call data to be a poor indicator of population size for the wood frog (*R. sylvatica*) (Stevens & Paszkowski 2004) and the natterjack toad (*Bufo calamita*) (Buckley & Beebee 2004). Buech and Egeland (2002b) recommend combining call surveys with netting and trapping to determine population sizes of frogs in small forest ponds. Statistical methods for estimating population sizes from call-survey data that allow for variation in detection probability are provided by Royle (2004).

Using multiple methods

All the methods described above have limitations in terms of efficiency for detecting amphibians at the species or individual level. These limitations are made very apparent when more than one method is used simultaneously in the same study. Table 7.3 summarises the results of several such studies. The overall conclusion is that using multiple methods yields the best results; this has been called the 'toolbox approach' (Olson *et al.* 1997).

Table 7.3. *Studies in which more than one method was used simultaneously, allowing the effectivenesses of methods to be compared*

Methods compared	Type of amphibian	Conclusion	Reference
Mark–recapture and removal sampling	Terrestrial salamanders	Removal sampling is good for short-term studies but does not sample animals living underground	Bailey *et al.* (2004a)
Call surveys and egg-mass counts	Breeding wood frogs (*Rana sylvatica*)	Call surveys predict egg production on some nights only	Stevens & Paszkowski (2004)
Dip-netting and funnel traps	Stream salamanders	Funnel traps are better at catching adults and catch more species; dip-netting is better for catching larvae	Willson & Dorcas (2003)
Transects, visual encounters, call surveys, trapping	Tropical forest frogs	Transects are best for long-term studies; good for sampling leaf-litter frogs	Rödel & Ernst (2004)
Dip-netting and enclosures	Aquatic amphibians	Timed dip-netting is more likely to detect rare species; enclosures are good for determining densities of common species	Wilson & Maret (2002)
Bottle traps and visual surveys	Aquatic larvae	Nocturnal visual surveys are better at detecting presence; they are also less intrusive	Jenkins *et al.* (2002)
Call surveys, netting and funnel traps, singly and together	Frogs and salamanders at forest ponds	All methods provide presence/absence data, but good estimates of population size require use of multiple methods	Buech & Egeland (2002b)
Drift fences and terrestrial funnel traps	Pond-breeding frogs and salamanders migrating to water	Funnel traps catch more animals, and more species, than drift fences do	Jenkins *et al.* (2003)
Call surveys and mark–recapture	Breeding green frogs (*Rana clamitans*)	Index of calling activity provides a good indicator of abundance	Nelson & Graves (2004)
Scan-searching, netting, double-observer estimation, mark–recapture	Frog tadpoles	Mark–recapture best; detection probabilities with other methods vary with turbidity of pool	Jung *et al.* (2002b)

Table 7.3. (*cont.*)

Methods compared	Type of amphibian	Conclusion	Reference
Visual encounter and quadrat sampling	Rainforest amphibians	Visual encounter detects more animals and more species, but quadrats are better for certain species and micro-habitats	Doan (2003)
Mark–recapture, transects and visual encounter	Terrestrial frogs (*Eleutherodactylus*)	Mark–recapture is accurate; other methods are imprecise	Funk *et al.* (2003)
Mark–recapture and removal sampling	Terrestrial salamanders	Mark–recapture is better; removal sampling detects only animals near the ground surface	Bailey *et al.* (2004a, 2004b)
Mark–recapture, drift fences and funnel traps	Newts (*Triturus*) at breeding ponds	Mark–recapture provides a more accurate estimate of population size	Weddeling *et al.* (2004)

Recording other data

Most census methods involved handling and keeping animals, if only briefly, and this provides a good opportunity to collect data that may be very valuable in studies of the long-term viability of amphibian populations in the face of decline. These include assessing female reproductive status (Reyer & Bättig 2004); measuring fluctuating asymmetry (McCoy & Harris 2003); recording deformities of the limbs (Ouellet 2000); and screening for the infectious disease chytridiomycosis (Briggs & Burgin 2003). Infectious diseases are a major cause of declines in amphibian populations throughout the world and, to prevent transmission of diseases between amphibian populations, it is important that anyone carrying out a census observes the DAPTF Fieldwork Code of Practice (www.open.ac.uk/daptf).

References

Anthony, B. P. (2002). Results of the first batrachian survey in Europe using road call counts. *Alytes* **20**, 55–66.

Arntzen, J. W., Goudie, I. B. J., Halley, J. & Jehle, R. (2004). Cost comparison of marking techniques in long-term population studies: PIT-tags versus pattern maps. *Amphibia–Reptilia* **25**, 305–315.

Bailey, L. L. (2004). Evaluating elastomer marking and photo identification methods for terrestrial salamanders: marking effects and observer bias. *Herpetological Review* **35**, 38–41.

Bailey, L. L., Simons, T. R. & Pollock, K. H. (2004a). Comparing population size estimators for plethodontid salamanders. *Journal of Herpetology* **38**, 370–380.

(2004b) Estimating site occupancy and species detection probability parameters for terrestrial salamanders. *Ecological Applications* **14**, 692–702.

Biek, R., Funk, C. W., Maxell, B. A. & Mills, L. S. (2002). What is missing in amphibian decline research: insights from ecological sensitivity analysis. *Conservation Biology* **16**, 728–734.

Briggs, C. & Burgin, S. (2003). A rapid technique to detect chytrid infection in adult frogs. *Herpetological Review* **34**, 124–126.

Buckley, J. & Beebee, T. J. C. (2004). Monitoring the conservation status of an endangered amphibian: the natterjack toad *Bufo calamita* in Britain. *Animal Conservation* **7**, 221–228.

Buech, R. R. & Egeland, L. M. (2002a). Efficacy of three funnel traps for capturing amphibian larvae in seasonal forest ponds. *Herpetological Review* **33**, 182–185.

(2002b). A comparison of the efficacy of survey methods for amphibians breeding in small forest ponds. *Herpetological Review* **33**, 275–280.

Camper, J. D. & Dixon, J. R. (1988). Evaluation of a microchip marking system for amphibians and reptiles. Research Publication 7100-159. Austin, Texas, Texas Parks and Wildlife Department.

Clarke, R. D. (1972). The effect of toe-clipping on survival in Fowler's toad (*Bufo woodhousei fowleri*). *Copeia*, 182–185.

Cooke, A. S. & Arnold, H. R. (2003). Night counting, netting and population dynamics of crested newts (*Triturus cristatus*). *Herpetological Bulletin* **84**, 5–14.

Crouch, W. B. & Paton, P. W. C. (2002). Assessing the use of call surveys to monitor breeding anurans in Rhode Island. *Journal of Herpetology* **36**, 185–192.

Daugherty, C. H. (1976). Freeze-branding as a technique for marking anurans. *Copeia*, 836–838.

Davies, N. B. & Halliday, T. R. (1979). Competitive mate searching in the common toad *Bufo bufo*. *Animal Behaviour* **27**, 1263–1267.

de Solla, S. R., Shirose, L. J., Fernie, K. J. *et al.* (2005). Effect of sampling effort and species detectability on volunteer based anuran monitoring programs. *Biological Conservation* **121**, 585–594.

Doan, T. M. (2003). Which methods are most effective for surveying rain forest herpetofauna? *Journal of Herpetology* **37**, 72–81.

Dodd, C. K. & Dorazio, R. M. (2004). Using counts to simultaneously estimate abundance and detection probabilities in a salamander community. *Herpetologica* **60**, 468–478.

Elmberg, J. (1989). Knee-tagging – a new marking technique for anurans. *Amphibia–Reptilia* **10**, 101–104.

Emlen, S. T. (1968). A technique for marking anuran amphibians for behavioral studies. *Herpetologica* **24**, 172–173.

Funk, W. C., Almeida-Reinoso, D., Nogales-Sornosa, F. & Bustamente, M. R. (2003). Monitoring population trends of *Eleutherdactylus* frogs. *Journal of Herpetology* **37**, 245–256.

Gibbons, J. W. & Semlitsch, R. D. (1981). Terrestrial drift fences and pitfall traps: an effective technique for quantitative sampling of animal populations. *Brimleyana* **7**, 1–16.

Gittins, S. P., Parker, A. G. & Slater, F. M. (1980). Population characteristics of the common toad (*Bufo bufo*) visiting a breeding site in mid-Wales. *Journal of Animal Ecology* **49**, 161–173.

Golay, N. & Durrer, H. (1994). Inflammation due to toe-clipping in natterjack toads (*Bufo calamita*). *Amphibia–Reptilia* **15**, 81–96.

Griffiths, R. A. (1985). A simple funnel trap for studying newt populations and an evaluation of trap behaviour in smooth and palmate newts, *Triturus vulgaris* and *T. helveticus*. *Herpetological Journal* **1**, 5–10.

Hagstrom, T. (1973). Identification of newt specimens (Urodela *Triturus*) by recording the belly pattern and a description of photographic equipment for such registration. *British Journal of Herpetology* **4**, 321–326.

Hairston, N. G. & Wiley, R. H. (1993). No decline in salamander (Amphibia: Caudata) populations: a twenty-year study in the southern Appalachians. *Brimleyana* **18**, 59–64.

Harvey, E. (2003). Evaluation of fluorescent marking techniques using cannibalistic salamander larvae. *Herpetology Review* **34**, 119–121.

Heyer, W. R., Donnelly, M. A., McDiarmid, R. W., Hayek, L.-A. C. & Foster, M. S. (eds.) (1994). *Measuring and Monitoring Biological Diversity. Standard Methods for Amphibians.* Washington, Smithsonian Institution Press.

Jenkins, C. L., McGarigal, K. & Gamble, L. R. (2002). A comparison of aquatic surveying techniques used to sample *Ambystoma opacum* larvae. *Herpetology Review* **33**, 33–35.

(2003). Comparative effectiveness of two trapping techniques for surveying the abundance and diversity of reptiles and amphibians along drift fence arrays. *Herpetological Review* **34**, 39–42.

Johnson, B. R. & Wallace, J. B. (2002). *In situ* measurement of larval salamander growth using individuals marked with acrylic polymers. *Herpetological Review* **33**, 29–32.

Joly, P. & Miaud, C. (1989). Tattooing as an individual marking technique in urodeles. *Alytes* **8**, 11–16.

Jung, R. E., Bonine, K. E., Rosenshield, M. L. *et al.* (2002a). Evaluation of canoe surveys for anurans along the Rio Grande in Big Bend National Park, Texas. *Journal of Herpetology* **36**, 390–397.

Jung, R. E., Dayton, G. H., Williamson, S. J., Sauer, J. R. & Droege, S. (2002b). An evaluation of population index and estimation techniques for tadpoles in desert pools. *Journal of Herpetology* **36**, 465–472.

Kurashina, N., Utsunomiya, T., Utsunomiya, Y., Okada, S. & Okochi, I. (2003). Estimating the population size of an endangered population of *Rana porosa brevipoda* Ito (Amphibia: Ranidae) from photographic identification. *Herpetological Review* **34**, 348–349.

MacKenzie, D. I., Nichols, J. D., Lachman, G. B. *et al.* (2002). Estimating site occupancy rates when detection probabilities are less than one. *Ecology* **83**, 2248–2255.

Marsh, D. M. & Goicochea, M. A. (2003). Monitoring terrestrial salamanders: biases caused by intense sampling and choice of cover objects. *Journal of Herpetology* **37**, 460–466.

Martof, B. S. (1953). Territoriality in the green frog, *Rana clamitans. Ecology* **34**, 165–174.

May, R. M. (2004). Ethics and amphibians. *Nature* **431**, 403.

McCarthy, M. A. & Parris, K. M. (2004). Clarifying the effects of toe clipping on frogs with Bayesian statistics. *Journal of Applied Ecology* **41**, 780–786.

McCoy, K. A. & Harris, R. N. (2003). Integrating developmental stability analysis and current amphibian monitoring techniques: an experimental evaluation with the salamander *Ambystoma maculatum. Herpetologica* **59**, 22–36.

Mullins, M. L., Pierce, B. A. & Gutzwiller, K. J. (2004). Assessment of quantitative enclosure sampling of larval amphibians. *Journal of Herpetology* **38**, 166–172.

Nace, G. W. & Manders, E. K. (1982). Marking individual amphibians. *Journal of Herpetology* **16**, 309–311.

Nelson, G. L. & Graves, B. M. (2004). Anuran population monitoring: comparison of the North American monitoring program's calling index with mark–recapture estimates for *Rana clamitans. Journal of Herpetology* **38**, 355–359.

Nishikawa, K. C. & Service, P. M. (1988). A fluorescent marking technique for individual recognition of terrestrial salamanders. *Journal of Herpetology* **22**, 351–353.

Olson, D. H., Leonard, W. P. & Bury, R. B. (eds.) (1997). *Sampling Amphibians in Lentic Habitats.* Olympia, Washington, Society for Northwestern Vertebrate Biology.

Ouellet, M. (2000). Amphibian abnormalities: current state of knowledge. In *Ecotoxicology of Amphibians and Reptiles,* ed. D. W. Sparling, G. Linder & C. A. Bishop. Pensacola, Florida, Society of Environmental Toxicology & Chemistry, pp. 617–661.

Pechmann, J. H. K., Scott, D. E., Semlitsch, R. D. *et al.* (1991). Declining amphibian populations: the problem of separating human impacts from natural fluctuations. *Science* **253**, 892–895.

Pellet, J. & Schmidt, B. R. (2005). Monitoring distributions using call surveys: estimating site occupancy, detection probabilities and inferring absence. *Biological Conservation* **123**, 27–35.

Perkins, D. W. & Hunter, M. L. (2002). Effects of placing sticks in pitfall traps on amphibians and small mammal capture rates. *Herpetological Review* **33**, 282–284.

Petranka, J. W., Smith, C. K. & Scott, A. F. (2004). Identifying the minimal demographic unit for monitoring pond-breeding amphibians. *Ecological Applications* **14**, 1065–1078.

Pierce, B. A. & Gutzwiller, K. J. (2004). Auditory sampling of frogs: detection efficiency in relation to survey duration. *Journal of Herpetology* **38**, 495–500.

Rafinski, J. N. (1977). Autotransplantation as a method of permanent marking for urodele amphibians (Amphibia, Urodela). *Journal of Herpetology* **11**, 241–242.

Reyer, H.-U. & Bättig, I. (2004). Identification of reproductive status in female frogs – a quantitative comparison of nine methods. *Herpetologica* **60**, 349–357.

Rödel, M.-O. & Ernst, R. (2004). Measuring and monitoring amphibian diversity in tropical forests. I. An evaluation of methods with recommendations for standardization. *Ecotropica* **10**, 1–14.

Royle, J. A. (2004). Modeling abundance index data from anuran calling surveys. *Conservation Biology* **18**, 1378–1385.

Schmidt, B. R. (2003). Count data, detection probabilities, and the demography, dynamics, distribution, and decline of amphibians. *Comptes Rendus Biologies* **326**, S119–S124.

 (2004). Declining amphibian populations: the pitfalls of count data in the study of diversity, distribution, dynamics, and demography. *Journal of Herpetology* **14**, 167–174.

Schmidt, B. R., Schaub, M. & Anholt, B. R. (2002). Why you should use capture–recapture methods when estimating survival and breeding probabilities: on bias, temporary emigration, overdispersion, and common toads. *Amphibia–Reptilia* **23**, 375–388.

Shirose, L. J., Bishop, C. A., Green, D. M. *et al.* (1997). Validation tests of an amphibian call count survey technique in Ontario, Canada. *Herpetologica* **53**, 312–320.

Skelly, D. K., Yurewicz, K. L., Werner, E. E. & Relyea, R. A. (2003). Estimating decline and distributional change in amphibians. *Conservation Biology* **17**, 744–751.

Stevens, C. E. & Paszkowski, C. A. (2004). Using chorus-size ranks from call surveys to estimate reproductive activity of the wood frog (*Rana sylvatica*). *Journal of Herpetology* **38**, 404–410.

Stuart, S. N., Chanson, J. S., Cox, N. A. *et al.* (2004). Status and trends of amphibian declines and extinctions worldwide. *Science* **306**, 1783–1786.

Veith, M., Lötters, S., Andreone, F. & Rödel, M.-O. (2004). Measuring and monitoring amphibian diversity in tropical forests. II. Estimating species richness from standardized transect censussing. *Ecotropica* **10**, 85–99.

Weddeling, K., Hachtel, M., Sander, U. & Tarkhnishvili, D. (2004). Bias in estimation of newt population size: a field study at five ponds using drift fences, pitfalls and funnel traps. *Herpetological Journal* **14**, 1–7.

Willson, J. D. & Dorcas, M. E. (2003). Quantitative sampling of stream salamanders: comparison of dip-netting and funnel trapping techniques. *Herpetological Review* **34**, 128–130.

Wilson, J. J. & Maret, T. J. (2002). A comparison of two methods for estimating the abundance of amphibians in aquatic habitats. *Herpetological Review* **33**, 108–110.

Wisniewski, P. J., Paull, L. M., Merry, D. G. & Slater, F. M. (1980). Studies on the breeding migration and intramigratory movements of the common toad (*Bufo bufo*) using Panjet dye-marking techniques. *British Journal of Herpetology* **6**, 71–74.

8 Reptiles

Simon Blomberg

Centre for Mental Health Research and School of Botany and Zoology,
Australian National University, Canberra, ACT 0200, Australia

Richard Shine

Zoology A08, School of Biological Sciences, University of Sydney, Sydney,
NSW 2006, Australia

Introduction

Most common survey methods employed to estimate the abundance of reptiles involve capturing individuals (Table 8.1). This is for two reasons: (a) reptiles tend to be mobile and/or shy and cryptic, so not all members of a population will be visible (and therefore amenable to counting by sight) at any one time; and (b) much more information can be obtained from an animal that has been captured than can be obtained from an animal that has simply been seen. For example, the animal may be weighed and measured, have its sex and reproductive condition determined, and have its parasite load assessed. An identifying mark may also be placed on the animal so that it can be re-identified, should it be recaptured at a later time. As well as providing advice on capturing reptiles, we also provide information on common techniques for marking individuals.

Reptiles are ectotherms. That is, they obtain their body heat from the external environment. This has major implications for any survey technique, in that weather conditions may greatly affect the activity and therefore the catchability of reptiles. The effect of weather can vary seasonally, as well as on a daily basis. This should be kept in mind when designing a survey programme.

Because it is unlikely that the whole population will be counted in any one census period (some individuals will be missed), statistical mark–recapture methods should generally be used to estimate population sizes/densities or survival probabilities for reptile populations (see Chapter 3).

Hand capturing

Hand capturing is used for small terrestrial snakes, lizards and tortoises.

Method

The simplest method used by field herpetologists to capture lizards (and other small terrestrial reptiles) is to search intensively in micro-habitats which they are known to frequent, and catch

Ecological Census Techniques: A Handbook, ed. William J. Sutherland.
Published by Cambridge University Press. © Cambridge University Press 1996, 2006.

Table 8.1. *A summary of methods suitable for various groups*

Method	Snakes	Lizards	Crocodilians	Turtles and tortoises	Page
Hand capturing	*	*	*	*	297
Noosing		+			301
Trapping	?	+	?	+	302
Marking individuals	*	*	*	*	305

* Method usually applicable, + method often applicable, ? method sometimes applicable. The page number for each method is given.

them by hand. Small lizards and snakes are found most easily by looking in potential shelter sites, for example by turning over rocks and logs, or by stripping bark from trees (for arboreal species). Sheets of scrap metal and derelict car bodies are often fruitful places to look (because the sun-warmed metal is very attractive to reptiles), though they could not be described as 'natural' habitats! Some herpetologists have standardised their 'catching effort' by placing their own consistently sized, regularly spaced sheets of tin in appropriate sunny positions, and looking under them in accord with some consistent schedule. A lightweight alternative to sheet metal is to use black plastic sheets (Kjoss & Litvaitis 2001). Sheets 1.5 m × 3 m × 0.1 mm can be pegged to the ground, and then checked underneath at regular time intervals. This method resulted in the capture of more smaller-bodied individuals, compared with those caught using funnel-box traps at the same site.

When 'turning cover', a pair of gardening gloves can be worn to prevent cuts to the hands, as well as bites and stings from various arthropods. If there are venomous snakes in the survey area, care should obviously be taken to avoid getting bitten. Logs and rocks should be turned towards the fieldworker, so that the log or rock is between the fieldworker and the snake.

A small hand-held torch powered by one or two penlight batteries is invaluable for looking into cracks or holes in search of reptiles. Alternatively, a mirror with a small hole in the centre to look through can be used. This has the advantage that shadows cast by the observer will not obscure the view.

When turning rocks or logs, a hand-held rake is useful for searching for small lizards and snakes that may have crawled under the leaf litter; however, one can improvise by using a small stick. Geckos often cling upside-down to rocks, so both the undersurface of rocks and the leaf litter should be examined where geckos occur. All logs and rocks should be replaced in their original position, so that disturbance to the habitat is minimised.

Some of these techniques may be particularly destructive to the habitat, both for the reptiles and for other organisms, so care should be taken to avoid causing undue disturbance. Care should also be taken when designing the survey to ensure that the survey method will not affect the habitat in such a way that subsequent survey estimates may be biased. Some species may increase in abundance in response to habitat disturbance, and some may decrease in abundance. This is especially important when censusing rare or endangered reptiles. One would certainly not want the survey method to cause the extinction of the study species, or any other species! Be sure to obtain any necessary permits from the appropriate wildlife and land-management authorities before you begin your survey.

Many lizards and snakes can also be counted or captured during their activity periods, rather than when they are hiding under cover. For many diurnal species, mid-morning is a good time to search, while the reptile is basking to elevate its body temperature. Walk slowly with the sun at your back, pausing frequently to scan suitable micro-habitats for your quarry. Nocturnal species can often be found by torchlight, in the same way as is often used for amphibians (see Chapter 7). Some species, such as geckos, are best detected by their dull eye shine (the reflection of the light source in the animal's eyes). This is easiest with a dull white light; the flashlight should be held close to the observer's eyes so that you are looking along the beam of light. Some geckos (e.g. *Diplodactylus maini*) fluoresce under ultraviolet (UV) light, and can be located using a small hand-held UV lamp (C. Dickman, personal communication).

Having located a lizard or a non-venomous snake, the easiest way to catch it is simply to pounce on it with an open, cupped hand, taking care not to crush it. Care should also be taken with species that practice tail autotomy (drop their tails). These species should not be held by the tail. Gloves can be useful when catching lizards or non-venomous snakes using this method, since even these animals can often give a nasty bite. Small, leaf-litter lizards can be particularly difficult to catch by hand. A technique that may be useful in this case is to scoop up the leaf-litter (containing the lizard) into a small plastic waste basket. The litter can then be searched and the lizard can easily be captured (Brattstrom 1996).

Lizards will often run to cover, such as a bush, when approached by an investigator. One technique that may be used to catch lizards in such situations is to surround the bush with small plastic pipes (closed at one end). Then the lizard can be flushed from the bush and (with luck), it will seek cover by running into one of the pipes, where it can be retrieved (Strong *et al.* 1993).

If a venomous snake has been located, it can be caught by pinning it behind the head, using a Y-shaped stick with some padding in the fork. The snake can then be picked up, with the neck held firmly (but not so tight as to choke the snake) between thumb and forefinger. Some larger species may also be picked up by the tail, and then pinned down by the neck. 'Snake tongs', large forceps manipulated by a trigger-grip, may also be useful when catching snakes. Sticks without padding should not be used to pin snakes, since they may hurt the snake. Fieldworkers wanting to catch venomous snakes should practise extensively on handling non-venomous snakes first, and should have a good knowledge of the necessary first-aid procedures. First-aid procedures differ for the various families of venomous snakes, so fieldworkers should be aware of the appropriate procedures for the bites of any snakes they are likely to encounter. No fieldworker should conduct studies of reptiles in remote areas on their own if dangerous snakes are known to be present.

Catching arboreal snakes presents different problems. Here, snake tongs are most useful. However, one technique that has been used successfully with brown tree snakes (*Boiga irregularis*) is to use a long forked stick (Engeman 1998). The stick should be 1–1.5 m in length, with the prongs of the fork of length 10–15 cm. Once the snake has been located, the fork is placed in the middle of the snake and rapidly twirled. The snake's response is to coil around the stick, with the result that the snake is wound onto the stick like 'winding spaghetti on a fork'. The snake can then easily be retrieved from the stick and restrained.

Once one has caught the lizard or snake, data can be taken from it immediately and it can be released, or it can be placed in a bag with a label for 'processing' at a later time. Cloth bags can be used for large specimens, and plastic bags can be used for small ones. If using plastic bags, make sure that the bag is inflated with air and contains some leaf litter for cover. More than one individual can be kept in a single bag; however, aggressive or cannibalistic species should be kept separately. Reptiles should not be kept in bags for more than a few hours, and they should never be left in the sun or inside a parked vehicle, because they will quickly succumb to heat exhaustion.

Terrestrial tortoises can be found and captured in the same way as lizards, but aquatic species (terrapins and turtles) obviously require different techniques. Freshwater species living in clear water can often be caught while snorkelling, although some soft-shelled turtles are too fast moving for even the strongest human swimmers. One alternative hand-capture technique for freshwater turtles is the use of a long-handled dip net from a canoe. It takes lots of practice, but it does work. Marine turtles are usually censused when females come ashore to lay eggs. Hatchling sea turtles may be censused when they leave the nest and migrate across the beach to the ocean. Male and immature sea turtles can be captured by hand by diving off a small powerboat; however, this method takes some practice.

Crocodilians (crocodiles and alligators) are usually hand-caught at night, because their eyes are highly reflective to the torch beam, allowing them to be located easily. Small crocodiles (length < 1.5 m) are simply jumped upon (beware of their larger relatives as you do so!), whereas larger animals are usually caught with small harpoons that lodge in the animal's scales. Some scientists catch alligators with heavy-duty fishing rods, simply casting a large hooked lure behind the alligator and retrieving it rapidly, so that the hooks lodge in the alligator's armour. Even small crocodilians can cause considerable lacerations if they are able to bite, so great caution and prior training are essential.

Advantages and disadvantages

Hand-capturing small lizards and snakes requires little specialised equipment, and many common species can be located and captured relatively quickly and easily. Retreat sites can be located and characterised. Hand-capturing is, however, labour-intensive. Suitable cover must be present if reptiles are to be captured from shelter sites. It is also difficult to standardise capture effort between fieldworkers. This method can also disturb the habitat considerably, if care is not taken to replace cover in its original position.

Biases

For terrestrial species, this method is biased towards species that use logs, rocks, bark, etc. for cover, and is not a good method for catching species that do not use these micro-habitats, for example burrowing species. Animals that live under rocks or logs that are too heavy to lift are also not sampled.

Some species may exhibit sex- or size-specific differences in catchability, which may be caused by sex- or size-specific differences in habitat use or activity patterns. However, it may be possible to alter a survey design to take such differences into account.

Noosing

Noosing is used to catch many lizards.

Method

Many lizards are most visible when they are active. However, they are wary when approached, and evade hand-capture by running away. Also, some lizards, such as varanids, iguanids and agamids, may sleep or bask in places that are difficult to reach, such as in the canopy of tall trees, or in burrows. In these situations, a noose may be used to capture lizards without having to approach the lizard too closely. Nooses consist of a long pole, with a loop of string at the tip, which can be tightened around the neck of a lizard and pulled tight in order to capture the animal. Nooses for small species may be constructed at short notice, using a small length of fishing line, or dental floss, and a stick. Heavy-duty nooses can be constructed from a long, strong fishing rod and some lightweight cord. These are especially useful for large agamids, iguanids and varanids.

A sophisticated rubber-band-powered noosing 'gun' may be constructed from a fishing-rod 'blank', with a noose made of fishing line attached to the tip. Fishing line is threaded through the hollow centre of the fishing rod and attached to a trigger grip. When the trigger is pulled, a rubber band tightens the noose (see Bertram & Cogger (1971) for construction details). Care must be taken to ensure that there is not too much tension on the noose, or the lizard may be injured.

To use a noose, the lizard must be approached slowly, and the noose slipped over its head. The lizard will frequently not be disturbed by this. In fact, many species will attempt to bite the end of the noose! When the noose is around the neck of the lizard, the string or fishing line should be quickly pulled tight. Lizards will frequently struggle violently, so the noose should be loosened and the lizard removed as soon as possible.

A related technique that works well for tropical skinks (for example, *Emoia*), and is the favoured technique for catching small lizards in the Pacific islands, involves attaching a small insect to a 'fishing rod' made from a stick and some fishing line. Once the lizard has a firm grasp on the insect, the lizard is flicked up into the air using the rod, and is caught with the free hand (Strong *et al.* 1993). The lizard can also be lured into a net or a collecting bucket using this technique (Durden *et al.* 1995).

Another technique that may be useful as an alternative to noosing is the use of a 'lizard grabber': a modified mechanic's bolt retriever (Witz 1996). Using this device has the advantage that lizards can be caught from behind; an advantage over the noose, which must be slipped over the lizard's head. This technique may be especially useful for lizards that are too shy or difficult to noose. A further alternative is to use a long stick, the tip of which is coated with mouse glue (Durtsche

1996). The sticky pad is simply touched onto the lizard, which becomes stuck and can then be retrieved. This method works best with small lizards (<10 g). Lizards can be removed from the pad using vegetable oil.

Advantages and disadvantages

This technique is good for catching active lizards and for catching lizards in places that are difficult to reach, such as at the entrances to burrows, or in trees. It is useful for species that are difficult to approach when active. Noosing works best with lizard species with distinct necks. Simple nooses can be easily constructed when needed. It is, however, labour-intensive and time-consuming, and thus faces similar problems to simple hand-capturing.

Biases

Noosing is biased towards active lizards that can easily be located by sight. It is not a useful technique for cryptic or burrowing species.

Trapping

Trapping is used to catch many small reptiles.

Method

For terrestrial reptiles, the most commonly used trap is a pitfall trap, consisting of a bucket sunk into the ground so that the lip is flush with the surface. A small layer of leaf-litter or some other cover should be provided for animals that fall in. This may also have the benefit of attracting arthropods to the trap, which can act as bait for lizards.

Cans, plastic tubes, or even milkshake cups can be used in place of buckets; the diameter and depth of the trap may affect the size and species composition of the reptiles that are trapped, so some experimentation with pitfall-trap design may be necessary. For example, James (1989, 1991) used buckets (350 mm in diameter and 450 mm deep) in a study of seven sympatric skink species. Adult body sizes ranged from 35 mm snout–vent length (SVL) to 100 mm SVL. Traps may be left in place permanently, or removed between census periods. If traps are left in place, lids should be fitted to the traps so that no animals can fall in between census periods. If pitfall traps are used in a forest or woodland, the traps can be filled with sticks so that any animal that enters the trap between censuses can crawl out by climbing up the sticks. In high-rainfall areas, pitfall traps should have drainage holes in the bottom, so that they do not fill with water. The drainage holes should not be so large, however, that animals can crawl through them and live under the trap!

Pitfall traps can be placed next to logs, rocks, or other habitat features that may be frequented by lizards or snakes. Alternatively, pitfall traps may be used in conjunction with drift fences to increase capture success. Drift fences are low fences (approximately 30 cm high) made of polythene, or some other cheap flexible material, and held erect by pegs. The bottom of the fence is dug a few centimetres into the soil to prevent animals burrowing underneath. Drift fences run along a trap line, between the traps, acting to guide reptiles into the traps. Reptiles that are crawling through the survey area come up against the fence and follow it along, looking for a way around the fence. They eventually come to a pit trap and fall into it. It is important to use a material for the drift fence that is difficult for lizards and snakes to climb. Nothing is so depressing as walking along a drift fence and seeing all the lizards that you wished to catch sitting on top of the fence, basking in the sun, and then jumping off and running away as you approach!

Traps should be checked and animals processed at least once per day during a trapping study. Traps may have to be checked more frequently in harsh environments (e.g. deserts) to avoid stress and mortality of trapped animals, or where trapping success is particularly high. Frequent checking is also essential if the local predators (birds, mammals, etc.) realise that you are providing a free meal and clear your traps before you do! The spacing and size of pit traps and the use of drift fences for designing sampling programmes have been discussed by Friend *et al.* (1989) and Morton *et al.* (1988). Check your traps carefully with a stick or a gloved hand before reaching in to pick up that lizard – there may also be a hidden snake, spider, or scorpion!

An alternative to pitfall traps that can be used in rocky areas where holes cannot be dug is the wire-funnel trap (Fitch 1987). Such a trap is made in the same shape as a lobster trap or fish trap, being basically a cylinder with a funnel at one or both ends. A drift fence leads the reptile into the funnel and thence into the trap. A wide range of sizes is possible, depending on the species to be caught. Large species require strong wire-mesh traps, whereas smaller animals can be caught in plastic bottles with the funnel attached to the lid. In place of funnel traps, aluminium-box traps such as those used for mammals may be used (Chapter 10). Meat, such as cat food, or fruit, such as bananas, may be useful as bait. Live bait, such as crickets, can also be used (Durden *et al.* 1995). Doan (1997) successfully used a modified Sherman trap baited with cracked hen eggs in a study of the large lizard *Tupinambis teguixin*. Trap design has been discussed in detail for snake species that pose medical or ecological risks (Rodda *et al.* 1999).

Baited funnel traps (Legler 1960) or drum nets are very effective for catching many species of freshwater turtles. The bait is usually meat or fish (cans of sardines are quick and easy to use), and the trap must be set so that part of it projects above the water. This gives trapped turtles a chance to breathe – otherwise, they will drown. An alternative is to use a trap that provides artificial basking sites for turtles; turtles bask on the basking sites and, when the investigator comes to check the trap, the turtles dive down under the water, where they are caught by a net (Petokas & Alexander 1979; Gianaroli *et al.* 2001).

Terrestrial tortoises are usually hand-captured or pitfall-trapped, but gopher tortoises can also be caught using a simple net trap at the burrow entrance (Bryan *et al.* 1991). When the tortoise emerges from its burrow and onto the net, a spring-loaded arm pulls the net up and around the tortoise in a purse arrangement. The trap takes 10–15 min to set, and results in minimal damage to the burrow or the tortoise.

Large and small crocodilians may be trapped using an assortment of snare techniques. Snare traps may be baited or unbaited, and traps may be placed on the shore of rivers (baited traps), or on trails that are regularly used by crocodilians (unbaited traps). See Mazzotti and Brandt (1988) and Kofron (1989) for construction details.

Arboreal lizards are particularly difficult to catch. One useful method for catching small, trunk-dwelling species is to wrap a length of netting around the trunk of the tree, above the lizard (Paterson 1998). The lizard can then be caught by hand on the tree trunk, or chased into the netting where it can become trapped. This technique works well for *Anolis distichus*, which is loath to run down the tree onto the ground. Larger arboreal lizards, such as varanids, can be captured using a snare attached to the side of a tree (Bennett *et al.* 2001). A barrier of vines and leaves is placed around the trunk of the tree, leaving a space for the lizard to pass, where the snare is to be set. Bennett *et al.* (2001) give construction details. For animals that live in tree hollows, a wire-mesh minnow trap can be placed over the hollow to catch the animal as it emerges (Zani & Vitt 1995).

A useful method that can be applied both for arboreal and for terrestrial species is to employ cardboard or wood boards coated with mouse glue. These act as sticky traps, and can be placed on the ground (perhaps associated with drift fences) to catch terrestrial species, or around the trunks of trees to catch arboreal species (Bauer & Sadlier 1992; Rodda *et al.* 1993). Animals can be removed from the traps using vegetable oil. It is important to check sticky traps regularly, since captured animals may be more exposed to heat stress if the trap is in the sun. Sticky traps may also leave animals exposed to predators. Sticky traps become less effective after a few days, so they may need to be replaced often. They also need to be replaced after rain. However, they are cheap to build and maintain. For shorter-term use, a strip of masking tape placed sticky-side up at a basking site, from whence you have just frightened a small lizard, may well capture the resident when it returns to recommence basking a few minutes later (Downes & Borges 1998).

One point to remember with any trap you set is that you have the responsibility to check the trap regularly. A lost or abandoned trap could go on catching (and killing) animals for many decades.

Advantages and disadvantages

Trapping often allows the capture of a large number of individuals, with comparatively little effort from the investigator. Trapping effort can be standardised in space and time, satisfying the assumptions of many statistical models for estimating survival and population size. Traps and drift fences may have to be constructed and installed at the survey site, which may take considerable time, effort and money. Traps have to be checked regularly, and should be closed or removed when not in use. Lost or abandoned traps may cause the death of animals at the survey site long after the study has concluded.

Biases

Only animals that are actively moving within the survey area will come into contact with traps or drift fences. Therefore, trap success will generally be very low in poor weather, when reptiles are inactive. Highly sedentary species will not be sampled.

Marking individuals

Marking can be used for all reptiles. Having captured a reptile, if a mark–recapture procedure is to be used to estimate the abundance of animals or their probability of survival over some time period, it is necessary to give individuals a unique mark so that they may be identified upon recapture. As with capture methods, the particular technique you adopt will depend on the natural history of the species studied, so some experimentation will probably be necessary, and is usually desirable.

Method

Marks may be applied to reptiles so that they last for short time periods, for longer time periods, or permanently. Marks can be applied using some sort of coding system, so a large number of combinations of marks can be used (Woodbury 1956).

Perhaps the simplest technique is to paint a mark on each animal, using paint (nail polish works well). A colour-coding system can easily be devised.

The most common technique that is used to mark small lizards is to remove toes from one or more feet using sharp fingernail scissors. Feet and toes can be numbered, and can yield a large number of combinations, depending on the coding system. Permanent marks can be applied to lizards or snakes by clipping scales and cauterising them with a small hand-held soldering iron. The scales will either not grow back, or will grow back in a different colour (usually white) if the underlying pigment layer has been destroyed. An alternative method is to mark animals using a hot or cold branding iron. Turtles and tortoises can be marked by cutting notches in the edge of the carapace, or by using metal or plastic tags attached to the flippers (for sea turtles). Crocodiles and alligators can be marked by removing tail scutes. Reptiles may also be tattooed using a small tattoo gun. Removing tissue requires a license in some countries, raises ethical questions and may cause public disquiet.

Passive-integrated-transponder (PIT) tags have recently become available for marking animals, and their use promises to be an effective method for permanently marking reptiles (Camper & Dixon 1988; Keck 1994). Tags can be inserted under the skin or in the abdominal cavity of small specimens. Each tag has a unique code that can be read using a hand-held scanner.

Advantages and disadvantages

Paint marking has the advantage that the animal is not physically harmed in any way. However, it may make animals more visible to predators. Also, paint will wear off after several days, so this technique is useful only for very-short-term studies.

The natural history of the species should be taken into account when deciding on an appropriate marking procedure. Toe-clipping would be a cruel and inappropriate procedure to use for highly arboreal lizards, or those with thick, fleshy toes. However, many terrestrial species seem to be unaffected by toe-clipping.

Apart from paint marking, the methods mentioned above all involve physically altering animals in some way. These methods result in a mark that is long lasting or permanent; however, care must be taken to ensure that animals are not harmed by the marking procedure. Permanent marking techniques, such as tattooing and toe-clipping, should be carried out only by experienced, trained herpetologists or under the supervision of a veterinarian. Fieldworkers should be aware of ethical considerations when censusing reptile populations, since accurate censusing requires interfering with animal populations to a certain extent. Study methods should be submitted to the appropriate animal-ethics committee for approval, when required by law.

Biases

Marks that are ambiguous or are lost can make the analysis of mark–recapture data very difficult. A major assumption of mark–recapture models is that marks are unique, and most models assume that marks are not lost over time. Another common assumption is that marking individuals does not affect their behaviour. If these assumptions are not met, population-size estimates and survival probabilities will be severely biased. Loss of marks can be corrected for by marking individuals twice, and then calculating the rate of loss. A better solution is to develop a better marking technique for your species of interest.

Acknowledgements

We thank M. Olsson and C. Dickman for useful comments on previous drafts of this chapter.

References

Bauer, A. M. & Sadlier, R. A. (1992). The use of mouse glue traps to capture lizards. *Herpetological Review* **23**, 112–113.

Bennett, D., Hampson, K. & Yngente, V. (2001). A noose trap for catching a large arboreal lizard, *Varanus olivaceus*. *Herpetological Review* **32**, 167–168.

Bertram, B. P. & Cogger, H. G. (1971). A noosing gun for live captures of small lizards. *Copeia*, 371–373.

Brattstrom, B. H. (1996). The "skink scooper:" a device for catching leaf litter skinks. *Herpetological Review* **27**, 189.

Bryan, T. W., Blankenship, E. L. & Guyer, C. (1991). A new method of trapping gopher tortoises (*Gopherus polyphemus*). *Herpetological Review* **22**, 19–21.

Camper, J. D. & Dixon, J. R. (1988). Evaluation of a microchip marking system for amphibians and reptiles. Research Publication 7100-159. Austin, Texas, Texas Parks and Wildlife Department, pp. 1–22.

Doan, T. M. (1997). A new trap for the live capture of large lizards. *Herpetological Review* **28**, 79.

Downes, S. J. & Borges, P. (1998). Sticky traps: an effective way to catch small terrestrial lizards. *Herpetological Review* **29**, 94.

Durden, L. A., Dotson, E. M. & Vogel, G. N. (1995). Two efficient techniques for catching skinks. *Herpetological Review* **26**, 137.

Durtsche, R. D. (1996). A capture technique for small, smooth-scaled lizards. *Herpetological Review* **27**, 12–13.

Engeman, R. M. (1998). An easy capture method for brown tree snakes (*Boiga irregularis*). *The Snake* **28**, 101–102.

Fitch, H. S. (1987). Collecting and life-history techniques. In *Snakes: Ecology and Evolutionary Biology*, ed. R. A. Seigel, J. T. Collins & S. S. Novak, New York, Macmillan, pp. 143–164.

Friend, G. R., Smith, G. T., Mitchell, D. S. & Dickman, C. R. (1989). Influence of pitfall and drift fence design on capture rates of small vertebrates in semi-arid habitats of Western Australia. *Australian Wildlife Research* **16**, 1–10.

Gianaroli, M, Lanzi, A. & Fontana, R. (2001). Utilizzo di trappole del tipo "bagno di sole artificiale" per la cattura di testuggini palustri. *Pianura* **13**, 153–155

James, C. D. (1989). Comparative ecology of sympatric scincid lizards (*Ctenotus*) in spinifex grasslands of central Australia. Unpublished Ph.D. Thesis. University of Sydney.

James, C. D. (1991). Population dynamics, demography and life history of sympatric scincid lizards (*Ctenotus*) in central Australia. *Herpetologica* **47**, 194–210.

Keck, M. B. (1994). Test for detrimental effects of PIT tags in neonatal snakes. *Copeia*, 226–268.

Kjoss, V. A. & Litvaitis, J. A. (2001). Comparison of two methods to sample snake communities in early successional habitats. *Wildlife Society Bulletin* **29**, 153–157.

Kofron, C. P. (1989). A simple method for capturing large Nile crocodiles. *African Journal of Ecology* **27**, 183–189.

Legler, J. M. (1960). A simple and inexpensive device for trapping freshwater turtles. *Proceedings of the Utah Academy of Science* **37**, 63–66.

Mazzotti, F. J. & Brandt, L. A. (1988). A method of live-trapping wary crocodiles. *Herpetological Review* **19**, 40–41.

Morton, S. R., Gillam, M. W., Jones, K. R. & Fleming, M. R. (1988). Relative efficiency of different pit-trap systems for sampling reptiles in spinifex grasslands. *Australian Wildlife Research* **15**, 571–577.

Paterson, A. (1998). A new capture technique for arboreal lizards. *Herpetological Review* **29**, 159.

Petokas, P. J. & Alexander, M. M. (1979). A new trap for basking turtles. *Herpetological Review* **10**, 90.

Rodda, G. H., McCoid, M. J. & Fritts, T. H. (1993). Adhesive trapping II. *Herpetological Review* **24**, 99–100.

Rodda, G. H., Sawai, Y., Chizsar, D. & Tanaka, H. (1999). *Problem Snake Management; the Habu and the Brown Treesnake*. Ithaca, New York, Comstock Publishing Associates.

Strong, D., Leatherman, B. & Brattstrom, B. H. (1993). Two new simple methods for catching small fast lizards. *Herpetological Review* **24**, 22–23.

Witz, B. W. (1996). A new device for capturing small and medium-sized lizards by hand: the lizard grabber. *Herpetological Review* **27**, 130–131.

Woodbury, A. M. (1956). Uses of marking animals in ecological studies: marking amphibians and reptiles. *Ecology* **37**, 670–674.

Zani, P. A. & Vitt, L. J. (1995). Techniques for capturing arboreal lizards. *Herpetological Review* **26**, 136–137.

9 Birds

David W. Gibbons and Richard D. Gregory

Royal Society for the Protection of Birds, The Lodge, Sandy, Bedfordshire SG19 2DL, UK

Introduction

Birds are among the easiest of animals to census. They are often brightly coloured, highly vocal at certain times of the year and relatively easy to see. They are also very popular, with the result that high-quality field guides are available in most parts of the world and there are many professionals and amateurs with a high level of identification skills. Because of this popularity, they are undoubtedly the most frequently surveyed of all taxonomic groups. The widespread involvement of volunteers in many schemes makes bird surveys an extremely cost-effective way of monitoring the overall health of the environment, as demonstrated by the inclusion of an indicator based on wild-bird population trends in the UK Government's list of headline indicators of sustainable development (Gregory et al. 2003, 2004c; Anon 2005; http://www.sustainable-development.gov.uk/indicators/headline/h13.htm) and among the European Union's structural and sustainability indicators (Gregory et al. 2005; http://epp.eurostat.cec.eu.int/portal/page?_pageid=1090,30070682,1090_33076576&_dad=portal&_schema=PORTAL).

Before you start counting

Before undertaking a survey you must decide on your objectives and plan accordingly. The temptation at this stage is often to be too ambitious, so careful thought should be given to your key objectives and priorities (see also Chapter 1). You may be interested in an inventory of a site, the population size (i.e. total or absolute numbers) of a species or set of species in a particular area, or a population index (i.e. relative numbers). In many instances, your aim may well be to estimate the total numbers of a particular species in an area. If, however, you are interested only in whether a population is increasing, decreasing or stable, then a population index is perfectly acceptable. Ideally, changes in a population index should be directly proportional to changes in population size; if the population halved then so would the index. Whether or not this assumption is valid, however, will not always be known and is difficult to ascertain. By obtaining repeated measures of population size or index over time, the population can be monitored. In general, monitoring with an index is more cost-effective and less demanding of time, and a reliable index may be preferable to a poor census.

Ecological Census Techniques: A Handbook, ed. William J. Sutherland.
Published by Cambridge University Press. © Cambridge University Press 1996, 2006.

Once you have decided on whether to obtain an index or a measure of total numbers, you need to consider the survey design. While this is covered in detail in Chapter 2, Bibby *et al.* (2000) and Gregory *et al.* (2004c), a very brief introduction is given here. The key issues at this stage revolve around understanding the subtle differences among precision, accuracy and bias, and how such factors influence survey design (Gregory *et al.* 2004c).

First, you must consider where to count by setting the survey boundaries. In many cases, these will be self-evident. If, for example, you wish to estimate the population size of a species in a particular habitat patch, island or country, then the survey boundaries may well be the boundary of the patch, island or country. However, the efficiency of your survey could be dramatically improved by ensuring that, even within that patch, island or country, you are not counting in entirely unsuitable areas. You could use existing information, such as from an atlas of bird distributions, to help you further refine your survey boundaries or, alternatively, you could do this using the species' known or likely habitat preferences.

Second, you must decide whether the survey is to be a complete census (all individuals counted), or a sample survey. For common and widespread species, it is usually preferable to survey representative samples of the population. For rare species, or those with clumped distributions, it may be more desirable and practical to count all individuals. Surveys can be a mix of census (for example, in the core of a species' range) and sample (at the edges). If you do plan to sample, you must decide on the sampling units and the sampling design. Units could, for example, be grid squares or habitat patches, and should be selected at random or stratified random to ensure that they are truly representative.

Third, once you have decided where to count, you must decide on how to count, i.e. the field census method. The rest of this chapter will help you with this decision by providing a range of options suitable for various species and conditions.

Finally, of course, you must report your results, although this chapter will not dwell on this subject.

Choice of census method

Table 9.1 lists the many methods that have been used for surveying birds. Though several methods are listed, there are broadly two types: those for species that are evenly distributed across the landscape; and those for species that are not (i.e. are highly clumped). Listing methods, territory mapping, point counts and transects, for example, are best for species that are evenly distributed (e.g. territorial species), whereas counts of colonies, roosts and flocks are best for species with clumped distributions. Do bear in mind, however, that the dispersion of a species may vary throughout the annual cycle. For example, seabirds often gather at breeding colonies in the spring, but are out at sea for much of the rest of the year. In contrast, many waterfowl are distributed widely across the landscape while breeding, but commonly congregate outside the breeding season and are then much easier to count. Different methods would thus be required for different seasons, though it is often much simpler to count birds when they are clumped than it is when they are dispersed.

Table 9.1. *A summary of methods suitable for various groups*

Method	Water-birds	Seabirds	Wading birds	Raptors	Game-birds	Near-passerines	Passerines	Page
Listing methods	?	?	?	+	+	+	+	311
Timed species counts	?	?	?	+	+	+	+	313
Territory mapping	+	*	+	+	+	+	*	314
Line transects	+	*	+	+	+	+	*	320
Point transects	?		?	?	?	+	*	324
Catch per unit effort	?					?	+	328
Capture–mark–recapture	?	?	?	?	?	?	?	330
Counting colonial nests	+	*		?		+	?	331
Counting roosts and flocks	*		*			?	?	335
Counting migrants				+			?	337
Indirect counts	+		?		+			339
Tape playback		+		+		?	?	341
Vocal individuality	?	?	?	?	?	?	?	342

* Method usually applicable, + method often applicable, ? method sometimes applicable. The page number for each method is given.

The methods outlined in Table 9.1 also differ in their ability to provide absolute or relative measures of abundance. Timed species counts and listing approaches, though efficient at providing rapid inventories of the species in a particular area, will yield information only on the relative abundance of species. Many of the other methods yield estimates of absolute abundance. Under most circumstances, census methods that involve catching birds are not preferred; most bird species are easier to see or hear than to catch, and catching involves extra investment in equipment and training. Occasionally, however, this is not the case and capture may be the only practical method.

Where to look for further information

Bibby *et al.* (2000) give an excellent and detailed review of techniques, while Bibby *et al.* (1998) discuss methods with particular reference to bird expeditions. Gregory *et al.* (2004c) cover survey design and general census techniques; Bennun & Howell (2002) present methods suitable for African forest birds; and Javed & Kaul (2002) do the same for Asian birds. Verner (1985) and Dawson (1985) review census methods, Koskimies & Väisänen (1991) describe those used in Finland, while the proceedings of the Asilomar Symposium (Ralph & Scott 1981) and of the European Bird Census Council conferences (Czech Stastný & Bejcek 1990, Hagemeijer & Verstrael 1994, Helbig & Flade 1999, Anselin 2004) contain many individual research papers on census and survey techniques. Gilbert *et al.* (1998) and Steinkamp *et al.* (2003) outline a variety of species-specific methods, Greenwood and Robinson (Chapters 2 and 3 in this volume) introduce much underlying theory, while Buckland *et al.* (2001) describe advanced methods for density estimation. Finally, Underhill and Gibbons (2002) outline the uses of bird-census information.

Listing methods

Listing methods are applicable to a wide range of species and habitats, but most widely used in tropical habitats. They are suitable for rapid assessments of poorly known areas. They can be used in population monitoring.

Method

Lists of birds recorded by birdwatchers are collected for a particular geographical area. Common species will occur on many lists, rare species on only a few. Thus the frequency of occurrence of species on lists, termed the 'reporting rate' by Harrison *et al.* (1997), is a crude measure of relative abundance.

Bart & Klosiewski (1989) compared the frequency of occurrence of birds at 50 counting stations (equivalent to 50 lists) with estimates of abundance from point counts (see below) at the same stations. Trends in species' populations were similar from the two methods, though those obtained from lists were about 40% lower when several individuals of a species were counted at stations. They concluded that such listing approaches were suitable for detecting change, but not necessarily the magnitude, and that listing would be preferable to counts only if they allowed many more observers to become involved; in practice this may often be the case because of the simplicity of listing. Roberts *et al.* (2004) have similarly shown that changes in frequency of occurrence on lists are a reasonable measure of changes in abundance over decadal time periods obtained from much more intensive territory-mapping and capture techniques.

Inevitably, the more effort that is put into generating each list, the longer that list is likely to be, making comparisons between areas with different levels of effort problematic. To overcome this, lists should ideally be produced for specific time periods, such as an hour or a day. Lists can also be constrained to a more precise geographical area. Hewish & Loyn (1989) found that producing species lists for 2-ha plots during 20-min periods appealed to observers because they felt that they were able to record all species present within the time period. The more lists that are produced, the more precise the reporting rates will be, so a reasonable number of lists, perhaps 15 or more, is required.

McKinnon lists (McKinnon & Phillips 1993) are a specific form of listing that records species on fixed-length lists rather than within fixed periods. To produce a McKinnon list, walk slowly around the study area listing the first *n* species encountered, where *n* could be, for example, 10, 15 or 20. List the names of all new species encountered and when *n* have been listed, start a new list and continue surveying until, again, *n* species have been encountered. Repeat this process until a reasonable number (>15) of lists has been produced. To give you an idea of what proportion of species present in your study area have been found, plot out the cumulative number of species recorded across the lists. This species-accumulation curve will begin to plateau when you have recorded a high proportion of the species present. As for other listing approaches, the relative abundance of each species is the proportion of lists on which it was recorded.

Advantages and disadvantages

The great appeal of listing methods is their simplicity. Counts of individuals are not needed, allowing more time to be spent on identification, which is particularly valuable for inexperienced observers, in poorly known regions, and in bird-rich habitats. They provide a simple measure of relative abundance, allowing indices to be compared between species and sites. However, the index produced will be most useful for moderately abundant species. Very common species will be recorded on all lists – and thus true variation in abundance of these species will be masked and trends dampened. Very rare species will be recorded on no, or few, lists, giving little variation in abundance between species and sites, often no better than recording their presence or absence. An advantage of the McKinnon lists over time-limited lists is that, because observers are not restricted to particular time periods, less skilled observers – who take longer to find and identify species – can still produce lists that will be comparable with those of more experienced observers. The more skilled observers will simply collect more lists. Data from such lists can be used to produce maps of distribution and geographical patterns of relative abundance. However, this approach has the considerable weakness that the index of abundance it produces is relative to that of other species (see Chapter 3).

Biases

Relative densities of vocal and highly detectable species will be overestimated. Unless told not to, observers may seek out rare species, thus inflating their proportional occurrence. Lists should be time-limited, or else abundances in areas with greater fieldwork effort (e.g. more accessible sites) will be higher than those in areas with lower effort. Abundances of species that flock will be underestimated, since the method is best suited to species that are evenly distributed across the landscape.

Timed species counts

This method is used for high-diversity communities, particularly tropical forests, and also for birds of savannah and semi-arid areas. It is suitable for rapid assessments of poorly known areas.

Method

Timed species counts (TSCs) are repeated species lists that yield indices of relative abundance and are a specific type of listing method. They are based on the simplistic assumption that, when birdwatching, on average common birds are noted first, whereas rare birds take longer to locate. The average time to first observation is thus a crude measure of abundance and can be used to make comparisons both between and within species.

In practice, counters divide the entire period of observation in the study area into a reasonably large number of shorter time periods (e.g. 6–12) within, say, 1 or 2 h. For example, walk slowly through the study area for 1 h and record the time or the block of time in which each species was first seen. Ignore subsequent observations of that species within the hour. If a species was recorded in the first 10-min interval it is allocated a score of 6, in the second a score of 5, in the third a score of 4 and so on. Score unrecorded species as 0. Repeat the 1-h count, e.g. 10–15 times, and ensure that the TSCs are spread well over the study area. For each species, calculate a mean score across all 1-h counts to give a relative measure of abundance (Pomeroy & Tengecho 1986; Pomeroy & Dranzoa 1997).

The method can be made increasingly complex by recording birds within set distance bands, e.g. up to and beyond 25 m from the observer (to remove biases due to noisy or conspicuous species) or within set height bands, e.g. above and below 3 m (to overcome problems with hard-to-detect understorey species that may be better censused by capture methods).

Measures of abundance for individual species from TSCs correlate well with those from line transects (Bennun & Waiyaki 1993; Pomeroy & Dranzoa 1997).

Advantages and disadvantages

Many of the advantages and disadvantages of general listing methods apply to TSCs and, again, this method has the major advantage of being quick. Common species are ignored once first seen and thus effort can be concentrated on finding less common species. A reasonably large area can be covered in the allocated time. The method is easy and does not require prior mapping or cutting of transect lines, and is a good way of rapidly evaluating the importance of sites. If, however, densities of common species are of interest then this is not a suitable method. Timed species counts provide only crude, relative indices of relative abundance. Maps of distribution and geographical patterns of abundance could be produced using this method.

Biases

Many of the biases of listing methods in general also apply to TSCs. Species vary greatly in their detectability (the probability of finding a species when it is present), and thus comparisons between species and sites need to be interpreted cautiously. Flocking species or those that aggregate (e.g. in fruiting trees) may have lower or higher indices than those that are dispersed more widely across the study area, depending on the counting behaviour of the observers.

Territory mapping

Territory mapping is used for territorial breeding species, e.g. some ducks, game birds and raptors, and most temperate passerines. It is widely used, but can be very time-consuming.

Method

During the breeding season, many species of birds are territorial and, in temperate areas, such behaviour is strongly synchronised. Males sing to defend their territories, nests are built within them, and the boundaries between territories are often clearly defined by disputes with neighbouring birds. The breeding territory can thus be used as a census unit, and territory (or spot) mapping, in which all signs of territory occupancy are marked on a large-scale map of a plot, can be used as an effective means to estimate absolute abundance. The aim of territory mapping is to determine how many territories of each species there are on a given plot. Standardised techniques are given in Kendeigh (1944), Enemar (1959), IBCC (1969), Marchant *et al.* (1990) and Bibby *et al.* (2000).

First, the study plot needs to be mapped at a scale of about 1:2500. To enable species' registrations to be located accurately on the map, any obvious features that can be easily identified during each visit (e.g. buildings, ponds, isolated trees, tracks, rides, hedges, etc.) should be marked on the map. The size of the plot should be such that a reasonable number of territories of each species is present. A realistic goal would be to ensure that there are five or more territories for half of the species on the plot (Terborgh 1989). Though these numbers may be minimal for statistical purposes, in practice they require a great deal of fieldwork. Plot size will vary with habitat, because bird density, diversity and conspicuousness vary with habitat too; 15–20 ha in closed habitats such as temperate woodland would be suitable (though perhaps half this in tropical forest), with 60–80 ha in open habitats (e.g. agricultural, moorland, grassland and steppe). If the census is being undertaken for long-term population monitoring, it might be important to ensure that the habitat is not strongly successional in nature.

Long, thin plots are unsuitable for territory mapping because the ratio of edge to area is high and many bird territories will overlap the plot boundary. Territories along the edge of the plot cause problems because it is often difficult to determine whether a particular territory belongs to the plot. Round or square plots are preferable. In addition, try to avoid using a species-rich feature of the landscape (e.g. a hedge) as a plot boundary, since this will serve to exacerbate any edge problems.

Several visits need to be made to each plot during the breeding season. In temperate regions, where breeding seasons are clearly defined, 5–10 visits per plot would be suitable; open habitats would be at the bottom end of this range, whereas woodland with high densities of birds would be at the top. The visits should be spread out throughout the season to ensure that both early- and late-breeding species are included in the censuses. In tropical areas with less clearly defined seasons, the number and timing of visits needs careful consideration; it is true to say that this method is less used in the tropics.

Many birds sing most during the first hour after dawn. In consequence, this period can be confusing in areas with high densities of birds, and is probably best avoided. Surveys should commence an hour after dawn and be completed well before mid-day because many species sing less and are less active later in the day. Temperate forest can be surveyed at a rate of about 5 ha per hour, tropical forests at about half this rate and more open habitats at about 20 ha per hour, depending on the precise objectives of the study. Each visit to a plot can thus be undertaken during a morning.

Prior to the start of the season, you will need to produce several copies of the plot map, one for each field visit and, ultimately, one for each species' map (see below). Large maps can be awkward to use in the field and are best attached to a clipboard. Cover maps with a large polythene bag if it is likely to rain. The plot should be covered at a slow walking pace with the route approaching within 50 m of every point on the plot. Each bird encountered is marked on the map using standard codes (Box 9.1). Evidence of nesting, such as nests, alarm calling and birds carrying nesting material or food, is particularly useful, as are simultaneous observations of different individuals of the same species (e.g. counter-singing or fighting males). Without these, the subsequent analysis of the maps (see below) is much less accurate (Tomiałojć 1980). It is necessary to work slowly and carefully to build up these records and to record inconspicuous species, though covering the plot too slowly may lead to unintentional double-counting of the same individuals. Mapping should extend slightly outside the study area to ensure that the territory boundaries of species at the edge of the plot are recorded (see 'Analysis of maps', below).

The territory-mapping method can be extended to cover a much larger geographical area simply by ignoring common species and mapping at a much bigger scale (e.g. 1 : 10,000; Robertson & Skoglund (1985)). By doing this, species that range over a much larger area, but are nevertheless territorial, can be censused (e.g. raptors).

Analysis of maps

At the end of the season all the information from the individual visit maps is transferred onto species' maps (one per species). Transfer registrations from the first-visit map and denote them by the letter 'A', those from the second-visit map by 'B' and so on. All the records of a particular

---

Box 9.1. **Activity codes for use in mapping censuses in Finland**

These activity codes have been developed from, and are very similar to, the mapping codes used by the British Trust for Ornithology in the UK. Most examples are for the chaffinch *Fringilla coelebs*. Some countries have standard codes for each species name (e.g. in the UK, CH = chaffinch).

 A chaffinch in song.

A chaffinch in song (exact location shown by the point).

 A chaffinch in song (location is not exact; the point where the observation was made is shown by the cross).

$\underline{\underline{F_{coe} \male}}$ A male chaffinch repeatedly giving alarm calls or other vocalisations (not song) thought to have strong territorial significance.

$\underline{F_{coe} \male}$ A male chaffinch calling.

$F_{coe} \male, \; F_{coe} \female, \; F_{coe}, \; F_{coe} \; 2\male 1\female, \; 3\, F_{coe} juv$

Chaffinch sight records, with age, sex, or number of birds if appropriate. Use F_{coe} $\male\female$ to indicate one pair of chaffinches, i.e. 2 F_{coe} $\male\female$ means two pairs together.

$F_{coe} \male^{f}$ A male chaffinch carrying food (or faeces).

$F_{coe} \female^{m}$ A female chaffinch carrying nest material.

$F_{coe}^{* \; 2E3N}$ An occupied nest of chaffinches, with two eggs (E) and three nestlings (N); *shows the location. Do not mark unoccupied nests, which are not of territorial significance by themselves.

$P_{maj}^{\boxplus \; 10E}$
Great Tit *Parus major* nesting in a specially provided site. Please remember to use this special symbol for a nest in a nest box.

F_{coe}^{*P} Chaffinch nest with a parent bird incubating or warming young.

$F_{coe} juv$ A chaffinch fledgling

$\cdot F_{coe}$ fam Juvenile chaffinches with parent(s) in attendance.

Movements of birds can be indicated by an arrow using the following conventions:

$F_{coe} \male$ A calling male chaffinch flying over (seen only in flight).

$F_{coe} \female$
$F_{coe} \female$ A female chaffinch moving between perches. The solid line indicates that it was definitely the same bird.

A singing chaffinch perched, then flying away (not seen to land).

A male chaffinch flying in and landing (first seen in flight).

A Siskin *Carduelis spinus* circling above the forest.

The following conventions indicate which registrations relate to different, and which to the same, individual birds. Their proper use will be essential for the accurate assessment of clusters.

Two chaffinches in song at the same time, i.e. definitely different birds. The hatched line indicates a simultaneous sighting/hearing of song and is of great value in separating territories.

The solid line indicates that the registrations definitely refer to the same bird.

The question-marked solid line indicates that the sightings/songs probably relate to the same bird. This convention is of particular use when your census route brings you back past an area already covered – it is possible to mark new positions of (probably the same) birds as those recorded before, without risk of double-recording. If you record birds without using the question-marked solid line, overestimation of territories will result.

No line joining the registrations – there is no assumption as to whether the records concern different birds, but, depending on the pattern of other registrations, they may be treated as if only one bird were involved (a question-marked dotted line indicates that the sightings/songs were almost certainly of different birds).

Two chaffinch nests occupied simultaneously, and thus belonging to different pairs. Only adjacent nests need to be marked in this way. Where they are marked without a line, it will be assumed that they were first and second broods, or a replacement nest following an earlier failure.

An aggressive encounter between two chaffinches; may be accompanied by notes on vocalisations.

species from the first visit are transferred to its species map, but with 'A' replacing the species code. The symbols from Box 9.1 are also incorporated on the species map. This is repeated for each of the visits until a map containing all of the registrations from all of the visits has been produced for each species.

The symbols on the species' maps should form clusters around which non-overlapping rings representing approximate territory boundaries can be drawn. Conventionally, at least two registrations are needed to define a cluster if there were 5–7 visits or three if there were 8–10 visits. To avoid including temporary migrants, records in the cluster must be from at least ten days apart. Simultaneous registrations indicate different individuals and should never be incorporated into the same cluster unless they are thought to be two adults of a pair. Records of nests can be counted as a cluster even in the absence of sufficient records of the adults.

Dealing with edge clusters, those that overlap the plot boundary, is problematical, and several analytical methods have been used. Treat all edge territories as belonging to the plot; include them if more than half of the registrations within the cluster lie in the plot; or use the proportion of a cluster's registrations that lie within the plot to calculate a fraction of a territory. The first method will overestimate densities.

Clusters can be difficult to differentiate and may overlap. For inconspicuous species, there may be few registrations per cluster and no simultaneous registrations. Thus, despite the existence of standard guidelines (IBCC 1969; Marchant 1983; Marchant *et al.* 1990; Bibby *et al.* 2000), analysis of species' maps can be subjective and requires experience, as well as time.

Though map analysis is generally undertaken at the end of the season, if it were undertaken during the season, fieldwork could be targeted at clarifying confusing situations. The species' maps could even be taken into the field.

Advantages and disadvantages

Territory mapping is very time-consuming and can thus be expensive and inefficient as a monitoring tool. The time commitment to cover one site using this method usually limits the total number of plots that can be covered. The alternative of covering more sites less intensively may be attractive if the aim is representative monitoring. At first sight, this would appear to be an accurate and precise method, but this is not always the case and one needs to be aware of the underlying assumptions being made. It is not suitable for species that are colonial or semi-colonial, for those that live in loose groups or whose territories are large relative to the study area, for species that sing for brief periods, or for species with complex mating systems. It can be used only when birds are territorial, and thus is largely suitable only for breeding birds, though migrant species may set up individual, rather than pair, territories on their wintering grounds (e.g. Rappole & Warner (1980) and Kelsey (1989), see 'Response to playback'). It does, however, yield a map of bird distributions that can be particularly useful for analysing fine-scale bird-habitat associations or in the management of an individual site. Because of the great amount of time spent in the field, the method is better buffered against environmental variation (e.g. weather and timing of visits in relation to a species' breeding cycle) than are other less time-consuming techniques, such as use of point counts and line transects. It allows density to be estimated directly (although it provides no measure of precision) and, despite its drawbacks, is still the favoured method for determining population sizes of territorial breeding species on moderate-sized plots of land in temperate regions. Mapping methods can also usefully be combined with nest finding, radiotelemetry, mist nesting etc. in research projects. Mapping has seldom been used in the tropics, largely because breeding is more asynchronous and many species have complex social behaviours.

Biases

In some species, unpaired birds sing more. For example, unpaired male ovenbirds *Seiurus aurocapillus* and Kentucky warblers *Oporonis formosus* sing 3.5 and 5.4 times more often than paired males, respectively (Gibbs & Wenny 1993). In the same study, all unpaired males were detected but only 50% of paired male ovenbirds and 65% of paired male Kentucky warblers were located over ten visits. Male sedge warblers *Acrocephalus schoenobaenus* cease to sing as soon as they have attracted a mate, whereas males of the congeneric reed warbler *Acrocephalus scirpaceus* continue to sing once mated (Catchpole 1973). These differences in behaviour can make it unclear whether the breeding population is being censused.

The method assumes that birds live in pairs in fixed, discrete and non-overlapping ranges, which is often not the case (e.g. polygynous species, polyterritorial species). Despite standard guidelines for map interpretation, there is nevertheless a good degree of subjectivity involved and this can lead to variation, which makes comparison between studies difficult. The method can be unreliable at high densities, if birds are not readily visible, if registrations are plotted inaccurately, or if it is difficult to obtain many simultaneous registrations.

Transects

There are two types of transect most commonly used in bird surveying; line transects and point counts (or point transects). Both are based on recording birds along a predefined route within a predefined survey unit. In the case of line transects, bird recording occurs continually, whereas for point transects it occurs at regular intervals along the route and for a given duration at each point. There are variations on this theme whereby birds are recorded to an exact distance (variable distance) or within bands (fixed width) from the transect point or line. While there are some important differences between the line and point transects, there are many practical and theoretical similarities.

Line and point transects are the preferred survey methods in many situations. They are highly adaptable and can be used in terrestrial and marine systems. They can be used to survey individual species, or groups of species. They are efficient in terms of the quantity of data collected per unit of effort expended, and for this reason they are particularly suited to monitoring projects. Both can be used to examine bird–habitat relationships (though generally in less detail than territory mapping), and both can be used to derive relative and absolute measures of bird abundance. Transects can usefully be supplemented and, to some degree, verified in combination with other count methods such as sound recording, mist netting and playback (e.g. Haselmayer & Quinn (2000) and Whitman *et al.* (1997), see 'Response to playback').

There are several issues to consider when using transects in the field. The recommended walking speed is particularly important for line transects, as are the counting instructions. A further important consideration is whether to use full distance estimation, i.e. estimating distances from the centre of the point count or from the transect line, to all birds heard or seen, or to use estimation within distance bands or belts. In the latter case, you need to decide on the specific distance bands. It is preferable to record some measure of the distance to each bird seen or heard because this provides a useful measure of bird detectability and allows species-by-species density estimation. Modern methods, known as *distance sampling* (i.e. employment of a variable circular plot or variable-width line transects), are used to analyse such data (Buckland *et al.* (2001); see 'Correcting for differences in detection probabilities'). Ideally, it is best to record the exact distance to birds, but, failing this, distance bands or belts can be used. As range-finders become increasingly affordable, they open the way for simple and accurate distance estimation, especially for single-species surveys.

The aim of transects is to record all birds identified by sight or sound with an estimate of distance when first detected. Distances should be estimated perpendicular to the transect line (rather than the distance from the bird to the observer), or from the point-count station. Birds that are seen flying over the census area (aerial species) are recorded separately because they cannot be included in standard density estimation. For such mobile species, it is best to make an estimate of their numbers along each section of transect, or at each point. If birds fly away as you are counting, record them from the point you first saw them. It is recommended that birds flushed as you approach a point-count station should be recorded from that point and included in the point-count totals (but you must make this plain in the write-up). Try to avoid double-counting the same individual birds at a point count or within a transect section by using careful observation and common sense. It is, however, correct to record what are likely to be the same individual birds when they are detected from subsequent point counts or transect routes. For surveys of breeding birds, between two and four visits to a site are recommended each season, depending on manpower, resources and specific objectives.

Line transects

Line transects are used for birds of extensive open habitats, e.g. shrub-steppe and moorland, offshore seabirds and waterbirds. This is a highly adaptable and efficient method.

Method

Line transects are undertaken by observers moving along a predetermined fixed route and recording the birds they see or hear on either side of that route. Line transects can be walked or driven on land, sailed at sea, or flown in the air. Because the observer needs to be able to move freely through the land, sea or air, transects are most suitable for large areas of continuous open habitat; but the method has proved highly adaptable.

First, the transect route(s) need to be chosen. Ensuring that their location is as random as possible is crucial to the success of the study. If, for example, a route were to follow a path, a hedge, a stream or a road, the results obtained could be markedly biased by the influence that these linear features might have on bird populations. A similar example would be of transect counts of fish-eating seabirds made from fishing trawlers that are, like the seabirds, actively seeking out fish stocks. The location of such transects could not be considered to be random, and would bias (probably inflate) any estimates made from the counts for particular species. The difficulty of randomly allocating routes due to access restrictions is a disadvantage of line transects. The Breeding Bird Survey in the United Kingdom (Gregory & Baillie 1998, 2004; Gregory 2000; Gregory *et al.* 2004a; Raven *et al.* 2005; http://www.bto.org/survey/bbsindex.htm) uses line transects located on a north–south or east–west axis within randomly allocated 1-km × 1-km grid squares of the

National Grid. In reality, few transects are able to follow the idealised route and some bias towards field boundaries, paths and roads may be inevitable. Birds are recorded in three distance bands from the transect line (0–25 m, 25–100 m, >100 m), or as in flight. There are three visits to each square each year: one to record land use and set up the counting route; and two to count the birds early in the morning, one early in the breeding season and one late. A similar model has been followed in the Republic of Ireland, Poland and Bulgaria. A transect could even be square, rectangular or triangular to ensure that the observer ends up at the starting point. Where maps are insufficient to plan a route precisely, it is a good idea to walk along compass bearings (Koskimies & Väisänen 1991), and generally to use a compass or GPS to aid navigation. Triangular line transects in forest, 2 km in length, have been used to survey capercaillie, *Tetrao urogallus*, in Scotland (Wilkinson *et al.* 2002).

The total length of transect route will vary depending upon the study in question. Practical considerations (e.g. time available to spend in the field, size of the area to be surveyed, study species or group of species) may well be most important. Ideally, split the total length down into several shorter lengths. These could either continue one on from the other or be wholly independent of one another. The latter may be more useful for analytical purposes since the separate lengths of transect may be considered statistically independent, or their autocorrelation modelled. In Finland, breeding birds are surveyed along rectangular transects with a total length of about 5 km (Koskimies & Väisänen 1991). In the UK Breeding Birds Survey, each of two 1-km transects is further subdivided into five 200-m lengths for more geographically precise recording of birds and habitats. The total length of transect will depend on your objectives and local conditions.

If several different transects are to be undertaken on a plot, they need to be sufficiently far apart to sample the birds appropriately. Sensible distances might be 150–200 m in closed habitats, but 250–500 m in open habitats, depending on your specific objectives. The distance between transects should be greater in open habitats because birds are more visible over greater distances and are more likely to flee greater distances from an observer.

Once the transect routes have been planned it is then necessary to decide how many visits are to be made to each route and what distance estimation is to be used. As for point counts (see below) it is sensible to repeat each transect one or more times to maximise the chance of recording all species since bird activity often varies across seasons (either because the species is absent, or because it becomes unobtrusive at certain times). Two counts per nesting season is usually the minimum for monitoring studies.

Methods for estimating density are very similar to those used with point counts (see Chapter 1 for details) and various approaches have been advocated (Järvinen & Väisänen 1975; Burnham *et al.* 1980; Bibby *et al.* 2000; Emlen 1977; Buckland *et al.* 2001). Simple indices of the number of birds recorded per unit length of transect obtained by counting birds, either up to an unlimited distance, or to a single fixed distance on either side of the transect, are unlikely to be reliable because birds are inevitably missed. Increasingly sophisticated methods under the banner of distance sampling are now commonly applied to transect data for birds and the necessary software is available freely (see 'Correcting for differences in detection probabilities').

In practice it may be sensible to map all bird records onto a schematic representation of the transect in your notebook, or onto a recording sheet. It might also help if distance bands were drawn onto this, too. Recording the birds in this manner means that a variety of different techniques can ultimately be used to analyse the data. If birds are recorded in separate distance bands, check that the distances can be reliably estimated. Try to standardise the rate of movement along the transect route; walking too fast misses birds, but walking too slowly may result in double-counting. A walking rate of 2 km per hour is reasonable in open habitats, though 1 km per hour would be more realistic in forest.

Line transects at sea

Away from nesting colonies, seabirds are frequently surveyed by transect from a ship (Tasker *et al.* 1984; Komdeur *et al.* 1992; Briggs *et al.* 1995; Bibby *et al.* 2000). Seabirds present particular problems because they are often recorded in flight, and their speed of flight in relation to the speed of the ship through the water and direction of flight relative to that of the ship influence the results markedly. In addition, some birds are attracted to boats, whereas others avoid them, and general viewing conditions are often difficult against the sea surface. The sophisticated methods used on land can generally not be replicated at sea, so surveys tend to focus on indices of relative abundance and have caveats attached. By standardising the count methods, it is at least possible to draw comparisons between different areas and different studies.

Two simple, standard methods for transects at sea have been advocated (Tasker *et al.* 1984; Webb & Durinck 1992). The first involves counts to one side of the ship during a set time period in a 300-m band (which is subdivided into 0–50-m, 50–100-m, 100–200-m and 200–300-m bands) or over 300 m. Birds are recorded perpendicular to the direction of the ship's course. By recording birds in distance bands, it is possible to generate detection functions and then apply adjustments specific to species and sea states. Frequently, two observers record the seabirds with counts every 10 min. The distance covered is calculated from the average ship speed in relation to the recording width and bird 'densities' are reported. The second method is the 'snapshot' approach in which observers estimate the number of flying birds and sitting birds in an imaginary box, for example, 300 m at right angles to the ship and as far ahead as all birds can reasonably be seen and identified. By taking a series of snapshots of this kind, it is possible to estimate the number of seabirds per unit area of sea. Validation of such figures from either method is in reality extremely difficult.

Line transects from the air

Waterfowl, some colonial species and seabirds are sometimes counted from the air by flying along transects of known length, width and direction (Komdeur *et al.* 1992). Though the use of a plane can be expensive, the speed of the plane, compared with that of a ship, does mean that the chances of double-counting the same birds, and thus overestimating density, are reduced. The flushing action of the plane can be helpful in some circumstances, although harmful disturbance must be avoided.

The width of the transect will vary with the particular application, but an overall width of about 200 m (100 m on either side of the plane) is sensible. The plane should be flown at a fixed and reasonably low (e.g. 50–100 m) altitude and at a moderate to low speed (e.g. 150 km per hour).

Table 9.2. *A comparison of line and point transects*

Line transects	Point transects
Suits extensive, open and uniform habitats	Suits dense habitats such as forest and scrub
Suits more mobile, large or conspicuous species and those that are easily flushed out	Suits more cryptic, shy and skulking species
Suits populations at lower density and more species-poor habitats	Suits populations at higher density and more species-rich habitats
Covers the ground quickly and efficiently, recording many birds	Time is *lost* moving between points, but counts give time to spot and identify shy birds
Double-counting of birds is a minor issue, since the observer is continually on the move	Double counting of birds is a concern within the counting period – especially for longer counts
Birds are less likely to be attracted to the observer	Birds may be attracted to the presence of observers at counting stations
Suited to situations where access is quite good and terrain easily worked	Suited to situations where access is restricted and difficult terrain
Can be used for bird-habitat studies	Better suited to bird-habitat studies
Errors in distance estimation have a smaller influence on density estimates (because the area sampled increases linearly from the transect line)	Errors in distance estimation can have a larger influence on density estimates (because the area sampled increases geometrically from the transect point)

In general, bird identification is more of an issue for aerial transects than for ship-based ones, and larger species are more readily surveyed. Identification becomes difficult if the plane is too high; and birds pass below too swiftly if it is too low. In practice, flight speeds may be too fast to count every individual bird and quick estimates of flock size are often needed. Bird density can be calculated from the number of birds counted and the overall area of transect covered (from its width, the speed of the plane and the time taken to complete the transect). The main disadvantages of the method are the expense, dependence on suitable flying and observation conditions, and restriction to larger, more visible species.

Advantages and disadvantages of line transects

Line transects can be undertaken at any time of year, on land, on sea or in the air. Line and point transects each have their own strengths and weaknesses (Table 9.2) and it is important that the methods are matched carefully with the survey objectives. Line transects are suited to large areas of homogeneous habitat, and are particularly useful where bird populations occur at low density. Because most birds are detected by song or call, a high level of observer experience is required. Estimates of density can be calculated using distance-sampling techniques. The area sampled by a line transect increases linearly away from the transect line; thus errors in detecting birds close to the observer and in distance estimation are less likely to bias density estimation than is the case with point counts. Random allocation of transect routes can be particularly difficult in some habitats and in some terrain. Because the observer is continually on the move, identification can be difficult and cryptic birds can easily be missed. The high costs of transects at sea can be reduced by observing from ships involved in other activities (though this may introduce some biases). Transects from the air are sometimes too quick to allow precise counts and the identification of some species, their age and sex.

Biases

Density estimation makes a number of assumptions; for example, it assumes that birds on the transect line are not missed (e.g. from walking too fast); that birds do not move before being detected (e.g. in response to the observer); that they are not counted twice along the same transect (walking too slowly); that distance is estimated without error; and that all observations are independent events (e.g. one bird is not detected because of the alarm calls of another). In practice, many of these assumptions will not be met and all may lead to bias. Because of the short length of time spent in the field, line transects can be markedly influenced by weather conditions. Ideally, counts should not be carried out in strong winds, rain or cold weather.

Point counts or point transects

These methods are used for highly visible or vocal species, often passerines, in a wide variety of habitats and are particularly suited to dense vegetation.

Method

A point count is a count undertaken from a fixed location for a fixed time. It can be undertaken at any time of year, and is not restricted to the breeding season. Point counts can be used to provide estimates of the relative abundance of each species or, if coupled with distance estimation, can yield absolute densities, too (Buckland *et al.* 2001).

Point-count stations (the position from which the count is made) should be laid out within the study plot either in a regular/systematic manner (e.g. on a grid) or in a random manner and stratified as appropriate. It is best not to place counting stations too close together. A sensible minimum distance is 200 m. If the distance between points is too great, however, too much time will be wasted travelling between the counting stations. A reasonably large number of point counts (more than 20) will be needed from each study plot; point counting is thus not a suitable technique for small study areas. Twenty counts can readily be made in a morning starting soon after dawn. An advantage of point counts over line transects is that it is often easier to approach and gain access to individual points than to establish transect lines, particularly when access and the terrain are problematic.

Wait for a set time, say a minute, before beginning to count at each station in order to allow the birds to settle down following your arrival. Count for a fixed time at each station. Ideally, this should be either 5 or 10 min, the actual duration depending on habitat and the bird communities present. If counts are too short, individuals are likely to be overlooked; if they are too long, some birds may be double-counted. Record all birds seen or heard. Endeavour to count each individual only once at each station. Most registrations will occur in the first few minutes, thus counting for too long can be inefficient. The time saved by counting for a shorter period can be used to count at more points, or to cut down the total time spent in the field. In areas with a very rich bird fauna or where species are hard to detect or identify, for example, in a tropical rainforest, it may be necessary to count for longer than 10 min.

In habitats with high densities of birds, it is easy to confuse different individuals, or to be uncertain whether you have already recorded a particular individual or not. A simple way of resolving this is to record their approximate positions in a notebook. This can be divided into four quarters, and birds recorded in these quarters (e.g. left and to the front, right and behind, etc.) marked accordingly by a species code. If you are counting in several different distance bands (see below), these could also be drawn as concentric circles around your central position.

If all that is required is an index of abundance, then count up to an unlimited distance, or only within an arbitrary range such as 30 m from the observer. However, such crude indices are generally not recommended and much can be gained by a little more effort. In practice, there are three ways of incorporating distance estimation into point counts to enable detectability to be assessed and density to be calculated. The simplest is to have two counting bands, and to record birds up to a fixed distance (e.g. 30 m in a forest, or 50 m in a more open habitat) and beyond that distance separately. Simple formulae (based on the possible manner in which detectability falls off with distance) can then be used to calculate the density of each species (see Chapter 3, Bibby *et al.* (1985) and (2000) and Buckland *et al.* (2001)). However, by recording birds in several distance bands, or to exact distances, a range of more sophisticated methods can be tested and applied (Buckland *et al.* 2001); such approaches are highly recommended and are much more efficient (Diefenbach *et al.* 2003).

It is often advisable to undertake at least two separate counts at each counting station. In temperate regions there is often one in the first half of the season and one in the second. This will not only ensure that both early and late breeders are recorded during the counts, but will, in part, also take into account seasonal variation in bird activity, since a species, though present, may be more detectable during one part of the season than during another; equally, summer migrants may arrive only part way through the season. In general, the maximum plot count for each species should be used in analyses of density. This increases the chance that all birds are detected at the centre of the count area, which is an important assumption in density estimation (see below). The maximum value need not be used if only relative indices are required, rather a mean value can be used. If several counts are made at each counting station, it can sometimes be difficult to relocate the precise counting station on subsequent visits. It is thus necessary to use a GPS or mark the locations of counting stations in a reasonably obvious manner (e.g. brightly coloured tape wrapped around a post or vegetation), particularly in habitats in which the vegetation is likely to grow rapidly between visits.

The North American Breeding Bird Survey (Robbins *et al.* 1986; Sauer *et al.* 2001; http://www.mbr-pwrc.usgs.gov/bbs/) uses 50 3-min counts at intervals of 0.8 km along randomly selected roadside routes. Birds are counted once at the height of the breeding season, starting 30 min after sunrise, and all birds heard and seen within 0.4 km of the road are recorded at each stop. Each route takes about 4–4½ h. No distance estimation is involved, thus all species are measured in terms of a relative index. Though it has gathered an enormous range of information and increased knowledge of bird populations hugely (2000 routes are counted each year and used to monitor about 230 species), the roadside nature of the scheme has led to problems in interpretation since habitat change along roads is unlikely to be representative of habitat change throughout North America as a whole. Such potential problems should always be considered if the census technique is to be used as the basis of a long-term monitoring scheme.

Advantages and disadvantages

Point counts are widely used to census songbirds. Line and point transects are compared with each other in Table 9.2. Point counts have been used, for example, to census waders, waterfowl and nocturnal birds in Finland (Koskimes & Väisänen 1991), parrots and hornbills in Indonesia (Marsden 1999) and the endemic bullfinch on the Azores (Bibby *et al.* 2000). Because most birds are detected by song or call, a high level of observing experience is required of observers. Counting stations are relatively easy to allocate randomly, which is not always the case for territory-mapping plots or line transects. Recording time is lost in moving between counting stations. Point counts are more suitable than transects where habitat is patchy, though much less so in open habitats where birds are likely to flee from the observer. Point counts are unsuitable for species that are easily disturbed or those that respond strongly to the presence of an observer. They are, however, very efficient for gathering large amounts of data quickly. Point counts can be used outside the breeding season.

Biases

Estimation of density assumes that all birds at the centre of the count area (i.e. where the observer stands) are recorded. This will not be the case if birds flee from the observer, or if the species is particularly skulking. Because the area sampled by a point count increases geometrically with distance from the observer, small errors in detecting birds close to the observer can seriously bias density estimates. Overestimates will occur if birds are attracted to the observer (Hilton *et al.* 2002). Where birds are highly mobile, the same bird may be recorded twice. Because of the short length of time spent in the field, point counts can be markedly influenced by weather conditions. Ideally, counts should not be carried out in strong winds, rain or cold weather.

Correcting for differences in detection probabilities

Method

Having carried out a survey of a species in a particular habitat, we often wish to compare our results with those of other similar studies. This is often easier said than done because to do so using the raw, or unadjusted, counts, you must assume that the probability of detecting birds is the same in each study. It is, however, unavoidable that some birds present in your study area will go undetected regardless of how well the survey is carried out. 'Detectability' is a key concept in wildlife surveys and needs to be assessed carefully. A comparison of unadjusted counts will be valid only if the numbers represent a constant proportion of the actual population present across space and time. This assumption is at best questionable and has been a matter of much debate (Buckland *et al.* 2001; Rosenstock *et al.* 2002; Thompson 2002).

The solution is to adjust counts to take account of detectability, for which various methods have been proposed (Thompson 2002). For example, the 'double-observer' approach uses counts from primary and secondary observers, who alternate roles, to model detection probabilities and adjust the counts (Nichols *et al.* 2000). The 'double-sampling' approach uses the findings from an intensive census at a subsample of sites to correct the unadjusted counts from a larger sample of sites (Bart & Earnst 2002). The 'removal model' assesses the detection probabilities of various species during the period of a point count and adjusts the counts accordingly (Farnsworth *et al.* 2002). Finally, 'distance sampling' models the decline in the detectability of species with increasing distance from an observer and corrects the counts appropriately.

Distance sampling is a specialised way of estimating bird densities from transect data and assessing the degree to which our ability to detect birds differs in various habitats and at various times (Buckland *et al.* 2001; Rosenstock *et al.* 2002; Thomas *et al.* 2005). The software and further information to undertake these analyses are freely available at http://www.ruwpa.st-and.ac.uk/distance. Distance sampling takes account of the fact that the number of birds we see or hear declines with distance from the observer. The shape of this decline, the distance function, differs among species, among observers and, importantly, among habitats. Birds in open grassland are detectable over much greater distances than are those in dense forest – even when they occur at the same densities. Distance sampling models the 'distance function' and estimates density taking into account both the birds that were observed and those that were present but were not detected.

Advantages and disadvantages

Distance sampling provides an efficient and simple way of estimating bird density from field data and corrects for differences in bird detectability under various circumstances (e.g. in different habitats or in the same habitat experiencing successional change). It allows for differences in conspicuousness between habitats and species (though not observers), enabling comparisons to be made between species, across habitats and at different times. These methods are strongly recommended. Density estimates improve with the number of birds recorded – a minimum of about 80 records is recommended. Such methods, however, demand a high degree of accuracy in distance estimation by observers and a series of statistical assumptions to be met. Some, if not many, of the underlying assumptions are likely to be broken in the field and it is often unclear how much bias has been introduced.

Biases

Distance sampling and related methods rely on a series of assumptions, which need careful evaluation in the field; and positive steps should be taken to lessen their effects (Buckland *et al.*

2001). The key assumptions of distance methods are that all the birds actually on the transect line or at the counting station are recorded (for cryptic and shy species this might not be true); that transect lines are randomly or systematically located; that birds do not move in response to the observer prior to detection; that distances are measured without error; and that the detection curve has a shoulder, i.e. detection rates are higher closer to the observer or transect line but fall away with distance (Buckland *et al.* 2001).

Capture techniques

Because many birds are visible and vocal, survey methods generally rely on observers seeing or hearing them. Species that live in dense understorey or forest canopy and are rarely seen or heard, however, can be censused by catching them in mist nets. There are situations in which capture methods can fulfil a useful monitoring role and deliver a great deal of extra information at the same time (Ralph & Dunn 2004).Two separate approaches are used; capture–mark–recapture (also known as mark–release–recapture, MRR) or catch per unit effort. The former allows estimation of population size the latter produces population indices.

Capture methods can be time-consuming and require substantial training to develop the skills necessary to catch, handle and mark birds safely. In many countries, a licence is needed before these techniques can be used. As a method of surveying birds, the return is poor in relation to the effort required, but there are other reasons to catch birds, for example to measure demographic parameters. Information on methods of capture and marking is given in Gosler (2004).

Catch per unit effort

This approach is used mainly for passerines, particularly of woodland, scrub and reed beds, but also for some riverine, dense-undergrowth and canopy species

Method

By placing standard lengths and types of mist nets in standard locations, for standard time periods under similar conditions, this method can be used to monitor changes in population level. Several schemes use capture per unit effort as a monitoring tool, the best known being the UK Constant Effort Sites (CES) scheme (Peach *et al.* 1996; http://www.bto.org/ringing/ringinfo/ces/index.htm), which has been adopted by an increasing number of European countries. The Monitoring Avian Productivity and Survival (MAPS) programme (Desante *et al.* 1993, 1999) is a similar initiative in North America.

These schemes differ from all the methods discussed so far in that demographic, as well as long-term population, information is collected. Thus, rather than simply documenting year-on-year changes in population level, these methods can help interpret such changes by highlighting whether productivity and survival are possible causes of population change.

The MAPS programme provides very specific detail on methods for participants (DeSante *et al.* 1993). Sites are chosen to fulfil the following requirements: they are at least 9 ha in size (though preferably up to 20 ha); contain a reasonable breeding population of birds; are away from areas where transient and migrant birds tend to congregate; and are in areas where active habitat management ensures that the habitat is held in lower successional stages. The number and length of nets set depends on the workforce available, though these parameters should remain constant from year to year. In the UK, where one or two people are working the site it is recommended that ten 12-m, 30-mm-mesh, four-tier, black, tethered, nylon mist nets are set uniformly in a 7–8-ha netting area within the whole plot, thus giving a density of about 1.25–1.5 nets per hectare. Nets are placed where capture efficiency is likely to be maximised, for example near water, along rides or at the edge of woodland. Nets must be set in exactly the same position each year, and are operated for six morning hours per day beginning at sunrise, for 12 days each, for periods about 10 days apart, from early May to late August. Netting starts when most spring migrants have already passed through and stops when autumn migrants begin to appear. No baits or playbacks are used to attract birds to the nets. All birds caught, including retraps, are identified, aged and sexed (see e.g. Pyle *et al.* (1987), Svensson (1992) and Baker (1993)) and all unringed birds ringed (banded).

It is essential to standardise catching time and the number of nets in a site. Simply calculating birds per 10 m of net per hour is insufficient, since doubling the number of nets in a site will not necessarily double the number of birds caught. For the same reason, catching for twice as long will not double the number of birds caught, particularly since this would mean more netting outside the period of peak activity.

Care must even be taken to standardise the mesh of the nets. Species weighing below 16 g are caught more frequently in 30-mm- than in 36-mm-mesh mist nets, whereas the reverse is true for those weighing above 26 g (Pardieck & Waide 1992). Thus, studies should not compare captures made with different mesh sizes.

Constant-effort ringing is often used to study birds of woodland and scrub, but is also good for surveying skulking rainforest-undergrowth species. It has also been used to monitor riverine species and was shown to correlate well with actual abundances of these species (Ormerod *et al.* 1988). It can also be used for canopy species, though, in this case, nets need to be raised many metres above the ground using pulleys or telescopic aluminium poles (Meyers and Pardieck 1993). A gunsling can be used for firing a line up to 45 m into the canopy and then using a pulley to pull up a mist net (Munn 1993).

Although MAPS and CES use mist nets, any accepted capture technique can be used, provided that effort is standardised (same number of traps, places, time periods, etc.).

The information obtained from such schemes has several uses. First, changes in the size of the adult population can be calculated. This can either be done in terms of a simple index based on the number of adult birds caught, in which case between-year comparisons are based on the number of individuals caught during the season irrespective of how many times each was caught, or it can be done using absolute population levels calculated from capture–recapture methods (see below and Chapter 3). Second, indices of post-fledging productivity can be calculated from the ratio of juveniles to adults caught late in the breeding season. Finally, adult survival-rates can be calculated from between-year retraps of ringed birds (Buckland & Baillie 1987; Peach *et al.* 1990; White & Burnham 1999). Because capture probability does not vary between years, survival-rate analyses are greatly simplified by constant-effort ringing. Such estimates are bound to be minimum because adults that survived between years, but did not return to the same site, are assumed to have died. Increasingly sophisticated methods have been developed to calculate survival rates (Clobert *et al.* 1987; Pollock *et al.* 1990; Lebreton *et al.* 1992; White & Burnham 1999; Nichols *et al.* 2004). First-year survivorship (the survival rate of young birds) cannot be so readily calculated, as the young birds often do not return to their natal site to breed.

Advantages and disadvantages

Unlike most other census methods, capture per unit effort provides information on productivity and survival. Constant-effort schemes are an excellent way of directing the efforts of numerous ringers who would otherwise ring in non-standardised and less useful ways. Long periods of training followed by application for a license are necessary before any ringing can be undertaken in some countries, and expensive specialist equipment is required. This makes it an unattractive method in many circumstances. In addition, it is time-consuming, sites are often chosen rather than randomly allocated, and habitat succession at sites can confuse the long-term picture. In consequence, it is not the most appropriate method for monitoring population levels and is usually able to cover only a very limited set of species. It can, however, be useful for censusing species that live in habitats within which observation is difficult (e.g. dense undergrowth, forest canopy and reed-beds). The method is best for species with high probabilities of retrapping (e.g. warblers).

Biases

Because of the constraints on finding somewhere suitable to perform constant-effort ringing, sites are rarely randomly allocated and thus between-year changes might not faithfully represent changes on a larger, e.g. regional or national, scale. Any change in methods between years, for example change in location, number and length of nets and their mesh size, could lead to bias, as could successional change to the habitats, which is likely in the typical range of habitats chosen. Some individuals of some species may become 'trap-shy' and thus will actively avoid being recaught.

Capture–mark–recapture

In principle this method can be used for a wide range of species, though in practice the method is rarely used as a census technique.

Method

If birds are caught and individually marked, for example with rings (bands), then the population size can be estimated from the ratio of marked to unmarked birds subsequently recaptured or resighted. This can readily be explained by example. Assume that 100 birds of a particular species were caught at a site, marked and released, and, a week later, 50 of the same species were caught (or seen), 25 of which had been marked on the first day. If the proportion of birds caught on the second date that were marked (25/50) is the same as that in the whole population, then the size of the population at that site is 200.

An array of models (see Chapter 1) has been developed to analyse data from capture–mark–recapture studies. The simplest, the Lincoln index (or Petersen method), assumes one capture and one recapture (or resighting) event only, and that the population is closed. The calculations are essentially those described above. Models that are more complex allow for multiple capture (resighting) events and for open populations. The latter types of model, generally known as Jolly–Seber models, provide information both on population size and on survival rates.

Though colour rings (bands) are the most common method of marking, several other techniques, such as the use of wing tags, neck collars, colour dyes and radio transmitters, are also available (see e.g. Bibby *et al.* (2000) and Gosler (2004)).

Advantages and disadvantages

In practice, capture–mark–recapture is rarely used to estimate population sizes of birds. For many species, it is difficult to catch a large enough sample and there is a host of sources of error. Since most species of bird are readily observable, other techniques are usually preferred. For some species, however, particularly skulking ones and those living in the forest canopy, it may be the only practical method.

Biases

Estimating population sizes using capture–mark–recapture requires that numerous assumptions are not broken. It assumes that birds mix freely within the population; that the population is closed and no birds enter or leave through births, deaths or movements; that marking does not affect the probability of recapture (or resighting); that marked birds have the same probability of survival as unmarked birds; and that marks do not fall off or become less visible. Although many of these assumptions will be broken, it is possible to minimise their influence on the results. If the first and second capture dates are close together, the study site well defined and the study undertaken outside the breeding and migration periods, then the population will more approximate a closed one and population estimates will be more reasonable.

Counting nests in colonies

This method is used for colonial nesters, particularly seabirds and waterbirds.

Method

About one-eighth of bird species nest in colonies and, as a consequence, are particularly easy to census during the breeding season, when that season is reasonably well defined. The technique adopted depends upon whether the colony is on a cliff face, or whether the species nests on the ground, in burrows, trees or bushes. Each is treated in turn below, and each method relies on discriminating occupied from unoccupied nests. Birkhead & Nettleship (1980), Lloyd *et al.* (1991), Gilbert *et al.* (1998), Mitchell *et al.* (2004) and Steinkamp *et al.* (2003)

provide details of methods for counting colonial seabirds, the latter also including colonial waterbirds. Two comprehensive manuals of methods for censusing seabirds, one for Britain and Ireland (Walsh *et al.* 1995), the other for Gough Island (Cuthbert & Sommer 2004), are also available.

Many colonial nesting species breed synchronously (although not all; and asynchronous breeding creates particular problems). This can be advantageous for censusing, since a high proportion of breeding birds will be at the colony during the same period. The best time to count is generally from midway through incubation to early in the nestling stage (Bullock & Gomersall 1981; Hatch & Hatch 1989). Any earlier and some clutches might not have been started; any later and some pairs may already have lost chicks and deserted the nest site, both leading to underestimates of the total breeding population. Other colonial species have a more protracted breeding season, sometimes with high rates of nest failure, e.g. the greater flamingo *Phoenicopterus rubber* (Green & Hirons 1988). For these, population estimation can be more difficult since at any time part of the population might be elsewhere, and the birds present during one visit might not necessarily be those present during another. Under such circumstances, individual marking of birds may be necessary.

Cliffs

Ideally, count from a position slightly above but opposite the colony. Do not count from immediately above the colony, since many nests will be missed. Several ornithologists have died studying seabirds and it is important to ensure that the counting position and access route to it are safe. Count pairs of birds or occupied nest-sites (or at least apparently occupied nest-sites). For some species that nest at very high densities (e.g. guillemots, *Uria aalge*), however, pairs of birds are difficult to count because this requires identification of all sites with eggs, young or incubating adults; this can take hours of observation. For such species, counts of individual birds are more effective. Subdivide large colonies into smaller units for ease of counting. Photographs can be used to divide the cliff into counting units and to provide a permanent record of them. Colony attendance can vary both with season and diurnally and should be taken into account in deciding when to count.

For some highly visible colonial nesters, particularly those that, like the gannet *Morus bassanus*, build substantial nests, it may be simpler to photograph the colony and to count nests directly from the photograph, or by projecting an image onto a large screen and counting from that. This method is particularly useful where there is no suitable position to count from, yet a photograph could be taken (e.g. from a boat or from the air). For some species, photography may be unsuitable since black and white seabirds are readily confused with guano and shadows.

Burrows

Burrow-nesting seabirds are particularly difficult to census, many of them return to land after dark, and burrows may be unoccupied or occupied by more than one pair. They are best censused by sampling, using random or stratified random quadrats or line transects (e.g. Brooke (1990), Gibbons & Vaughan (1998), Ratcliffe *et al.* (1998) and Mitchell *et al.* (2004); see also Chapter 2) and then counting the number of burrows that appear occupied. Circular sampling plots are easy to use in practice, since a fixed length of rope tied to a stake will give a fixed size of plot. A rope and stake are also easy to carry into the field. Occupied burrows can often be recognised by a range of features such as feathers, excavated earth, droppings, broken eggshells and smell (especially when young are present) or by planting matches or toothpicks around the entrance to the burrow and seeing whether these get knocked over, but beware pre-breeding birds that are prospecting for nest sites. An endoscope (optical fibrescope) or miniature nest camera on a flexible pole can also be used to examine the nest contents, though if the burrow contains too many bends this may be time-consuming or impossible. It is often useful to play recordings of the call; if a response is elicited then the burrow is occupied, though not all birds will necessarily respond (see 'Response to playback', below). Digging down to the nest to expose the nest chamber is discouraged. For nocturnal species, it may be useful to play recordings of the call to elicit a response (James & Robertson 1985; Ratcliffe *et al.* 1998; Mitchell *et al.* 2004). As digital nest-camera technology and recording improves, it is increasingly possible to use remote triggering and digital recording to monitor nest burrows in an unobtrusive and highly efficient manner.

One perennial problem with censusing burrow-nesting species is that it is necessary to distinguish the burrows of various species and to exclude those of mammals. This is not always straightforward. The simplest way of overcoming this is to survey in places, or at times of year, when only the species under study is present. This will clearly not always be practical, and should the use of endoscopy, nest cameras or playback prove impossible, it may be necessary to develop more sophisticated techniques, such as that of Alexander & Perrins (1980), which is based on mark–recapture.

Colonies of ground-nesting species

Many species of seabirds, e.g. gulls, terns, penguins and albatrosses, nest in colonies on the ground. If the colony is small (fewer than 200 pairs) and can easily be viewed, then the number of nests may be counted directly. For larger colonies, it is probably sensible to subdivide the colony and count each section separately. Old nests, which can readily be identified by the lack of a white coat of faeces, should not be included in the counts.

Counts should be carried out at the time of year when adults are most likely to be on the nest (usually from mid-incubation to soon after hatching) and during the time of day when attendance is most stable. This will probably vary between species and colonies, but, as a rule, avoid early morning and evenings.

Particularly extensive colonies are probably best censused using line transects or quadrats (e.g. Thompson and Rothery (1991); see Chapter 3). To do this, first map the colony boundaries and calculate the overall area of the colony. When using transects, define their location, mark them on the ground with string, walk their length and count all occupied nests up to a set distance (e.g. 1 m) from the transect line. Do not count the same nest twice. Alternatively, if using quadrats locate them at random within the colony, or at equal distances along a transect, and count all occupied nests within each quadrat. It is simple to calculate the colony size from the total area of the colony and the total number of occupied nests in, and area of, all transects or quadrats. Rather than simply counting all nests within fixed-width transects, all nests seen from the transect could be counted, their distance from the transect line measured and distance-sampling methods (see above and Chapter 3) used to estimate densities and thus total numbers of nests.

Any technique that forces adults off their nests is traumatic both for birds and for the observer. It is essential to ensure that disturbance is kept to a minimum and adults should not be kept off the nest for more than 30 min, ideally less. Colonies should not be disturbed when it is very wet, cold or hot, for fear of causing egg and chick losses. Care should be taken to ensure that chicks do not run off, predators do not take advantage of the disturbance and eggs and young are not trampled. Such techniques are probably best not used near public areas.

Rather than walking through colonies to count nests, a more rapid and less invasive technique for small colonies is the flush count (Steinkamp *et al.* 2003). In this approach, all birds in the colony are flushed into the air with a loud noise and all flying birds counted. Using this method, the colony size can be estimated only if the relationship between flying birds and breeding pairs is known. For Arctic terns *Sterna paradisaea* three flushed birds correspond to two breeding pairs (Bullock & Gomersall 1981; Whilde 1985; Bibby *et al.* 2000), though this relationship will vary among species. An obvious concern with this approach is disturbance and nest abandonment at times when birds are particularly sensitive (e.g. during colony settlement early in a breeding season).

Tree colonies

Many herons, egrets, storks and spoonbills nest in dense colonies in trees, though species from several other groups (e.g. crows and weaver birds) do so as well. For those which nest in deciduous trees reasonably early in the year, nests are best counted before the leaves have completely emerged, or else the nests will become obscured. Occupied nests can often be identified by the presence of fresh nesting material, droppings in the nest, or underneath, and incubating or attendant adults or chicks calling in the nest. Alternatively, it may be necessary to use a mirror or miniature video camera on the end of a long telescopic pole to see into the nest cup. Many herons, egrets, storks and spoonbills are sensitive to disturbance and for these species it is probably not a good idea to visit until the colony is well established, once egg-laying has commenced. Even then, extreme care should be taken to ensure minimal disturbance.

Fortunately, many tree-nesting species are highly visible from a distance, so, provided that a suitable vantage point can be found, sensible nest counts are reasonably straightforward. Observation from custom-built tower hides is often best. Where a suitable ground-based vantage point cannot be found, aerial counts can be undertaken, finances permitting. Aerial counts of great blue herons *Ardea herodias* recorded 87% of the ground total while aerial photographs recorded 83% (Gibbs *et al.* 1988). Aerial methods were considered less disruptive than ground counts, were precise and were highly repeatable. The least disruptive method used, however, was a ground count of used nests after the breeding season.

Advantages and disadvantages

The biggest advantage of this method is that counts are undertaken at a time of year when the species is highly clumped and thus can be counted in a very cost-effective manner. At other times of year these species may well be spread over a very much larger area and are consequently very difficult to census. The disadvantages are that it is suitable only for breeding birds and that care has to be taken to keep disturbance to a minimum and to count all nests present.

Biases

Colony attendance can vary both diurnally and throughout the nesting season and this must be taken into account. Poor vantage points for counting can lead to unknown biases and probable under-recording.

Counting roosts

This method is used for communally roosting species, particularly waders, many wildfowl, parrots and some passerines.

Method

Many species of birds roost communally either during the night or, among coastal species, at high tide when their feeding grounds are covered. Birds are highly clumped at roosts and thus can be efficiently censused at this time. Many species roost only in the non-breeding season, though some species (e.g. colonial corvids such as the jackdaw *Corvus monedula*) roost during the breeding season as well; in this case males go to roost whilst females incubate.

When roosts are small and easily viewed, birds can be counted at the roost. When they are large or hidden, for example in trees or on rooftops, it is best to count flocks of birds (see 'Counting flocks', below) entering or leaving the roost. This is particularly the case at dusk when flocks of birds coming to roost may be visible against the sky, but are invisible once on the ground or in the trees.

For some estuarine species where alternative feeding areas such as salt marsh are available once the mudflats have been covered by the incoming tide, accurate counts can be obtained during very high tides when the sea covers all potential feeding sites. Waders are best counted at high-tide roosts during the 2 h either side of the highest spring tide of the month. If counting over a large area encompassing several roosts, count all roosts simultaneously, since individuals may move between roost sites. Pithon and Dytham (1999) used teams of volunteers to census ring-necked parakeets *Psittacula krameri* using simultaneous counts at all known roosts.

Roosts can be counted only once located. However, since many roost sites are traditional they are often well known. Unknown roost sites can be located by following the flight paths of flocks of birds as dusk, or high tide, approaches. Some coastal species, however, can roost on agricultural land up to 1 km inland.

Advantages and disadvantages

This is by far the easiest way of counting many species that are widely dispersed at other times, and is particularly useful outside the breeding season.

Biases

Some roosts are enormous and contain several million birds (e.g. of starlings *Sturnus vulgaris* and quelea *Quelea quelea*). As when counting flocks (see below), numbers of birds in large roosts are probably underestimated, especially if the species is small.

Counting flocks

This technique is used for flocking species, particularly waders, wildfowl and some passerines.

Method

Where the flock is of no more than a few hundred birds, all can be counted directly from a suitable vantage point through binoculars or a telescope. This is easy with large birds but becomes progressively more difficult with larger numbers and smaller birds at greater distances.

For small flocks, say fewer than 500 birds, count individual birds. With large numbers of birds, or with mobile flocks, however, count in tens, twenties or even greater numbers rather than counting individual birds, and estimate what proportion of the flock each represents. Use landmarks to divide large flocks on the ground into smaller groups. When flocks are particularly dense, try to count them from above if possible. However, take care not to disturb the birds, by keeping at a distance and counting from a concealed location. Try to ensure that the sun is behind you when counting, since this will improve your view of the flock, allowing you to discriminate among species and to make a more accurate census.

Advantages and disadvantages

This method is of most use outside the breeding season. Unlike roosts, the location of flocks is often not traditional and may respond to short-term changes in food resources. Flocks may be composed of several species, so it is necessary to get close enough to ensure that you can count each species separately without disturbing the flock. When searching an area systematically for flocks the data obtained can be markedly non-normal in distribution (e.g. many areas with no birds, and thus zero counts, and a few areas with very large counts); such data can present problems during statistical analyses.

Biases

Prater (1979) has shown that observers generally overestimate sizes of small flocks (a few hundred birds) and underestimate sizes of large flocks (a few thousand birds), and that there is a high degree of between-observer variability. Rapold *et al.* (1985) also document large observer errors in the estimation of flock sizes. To help overcome this, obtain repeated counts of a flock and, where possible, get others to count the same flock, too, for comparison. Flocks tend to bunch in the centre so that different parts of the flock covering similar areas will not contain the same number of birds.

Counting migrants

This method is used for migrating raptors, storks, pelicans and some passerines.

Method

Diurnal migrants

Several species of large diurnal migrants pass through bottlenecks on their migration routes. Raptors and storks on migration in Europe and the Middle East concentrate at narrow sea cross-ings, and birds at many of the best-known sites, such as the Straits of Gibraltar and the Bosphorus (Turkey), are routinely counted. During the autumn in North America, many hawks pass through migration funnels along the barrier islands off the coast and over ridgetops in the Appalachians and down through central America. Because these species are often widely dispersed at other times of year, often over huge areas with sparse human populations, counting birds at bottle-necks, particularly when they are limited in number, can be an extremely efficient census method (Meyburg & Chancellor 1994).

Though the most complete counts are made over the entire migration period, 80%–90% of some raptor species pass through a bottleneck during a 2–3-week time window, the dates of which are often well known. Teams of observers position themselves at vantage points (often high ground, with a wide field of view) 6–8 km apart across the breadth of the flyway. The number of teams will depend on the breadth of the flyway, though 6–8 km is about the optimal distance to avoid different teams counting the same birds. Each team should have a few observers, including someone skilled in raptor identification, and another member to take notes. Bildstein & Zalles (1995) give further detail on raptor-counting methods, but essentially this is done by observers methodically scanning the horizon, then moving their binoculars up one field of view, and repeating this several times. The number of gliding birds (not those circling in thermals) passing per hour is counted. It is useful if one observer counts to the north, one to the south and one overhead. Ideally, the teams should communicate with radio transmitters to avoid duplicate counting. Where numbers become too great to count, it may be sensible to photograph the passing flock, project the image onto a screen, and count the dots (Smith 1985). However, this can be slow and the same birds may be duplicated on different images. Well-coordinated raptor counts, repeated from year to year, are used to monitor raptors in North America (Lewis & Gould 2000).

Nocturnal migrant songbirds

A large proportion of migrants travel at night and some nocturnal migrant songbirds call to one another to keep in touch. These contact calls are generally specific to the species, though it is often hard to hear them against a background of other noise. Using sensitive microphones and customised software it is possible to distinguish automatically among species and to count the number of each species passing overhead at night (Evans 1994; Evans & Rosenberg 2000; http://www.birds.cornell.edu/brp/idex.html). Currently the technology does not count birds that fly above 1000 m and, for monitoring purposes, it assumes that a constant proportion of each species calls as they fly; whether or not this is the case is unknown.

An alternative approach to studying nocturnal migrants is to observe them through a telescope as their silhouettes pass across the Moon's disc at night (Lowery & Newman 1966; Alerstam 1990). With this Moon-watching method, birds within the narrow cone of sky between the observer and the Moon can be seen and counted. Migration intensity can be determined by calculation. Though the method requires cloud-free weather, the smallest songbirds can be detected at a distance of 2 km with a 20 × telescope.

Radar has been used to ascertain migration routes as well as to calculate the size and speed of migrating flocks, though rarely their specific identity. It has even been used to estimate seabird numbers (Burger 1997). Useful summaries of the uses of radar in ornithology are given in Eastwood (1967) and Alerstam (1990); however, since the method requires access to extremely sophisticated equipment it is beyond the reach of most ornithologists.

Advantages and disadvantages

A large proportion of the population of some migrants passes through bottlenecks, thus allowing a cost-efficient method of counting these species. Identification of species can be difficult, particularly for high-flying and nocturnal migrants. Large numbers of well-coordinated personnel are needed to count migrating raptors.

Biases

Weather conditions can cause raptor streams to change position, and thus flight paths can be missed. Differing levels of observer experience can lead to bias through misidentification. Double counting by observers who are spaced out to count over a broad front can lead to overestimates of population size. At mid-day, some raptors may migrate at heights invisible to the naked eye and, in general, the uncertainty over heights at which individual species migrate makes calculations of migration intensity difficult. Some migration bottlenecks may have been overlooked.

Indirect methods of censusing

Indirect methods are used for censusing wildfowl and gamebirds (droppings) and to detect the presence of elusive, nocturnal, ground-living species (footprints).

A variety of indirect methods can be used to detect the presence, and in some cases abundance, of species; two are considered here – measurements of droppings and footprints.

Dropping counts

Method

Determining the distribution of feeding wildfowl can be very time-consuming since they often feed in flocks that frequently move between sites. In some instances, daily observations are needed to determine which sites a flock visits. A simple way of overcoming this is to count the density of droppings at each site (Owen 1971). A single count can give a relative measure, but a much more accurate measure is obtained if plots are cleared on a regular basis.

In order to do this it is first necessary to determine how long the droppings last before they disintegrate and become either indistinguishable from one another, or completely unidentifiable. Fresh droppings can be marked with bamboo stakes and then revisited over a period of days to determine how long they last prior to disintegration. Inevitably, they will disintegrate much more quickly in the rain and when subjected to trampling. Bamboo stakes are then placed at random locations (20 stakes gives a good sample if the species is reasonably common) throughout the feeding site. All droppings within a given radius of each bamboo stake are then removed. The simplest way to measure this radius in the field is to use a length of string tied to the bamboo stake. If a spoon is tied to the other end, it can be used to 'flick' the droppings off the circular plot.

The area is then revisited at an interval such that all droppings produced in the interim period will still be visible, and the numbers present in the set radius around each of the randomly allocated stakes are counted. The mean number of droppings produced per unit area per day can then be calculated from the number of droppings counted, the number of days between clearing and counting, and the area of each circular sample plot. The whole procedure can then be repeated if required.

These data provide only a relative measure of the extent to which various sites are used, though they can be converted to the number of bird-days by estimating the dropping rate of the species. This involves watching an individual bird's bottom for a period of 10 or 15 min. If the bird turns out of sight or the view becomes blocked, switch observations to another individual. Intake rate may vary with position in the flock; thus, to determine a mean dropping rate, observations of birds throughout the flock should be made.

Gamebirds also have persistent and recognisable droppings. This is particularly useful for surveying elusive forest pheasants. Recording the presence/absence of droppings during a timed search and counting the number of droppings found along transects are frequently used methods.

Advantages and disadvantages

This is a very useful way of censusing elusive forest pheasants and determining site usage by wildfowl remotely. Some sites may be visited by several species, and distinguishing among their droppings may be difficult.

Biases

Heavy rain and trampling can cause droppings to disintegrate, which makes counting more difficult.

Footprints and tracking strips

Method

This method has been used to detect the presence of the ground-dwelling Jerdon's courser *Rhinoptilus bitorquatus*. Because it is nocturnal and lives in a densely wooded habitat, the species is very elusive and its presence was detected from footprints left in tracking strips (Jeganathan *et al.* 2002). These tracking strips were 5-m lengths of fine-grained soil about 2 cm deep. Birds that ran across these strips left tracks of footprints, and, to find out which species left which tracks, automatic cameras were placed at the ends of some. By measuring the shape of the footprints, either from casts or photographs, the distinctive tracks of Jerdon's courser were determined, allowing them to be separated from those of several similar species in the area. In areas where the species was known to occur, its tracks were recorded on about one in 30 nights. Calculations suggested that, if a grid of 15 strips were checked for about a month, Jerdon's courser would be very likely to be detected if present.

Advantages and disadvantages

This is a time-consuming method. It requires the preparation of tracking strips and consideration of the best type of substrate for capturing footprints; this may well require trial-and-error assessments of the soil types locally available. It is necessary to distinguish among the tracks of various, sometimes closely related, species with similar tracks. To do this it is important to determine which species left which track and this is best done with automatic cameras, fired when a bird walks across the strip. Inclement weather can destroy tracking strips, as can cattle or sheep, so strips may require frequent checking. It is difficult to estimate numbers of birds rather than their simple presence. The method is, however, non-disruptive – observers are not present when birds are active – and may be the only method to detect the presence of such cryptic species.

Biases

Misidentification of tracks, particularly among closely related species, could lead to errors in detecting whether a species is present or not.

Response to playback

This approach is used for many species that cannot easily be seen or heard in their breeding or non-breeding habitat.

Method

Some species of birds are notoriously difficult to see or hear, but will respond to a recording of their song or call. Examples of such species are those that have a skulking behaviour; those that live in dense habitats at certain times of the year; those that are nocturnal or crepuscular; and those that nest down burrows. Recordings of the songs and calls of many species are now available commercially, and can be copied onto tape or digital media. Ideally, use a tape loop or digital equivalent, so the song will continue to be broadcast until the player is switched off. If you are using tape playback, and a tape loop is not available, start recording at the beginning of the tape, so that it can easily be wound back to the start of the recording. The song can be broadcast from a hand-held loudspeaker or from one mounted on a vehicle. It is often sensible to combine passive recording with playback to increase the chance of detecting birds (frequently half the time is given over to passive recording and half to playback and listening for a response). One problem with the use of playbacks is that some individuals or species may habituate themselves (cease to respond) to the playback if it is used too frequently.

Playback can be used alongside other census methods, for example, during territory mapping or transects. Wotton *et al.* (2002) used tape playback alongside a transect method to increase their chances of detecting ring ouzels, *Turdus torquatus*, in a survey of the population in the British uplands. In this open and rocky terrain ouzels can be surprisingly difficult to detect. Playback accounted for about a third of all territories located, suggesting that it was successful in eliciting responses, but the response rate varied both seasonally and geographically. The new method set a baseline against which future surveys could be compared, but fell short of establishing a full calibration of response rates against known numbers. Similarly, a combination of playback and point-count observations has been proposed as a standard survey protocol for burrowing owls, *Athene cunicularia*, in North America (Haug & Didiuk 1993; Conway & Simon 2003). The method has also proved effective in assessing populations of North American wood warblers (*Parulini*) on their wintering grounds in Central America (Holmes *et al.* 1989; Lynch *et al.* 1985; Sliwa & Sherry 1992; Graves 1996). At this time of year, although these birds often establish territories, they do so in dense and sometimes remote habitats and can be highly inconspicuous. Playback greatly increases the probability of detecting a species when it is present and, if used in a standardised manner, allows the population to be monitored.

The results from census work involving playback, however, need careful interpretation. If the aim is simply to determine whether a given species is present in an area, then playback may simply increase the chance of finding it, as described above. If, however, the aim is to estimate the population size or to produce a population index, then more care is needed. To generate a reliable population index the probability of birds responding to playback needs to be held constant. This can be helped, for example, by standardising the manner in which the call is played (same volume, recording, playback length, time of day, season, etc.) and ensuring that the call is not played to any one individual too frequently, causing it to habituate and respond less frequently. Playback has been used widely for monitoring populations of marsh birds, owls and raptors (Fuller & Mosher 1981; Mosher *et al.* 1990; Lor & Malecki 2002; Newton *et al.* 2002; Conway 2003). Detailed studies of black rails, *Laterallus jamaicensis*, in North America, however, illustrate some of the difficulties in using this technique to estimate population size and monitor populations (Legare *et al.* 1999; Conway *et al.* 2004). Response to playback varies with the sex of the bird, month, year and temperature (Legare *et al.* 1999) and yet vocal surveys provide the only practical way to census these birds and standardised procedures have been proposed for monitoring purposes (Conway *et al.* 2004).

Estimating absolute population size from playback is more complex because the probability of the average bird in the population responding to playback needs to be estimated. Detailed additional work will often be required to learn more about when and where birds will response and at what rates. The most recent census of seabirds in Britain and Ireland, for example, relied heavily on tape playback for three species of burrow-nesting procellariiformes: storm petrels, *Hydrobates pelagicus*; Leach's petrels, *Oceanodroma leucorhoa*; and Manx shearwater, *Puffinus puffinus* (Mitchell *et al.* 2004). Additional calibration work was undertaken to estimate the response rate, which can vary between the sexes, within and among colonies, and even across years (Ratcliffe *et al.* 1998; Mitchell *et al.* 2004).

Advantages and disadvantages

Skulking, secretive and nocturnal species that would otherwise be overlooked can be located and population indices or population estimates produced. Careful and highly standardised use of playback is essential. Great attention needs to be paid to survey protocols and the methods need to be published in full to allow others to replicate them in the future and make comparisons. Some species and individuals may rapidly habituate themselves to playback. Variation in response rate needs careful assessment and will often require research that is more intensive.

Biases

Care should be taken to ensure that playbacks are broadcast for set durations, at a standard volume and under set conditions (e.g. time of day, weather), otherwise responses will vary. Ensure that the precise use of playbacks is noted, otherwise it may be impossible to repeat the survey. This has particularly been the case when playback has been used with constant-effort mist netting; sometimes tapes have been used, sometimes not, and use has often gone unrecorded. Such variation makes comparison of the results very difficult.

Vocal individuality

This approach is used for rare species that are difficult to see or capture in their preferred habitats.

Method

The songs and calls of many bird species are individually unique and often identifiable, if not by ear, then from a sonogram analysis, allowing them to be used for monitoring (Saunders & Wooller 1988). Sound recordings of birds are widely available, for example, from the British Library's sound archive (http://www.bl.uk/collections/sound-archive/wild.html), or from similar sources, and constitute a highly useful resource. Acoustically distinct calls of this kind have considerable potential in monitoring and conservation, particularly for birds that occur in dense vegetation or are otherwise difficult to observe, but this potential has not always been realised (McGregor *et al.* 2000). The method involves recording songs or calls with a directional microphone and examining sound spectrograms using freely available software. Spectrograms from an individual bird are often recognisable by eye and discrimination can be formalised using statistical techniques. The songs or calls of individuals are recorded onto tape or digital medium, preferably with a directional microphone, and sound spectrograms produced using readily available computer software such as RAVEN (http://www.birds.cornell.edu/brp/raven/Raven.html). The spectrograms of individuals can then be visually separated, either by a panel of observers or by the more time-consuming, but more rigorous, approach of measuring the duration and frequency of each component of the spectrogram, and using discriminant function analysis to distinguish among individuals (Gilbert *et al.* 1994).

Work on European bitterns, *Botaurus stellaris*, denizens of dense reed-beds, has shown that their booming calls are individually distinct. This has allowed their numbers to be tracked precisely and year-to-year survival to be estimated (Gilbert *et al.* 2002). Knowledge gained using this method greatly increased the researchers' understanding of bittern behaviour and it is hard to see how this could have been achieved using other methods. In a study of the corncrake, *Crex crex*, which is highly cryptic and calls from rank and dense habitat mainly at night, information gained from vocalisations increased census estimates in some areas by 20%–30% (Peake & McGregor 2001). This study also showed that males called less frequently than had previously been thought. The churring call of the male European nightjar *Caprimulgus europaeus*, a mainly nocturnal and mobile species, has also been shown to differ among individuals (Rebbeck *et al.* 2001). The pulse rate of calls and the phase lengths together allow the identification of the great majority of males. Interestingly, males were shown to move some distance within a breeding season, but they returned to the same territory year after year. In each case, quantitative rules were developed to help discriminate one bird from another, but this is not always straightforward and, in some cases, ambiguity remains. One can also apply capture–mark–recapture methods to resightings based on vocalisations to estimate population size. In contrast to the successful studies described above, while the calls of black-throated diver, *Gavia arctica*, were found to be distinct, their calls proved to be too infrequent and too difficult to record to make the method viable (McGregor *et al.* 2000).

Advantages and disadvantages

This is sometimes the only possible method and produces minimal disturbance. It is often advisable to pilot the method before embarking on a full study because there is no guarantee that it will work in all situations. It is non-intrusive; this might be particularly useful in studying rare and endangered species. The disadvantages are that it requires a good deal of hard work to collect high-quality recording of birds that often live at low densities across widely scattered sites and to analyse and distinguish among calls. Ideally, one also needs an independent means of identification, such as marking or radio tracking, to corroborate the findings. It requires specialist and quite expensive equipment; it often only tells us about breeding males; and it can be very time-consuming unless the analysis is automated (Rebbeck *et al.* 2001). For some species, the calls of individuals may vary between years, even though they are consistent within years.

Biases

Only calling or singing birds (mostly males) can be censused by this method. Females, immature males, non-breeding males and possibly males at low densities may well be missed since they may vocalise less frequently.

References

Alerstam, T. (1990). *Bird Migration*. Cambridge, Cambridge University Press.

Alexander, M. & Perrins, C. M. (1980). An estimate of the numbers of shearwaters on the Neck, Skomer, 1978. *Nature in Wales* **17**, 43–46.

Anon 2002. *The UK Sustainable Development Strategy*. London, Department for Environment, Food and Rural Affairs.

Anselin, A. (ed.) (2004). *Bird Numbers 1995, Proceedings of the International Conference and 13th Meeting of the European Bird Census Council, Pärnu, Estonia. Bird Census News* **13**, 1–198.

Baker, J. K. (1993). *Identification Guide to European Non-passerines*. Thetford, British Trust for Ornithology.

Bart, J. & Klosiewski, S. P. (1989). Use of presence–absence to measure changes in avian density. *Journal of Wildlife Management* **53**, 847–852.

Bart, J. & Earnst, S. (2002). Double sampling to estimate density and population trends in birds. *The Auk* **119**, 36–45.

Bennun, L. & Howell, K. (2002). Birds. In *African Forest Biodiversity: A Field Survey Manual for Vertebrates*, ed. G. Davies & M. Hoffmann. Oxford, Earthwatch, Europe, pp. 121–153.

Bennun, L. A. & Waiyaki, E. M. (1993). Using timed species counts to compare avifaunas in the Mau Forests, south-west Kenya. In *Birds and the African Environments: Proceedings of the Eighth Pan-African Ornithological Congress*, ed R. T. Wilson. *Annales du Musée Royal de l'Afrique Centrale (Zoologie)* **268**, 366.

Bibby, C., Jones, M. & Marsden, S. (1998). *Expedition Field Techniques*. London, Royal Geographical Society. http://home.eclions.net/scno.Birdsurvey.pdf.

Bibby, C. J., Burgess, N. D. & Hill, D. A. (2000). *Bird Census Techniques*, 2nd edn. London, Academic Press.

Bibby, C. J., Phillips, B. N. & Seddon, A. J. (1985). Birds of restocked conifer plantations in Wales. *Journal of Applied Ecology* **22**, 619–633.

Bildstein, K. L. & Zalles, J. I. (eds.) (1995). *Raptor Migration Watch-site Manual*. Kempton, Pennsylvania, Hawk Mountain Sanctuary Association.

Birkhead, T. R. & Nettleship, D. N. (1980). *Census Methods for Murres Uria Species: A Unified Approach.* Ottawa, Canadian Wildlife Service.

Briggs, K. T., Tyler, W. B., & Lewis D. B. (1985). Comparison of ship and aerial surveys of birds at sea. *Journal of Wildlife Management* **49**, 405–411.

Brooke, M. (1990). *The Manx Shearwater.* London, Poyser.

Buckland, S. T. & Baillie, S. R. (1987). Estimating bird survival rates from organised mist-netting programmes. *Acta Ornithologica* **23**, 89–100.

Buckland, S. T., Anderson, D. R., Burnham, K. P., Laake, J. L. & Borchers, D. L. (2001). *Introduction to Distance Sampling: Estimating Abundance of Biological Populations.* Oxford, Oxford University Press.

Bullock, I. D. & Gomersall, C. H. (1981). The breeding population of terns in Orkney and Shetland in 1980. *Bird Study* **28**, 187–200.

Burger, A. E. (1997). Behaviour and numbers of marbled murrelets measured with radar. *Journal of Field Ornithology* **68**, 208–223.

Burnham, K. P., Anderson, D. R. & Laake, J. L. (1980). Estimation of density from line transect sampling of biological populations. *Wildlife Monographs* **72**, 1–200.

Catchpole, C. K. (1973). The functions of advertising song in the sedge warbler *Acrocephalus schoenobaenus* and the reed warbler *A. scirpaceus. Behaviour* **46**, 300–320.

Clobert, J., Lebreton, J. D. & Allaine, D. (1987). A general approach to survival estimation by recaptures or resightings of marked birds. *Ardea* **75**, 133–142.

Conway, C. J. (2003). *Standardized North American Marshbird Monitoring Protocols.* Tucson, Arizona, U.S. Geological Survey, Arizona Cooperative Fish and Wildlife Research Unit.

Conway, C. J. & Simon, J. C. (2003). Comparison of detection probability associated with burrowing owl survey methods. *Journal of Wildlife Management* **67**, 501–511.

Conway, C. J., Sulzman, C. & Raulston, B. E. (2004). Factors affecting detection probability of California black rails. *Journal of Wildlife Management* **68**, 360–370.

Cuthbert, R. J. & Sommers, E. S. (2004). *Gough Island Bird Monitoring Manual.* RSPB Research Report, 5. Sandy, RSPB.

Dawson, D. G. (1985). A review of methods for estimating bird numbers. In *Bird Census and Atlas Studies,* ed. K. Taylor, R. J. Fuller & P. C. Lack. Tring, British Trust for Ornithology, pp. 27–33.

DeSante, D. F., Burton, K. M. & Williams, O. E. (1993). The Monitoring Avian Productivity and Survivorship (MAPS) program second (1992) annual report. *Bird Populations* **1**, 1–28.

DeSante, D. F., O'Grady, D. R. & Pyle, P. (1999). Measures of productivity and survival derived from standardized mist-netting are consistent with observed population changes. *Bird Study* **46**, S178–S188.

Diefenbach, D. R., Brauning, D. W. & Mattice, J. A. (2003). Variability in grassland bird counts related to observer differences and species detection rates. *Auk* **120**, 1168–1179.

Eastwood, E. (1967). *Radar Ornithology.* London, Methuen.

Emlen, J. T. (1977). Estimating breeding season bird densities from transect counts. *Auk* **94**, 445–468.

Enemar, A. (1959). On the determination of the size and composition of a passerine bird population during the breeding season. *Vår Fågelvärld*, Supplement **2**, 1–114.

Evans, W. R. (1994). Nocturnal flight call of Bicknell's thrush. *The Wilson Bulletin* **106**, 55–61.

Evans, W. R. & Rosenberg, K. V. (2000). Acoustic monitoring of night-migrating birds: a progress report. In *Strategies for Bird Conservation: The Partners in Flight Planning Process. Proceedings of the 3rd Partners in Flight Workshop,* eds. R. Bonney, D. N. Pashley, R. J. Cooper & L. Niles. Cape May, New Jersey, Cornell Laboratory of Ornithology. http://www.birds.cornell.edu/pifcapemay.

Farnsworth, G. L., Pollock, K. H., Nichols, J. D. *et al.* (2002). A removal model for estimating detection probabilities from point count surveys. *Auk* **119**, 414–425.

Fuller, M. R. & Mosher, J. A. (1981). Methods of detecting and counting raptors. *Studies in Avian Biology* **6**, 235–246.

Gibbons, D. W. & Vaughan, D. (1998). The population size of Manx shearwater *Puffinus puffinus* on 'The Neck' of Skomer Island: a comparison of methods. *Seabird* **20**, 3–11.

Gibbs, J. P. & Wenny, D. G. (1993). Song output as a population estimator: effect of male pairing status. *Journal of Field Ornithology* **64**, 316–322.

Gibbs, J. P., Woodward, S., Hunter, M. L. & Hutchinson, A. E. (1988). Comparison of techniques for censusing great blue heron nests. *Journal of Field Ornithology* **59**, 130–134.

Gilbert, G., Gibbons, D. W. & Evans, J. (1998). *Bird Monitoring Methods – A Manual of Techniques for Key UK Species*. Sandy, RSPB.

Gilbert, G., McGregor, P. K. & Tyler, G. (1994). Vocal individuality as a census tool: practical considerations illustrated by a study of two rare species. *Journal of Field Ornithology* **65**, 335–348.

Gilbert, G., Tyler, G. A. & Smith, K. W. (2002). Local annual survival of booming male great bittern *Botaurus stellaris* in Britain, in the period 1990–1999. *Ibis* **144**, 51–61.

Gosler, A. (2004) Birds in the hand. In *Bird Ecology and Conservation; a Handbook of Techniques*, ed. W. J. Sutherland, I. Newton & R. E. Green. Oxford, Oxford University Press, pp. 85–118.

Graves, G. R. (1996). Censusing wintering populations of Swainson's warblers: surveys in the Blue Mountains of Jamaica. *Wilson Bulletin* **108**, 94–103.

Green, R. E. & Hirons, M. G. J. (1988). Effects of nest failure and spread of laying on counts of breeding birds. *Ornis Scandinavica* **19**, 76–78.

Gregory, R. D. (2000). Development of breeding bird monitoring in the United Kingdom and adopting its principles elsewhere. *The Ring* **22**, 35–44.

Gregory, R. D. & Baillie, S. R. (1998). Large-scale habitat use of some declining British birds. *Journal of Applied Ecology* **35**, 785–799.

   (2004). Survey design and sampling strategies for breeding bird monitoring. *Bird Census News* **13**, 19–31.

Gregory, R. D., Baillie, S. R. & Bashford, R. I. (2004a). Monitoring breeding birds in the United Kingdom. *Bird Census News* **13**, 101–112.

Gregory, R. D., Gibbons, D. W. & Donald, P. F. (2004b). Bird census and survey techniques. In *Bird Ecology and Conservation; a Handbook of Techniques*, ed. W. J. Sutherland, I. Newton & R. E. Green. Oxford, Oxford University Press, pp. 17–56.

Gregory, R. D., Noble, D. A. & Custance, J. (2004c). The state of play of farmland birds: population trends and conservation status of farmland birds in the United Kingdom. *Ibis* **146** (Suppl. 2), 1–13.

Gregory, R. D., Noble, D., Field, R. *et al.* (2003). Using birds as indicators of biodiversity. *Ornis Hungarica* **12–13**, 11–24.

Gregory, R. D., van Stien, A. J., Vorisek, P. *et al.* (2005). Developing indicators for European birds. *Philosophical Transactions of the Royal Society London* B **360**, 269–288.

Hagemeijer, W. & Verstrael, T. (eds.) (1994). *Bird Numbers, 1992. Distribution, Monitoring and Ecological Aspects*. Beek-Ubbergen, SOVON.

Harrison, J. A., Allan, D. G., Underhill, L. G. *et al.* (eds.) (1997) *The Atlas of Southern African Birds*. Johannesburg, BirdLife South Africa.

Hatch, S. A. & Hatch, M. A. (1989). Attendance patterns of Murres at breeding sites: implications for monitoring. *Journal of Wildlife Management* **53**, 43–493.

Haselmayer, J. & Quinn, J. S. (2000). A comparison of point counts and sound recording as bird survey methods in Amazonian southeast Peru. *Condor* **102**, 887–893.

Haug, E. A. & Didiuk, A. B. (1993). Use of recorded calls to detect burrowing owls. *Journal of Field Ornithology* **64**, 188–194.

Helbig, A. J. & Flade, M. (1999). *Bird Numbers 1998; where Monitoring and Ecological Research Meet*. *Die Vogelwelt*, Supplement.

Hewish, M. J. & Loyn, R. H. (1989). *Popularity and Effectiveness of Four Survey Methods for Monitoring Populations of Australian Land Birds*. Victoria, Royal Australasian Ornithologists' Union.

Hilton, G. M., Atkinson, P. W., Gray, G. A. L., Arendt, W. J. & Gibbons, D. W. (2002). Rapid decline of the volcanically threatened Montserrat oriole. *Biological Conservation* **111**, 79–89.

Holmes, R. T., Sherry, T. W. & Reitsma, L. (1989). Population structure, territoriality and overwinter survival of two migrant warbler species in Jamaica. *Condor* **91**, 545–561.

IBCC (1969). Recommendations for an international standard for a mapping method in bird census work. *Bird Study* **16**, 248–255.

James, P. C. & Robertson, H. A. (1985). The use of playback recordings to detect and census nocturnal burrowing seabirds. *Seabird* **7**, 18–20.

Järvinen, O. & Väisänen, R. A. (1975). Estimating relative densities of breeding birds by the line transect method. *Oikos* **26**, 316–322.

Javed, S. & Kaul, R. (2002). *Field Methods for Bird Surveys*. New Delhi, Bombay Natural History Society, Deparment of Wildlife Sciences, Aligarh Muslim University and World Pheasant Association.

Jeganathan, P., Green, R. E., Bowden, C. G. R. *et al.* (2002). Use of tracking strips and automatic cameras for detecting critically endangered Jerdon's coursers *Rhinoptilus bitorquatus* in scrub jungle in Andhra Pradesh, India. *Oryx* **36**, 182–188.

Kelsey, M. G. (1989). A comparison of the song and territorial behaviour of a long distance migrant, the marsh warbler *Acrocephalus palustris*, in summer and winter. *Ibis* **131**, 403–414.

Kendeigh, S. C. (1944). Measurement of bird populations. *Ecological Monographs* **14**, 67–106.

Komdeur, J., Bertelsen, J. & Cracknell, G. (1992). *Manual for Aeroplane and Ship Surveys of Waterfowl and Seabirds*. IWRB Special Publication 19. Slimbridge, IWRB.

Koskimies, P. & Väisänen, R. A. (eds.) (1991). *Monitoring Bird Populations*. Helsinki, Finnish Museum of Natural History.

Lebreton, J., Burnham, K. P., Clobert, J. & Anderson, D. R. (1992). Modelling survival and testing biological hypotheses using marked animals: a unified approach with case studies. *Ecological Monographs* **62**, 67–118.

Legare, M. L., Eddleman, W. R., Buckley, P. A. & Kelly, C. (1999). The effectiveness of tape playback in estimating black rail density. *Journal of Wildlife Management* **63**, 116–125.

Lewis, S. A. & Gould, W. R. (2000). Survey effort effects on power to detect trends in raptor migration counts. *Wildlife Society Bulletin* **28**, 317–329.

Lloyd, M. C., Tasker, M. L., & Partridge, K. (1991). *The Status of Seabirds in Britain and Ireland*. London, Poyser.

Lor, S. & Malecki, R. A. (2002). Call–response surveys to monitor marsh bird population trends. *Wildlife Society Bulletin* **30**, 1195–1201.

Lowery, G. H. & Newman, R. J. (1966). A continent-wide view of bird migration on four nights in October. *Auk* **83**, 547–586.

Lynch, J. F., Morton E. S. & Van der Voort, M. E. (1985). Habitat segregation between the sexes of wintering hooded warblers (*Wilsonia citrina*). *Auk* **102**, 714–721.

Marchant, J. H. (1983). *BTO Common Bird Census Instructions*. Tring, British Trust for Ornithology.

Marchant, J. H., Hudson, R., Carter, S. P. & Whittingham, P. (1990). *Population Trends in British Breeding Birds*. Tring, British Trust for Ornithology.

Marsden, S. J. (1999). Estimation of parrot and hornbill densities using a point count distance sampling method. *Ibis* **141**, 377–390.

McGregor, P. K., Peake, T. M. & Gilbert, G. (2000). Communication behaviour and conservation. In *Behaviour and Conservation*, ed. M. Gosling & W. J. Sutherland. Cambridge, Cambridge University Press, pp. 261–280.

McKinnon, J. & Phillips, K. (1993). *A Field Guide to the Birds of Borneo, Sumatra, Java and Bali*. Oxford, Oxford University Press.

Meyburg, B. U. & Chancellor, R. D. (1994). *Raptor Conservation Today*. Berlin, Pica Press, for the World Working Group on Birds of Prey and Owls.

Meyers, J. M. & Pardieck, K. L. (1993). Evaluation of three elevated mist-net systems for sampling birds. *Journal of Field Ornithology* **64**, 270–277.

Mitchell, I. P., Newton, S. F., Ratcliffe, N. & Dunn, T. E. (2004). *Seabird Populations of Britain and Ireland: Results of the Seabird 2000 Census (1998–2002)*. London, Poyser.

Mosher, J. A., Fuller, M. R., & Kopenny, M. (1990). Surveying woodland raptors by broadcast of conspecific vocalisations. *Journal of Field Ornithology* **61**, 453–461.

Munn, C. A. (1993). Tropical canopy netting and shooting lines over tall trees. *Journal of Field Ornithology* **64**, 454–463.

Newton, I., Kavanagh, R., Olsen, J. & Taylor, I. (2002). *Ecology and Conservation of Owls*. Collingwood, Victoria, CSIRI Publishing.

Nichols, J. D., Hines, J. E., Sauer, J. R. *et al.* (2000). A double-observer approach for estimating detection probability and abundance from avian point counts. *The Auk* **117**, 393–408.

Nichols, J. D., Kendall, W. L. & Runge, M. C. (2004). Estimating survival and movement. In *Bird Ecology and Conservation; a Handbook of Techniques*, ed. W. J. Sutherland, I. Newton & R. E. Green. Oxford, Oxford University Press, pp. 119–140.

Ormerod, S. J., Tyler, S. J., Pester, S. J. & Cross A. V. (1988). Censusing distribution and population of birds along upland rivers using measured ringing effort: a preliminary study. *Ringing and Migration* **9**, 71–82.

Owen, M. (1971). The selection of feeding sites by white-fronted geese in winter. *Journal of Applied Ecology* **8**, 905–917.

Palmeirim, J. M. & Rabaça, J. E. (1993). A method to analyse and compensate for time-of-day effects on bird counts. *Journal of Field Ornithology* **65**, 17–26.

Pardieck, K. & Waide, R. B. (1992). Mesh size as a factor in avian community studies using mist nests. *Journal of Field Ornithology* **63**, 250–255.

Peach, W. J., Buckland, S. T. & Baillie, S. R. (1990). Estimating survival rates using mark–recapture data from multiple ringing sites. *The Ring* **13**, 87–102.

(1996). The use of constant effort mist-netting to measure between-year changes in the abundance and productivity of common passerines. *Bird Study* **43**, 142–156.

Peake, T. M. & McGregor, P. K. (2001). Corncrake *Crex crex* census estimates: a conservation application of vocal individuality. *Animal Behaviour and Conservation* **24**, 81–90.

Pithon, J. A. & Dytham, C. (1999). Census of the British ring-necked parakeet *Psittacula krameri* population by simultaneous counts of roosts. *Bird Study* **46**, 112–115.

Pollock, K. H., Nichols, J. D., Brownie, C. & Hines, J. E. (1990). *Statistical Inference for Capture–Recapture Experiments*. Bethesda, Maryland, Wildlife Society.

Pomeroy, D. & Dranzoa, C. (1997). Methods of studying the distribution, diversity and abundance of birds in East Africa – some quantitative approaches. *African Journal of Ecology* **35**, 110–123.

Pomeroy, D. & Tengecho, B. (1986). Studies of birds in a semi-arid area of Kenya III – the use of 'timed species counts' for studying regional avifaunas. *Journal of Tropical Ecology* **2**, 231–247.

Prater, A. J. (1979). Trends in accuracy of counting birds. *Bird Study* **26**, 198–200.

Pyle, P., Howell, S. N. G., Yunick, R. P. & DeSante, D. F. (1987). *Identification Guide to North American Passerines*. Bolinas, California, Slate Creek Press.

Ralph, C. J. & Scott, J. M. (eds.) (1981). *Estimating Numbers of Terrestrial Birds. Studies in Avian Biology,* 6. Las Cruces, Cooper Ornithological Society.

Ralph, C. J. & Dunn, E. H. (eds.) (2004). *Monitoring Bird Populations Using Mist Nets*. Studies in Avian Biology No. 29, Pennsylvania, Cooper Ornithological Society.

Rapold, C., Kersten, M. & Smith, C. (1985). Errors in large scale shorebird counts. *Ardea* **73**, 13–24.

Rappole, J. H. & Warner, D. (1980). Ecological aspects of migrant bird behaviour in Veracruz, Mexico. In *Migrant Birds in the Neotropics: Ecology, Behavior, Distribution and Conservation*, ed. A. Keast & E. S. Morton. Washington, Smithsonian Institution Press, pp. 353–395.

Ratcliffe, N., Vaughan, D., Whyte, C. & Shepherd, M. (1998). Development of playback census methods for storm petrels *Hydrobates pelagicus*. *Bird Study* **45**, 302–312.

Raven, M. J., Noble, D. G. & Baillie, S. R. (2005). *The Breeding Bird Survey 2004*. BTO Research Report 403. Thetford, British Trust for Ornithology.

Rebbeck, M., Corrick, R., Eaglestone, B. & Stainton, C. (2001). Recognition of individual European nightjars *Caprimulgus europaeus* from their song. *Ibis* **143**, 468–475.

Robbins, C. S., Bystrak, D. & Geissler, P. H. (1986). *The Breeding Bird Survey: Its First Fifteen Years, 1965–1979*. Washington, United States Department of the Interior, Fish and Wildlife Service.

Robertson, J. G. M. & Skoglund, T. (1985). A method for mapping birds of conservation interest over large areas. In *Bird Census and Atlas Studies*, ed. K. Taylor, R. J. Fuller & P. C. Lack. Tring, British Trust for Ornithology, pp. 67–72.

Rosenstock, S. S., Anderson, D. R., Giesen, K. M., Leukering, T. & Carter, M. F. (2002). Landbird counting techniques: current practices and an alternative. *Auk* **119**, 46–53.

Sauer, J. R., Hines, J. E. & Fallon, J. (2001). *The North American Breeding Bird Survey, Results and Analysis 1966–2000*. Version 2001.2. Laurel, Maryland, USGS. Patuxent Wildlife Research Center.

Saunders, D. A. & Wooler, R. D. (1988). Consistent individuality of voice in birds as a management tool. *Emu* **88**, 25–32.

Sliwa, A. & Sherry, T. W. (1992). Surveying wintering warbler populations in Jamaica: point counts with and without broadcast vocalizations. *Condor* **94**, 924–936.

Smith, N. G. (1985). Dynamics of the transithmian migration of raptors between Central and South America. In *Conservation Studies on Raptors*, ed. I. Newton & R. D. Chancellor. Cambridge, International Council for Bird Preservation, pp. 271–290.

Stastný, K. & Bejček, V. (eds.) (1990). *Bird Census and Atlas Studies. Proceedings of the XIth International Conference on Bird Census and Atlas Work*. Prague, Institute of Systematic and Ecological Biology.

Steinkamp, M., Peterjohn, H., Bryd, V., Carter, H. & Lowe, R. (2003). Breeding season survey techniques for seabirds and colonial waterbirds throughout North America. http://www.im.nbs.gov/cwb/manual/.

Svensson, L. (1992). *Identification Guide to European Passerines*. Stockholm, Fingraf AB.

Tasker, M. L., Hope Jones, P., Dixon, T. & Blake, B. F. (1984). Counting seabirds at sea from ships: a review of methods employed and a suggestion for a standardised approach. *Auk* **101**, 567–577.

Terborgh, J. (1989). *Where Have All the Birds Gone?* Princeton, Massachusetts, Princeton University Press.

Thomas, L., Laake, J. L., Strindberg, S. *et al.* (2005). Distance 5.0. Research Unit for Wildlife Population Assessment University of St Andrews, UK http://www.ruwpa.st.and.ac.uk/distance/.

Thompson, K. R. & Rothery, P. (1991). A census of black-browed albatross *Diomeda melanophrys* population on Steeple Jason Island, Falkland Islands. *Biological Conservation* **56**, 39–48.

Thompson, W. L. (2002). Towards reliable bird surveys: accounting for individuals present but not detected. *Auk* **119**, 18–25.

Tomiałojć, L. (1980). The combined version of the mapping method. In *Bird Census Work and Nature Conservation*, ed. H. Oelke. Göttingen, Dachverband Deutscher Avifaunisten, pp. 92–106.

Underhill, L. & Gibbons D. (2002). Mapping and monitoring bird populations; their conservation uses. In *Conserving Bird Biodiversity; General Principles and Their Application*, ed. K. Norris & D. Pain. Cambridge, Cambridge University Press, pp. 34–60.

Verner, J. (1985). Assessment of counting techniques. In *Current Ornithology*, Volume 2, ed. R. F. Johnson. New York, Plenum Press, pp. 247–302.

Walsh, P. M., Halley, D. J., Harris, M. P. *et al.* (1995). *Seabird Monitoring Handbook for Britain and Ireland*. Peterborough, JNCC.

Webb, A. & Durinck, J. (1992). Counting birds from ships. In *Manual for Aeroplane and Ship Surveys of Waterfowl and Seabirds*, ed. J. Komduer, J. Bertelsen & G. Cracknell. IWRB Special Publication 19. Copenhagen, IWRB/JNCC/Ornis Consult A/S Ministry of the Environment.

Whilde, A. (1985). The 1984 all-Ireland tern survey. *Irish Birds* **3**, 1–32.

White, G. C. & Burnham, K. P. (1999) Program MARK: survival estimation from populations of marked animals. *Bird Study* **46**, S120–S139.

Whitman, A. A., Hagan, J. M., & Brokaw, N. V. L. (1997). A comparison of two bird survey techniques used in a subtropical forest. *Condor* **99**, 955–965.

Wilkinson, N. I., Langston, R. H., Gregory, R. D., Gibbons, D. W. & Marquiss, M. (2002). Capercaillie *Tetrao urogallus*, abundance and habitat use in Scotland, in winter 1998–99. *Bird Study* **49**, 177–185.

Wotton, S. R., Langston R. H. W. & Gregory, R. D. (2002). The breeding status of the ring ouzel *Turdus torquatus* in the UK in 1999. *Bird Study* **49**, 26–34.

10 Mammals

Charles J. Krebs

Department of Zoology, University of British Columbia, Vancouver, B.C., Canada V6T 1Z4

Introduction

Census methods for mammals depend critically on the size of the species and its natural history. If species are diurnal, common and highly visible, the census problem is relatively simple. If species are nocturnal, rare and difficult to detect, the census problems are most difficult. As in all ecological census work, you need to decide the purpose of the study and the level of precision you require. Higher precision bears costs in time and money, and methods that lead to higher precision might not be practical for some species within a finite budget.

A sequence of decisions to facilitate the choice of methods for a mammal census is outlined in Figure 10.1 (Table 10.1). Just because many studies of a particular species or group of species have used a particular method does not mean that you must use this method for your study. Many studies have not used the best methods in the past, and there is no reason to continue using sub-optimal techniques that waste time and money.

Total counts

The simplest way to determine how many individuals of a particular species of mammal live in an area is to count all of them. This census method we might consider the Holy Grail of mammal-census methods, yet it can hardly ever be achieved. Total counts can be done on large mammals in restricted areas (Bookhout 1994), but one should always be sceptical of the accuracy of total counts, since in most cases to date there is a negative bias – estimated numbers are less than actual numbers. Grey whales have been counted since 1975 as they migrate south along the coast of central California and, while one might think that a party of observers could count all the whales moving close to shore, in fact they recorded only 79% of the whales (Rugh *et al.* 1990). The message is clear: if you think you can do a total count, check your assumption by double-counting with two independent sets of observers.

Method

If numbers are small, you can count isolated blocks of individuals, but if they are larger you will need to photograph the groups and count from the photos. This method can be used with

Ecological Census Techniques: A Handbook, ed. William J. Sutherland.
Published by Cambridge University Press © Cambridge University Press 1996, 2006.

Start

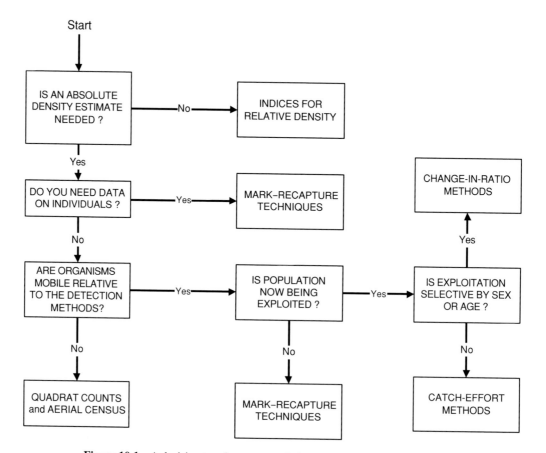

Figure 10.1   A decision tree for census techniques for mammals. It is important as a first step to have clear objectives for your census, and then to pick the best methods to achieve these objectives. Within each box of methods there are many alternative procedures (see Chapter 2). (Modified after Caughley (1977).)

helicopter or airplane counts of large mammals like elephants, reindeer and red deer, if visibility is excellent. Helicopters, although expensive, are particularly useful since they can hover until the count has been completed.

Biases

Few total counts have escaped the negative bias of undercounting what is actually there. There is unfortunately not a consistent undercounting bias that can be used for corrections of the raw counts. Observers differ dramatically in their ability to see even large mammals, so care must be taken to use the same observers whenever possible. Trees and shrubs hide some individuals, whereas others are spooked by the helicopter or airplane and may be counted twice.

Table 10.1. *A summary of methods suitable for various groups*

	Carnivores	Sea mammals	Primates	Ungulates	Bats	Rodents	Rabbits, hares and pikas	Insectivores and elephant shrews	Endentates	Page
Nesting or resting structures			+			+				354
Bat roosts and nurseries					*					354
Line transects	?	+	+	*	?	?	+		+	356
Aerial surveys	?	*	+	*						358
Individual recognition	?	+	+	+						359
Counting calls	+	?	+		+					360
Trapping	?			?	?	*	*	*		360
Counting dung	+			*	?	+	+		?	363
Feeding signs for herbivores	+		?	?	?	+	?		?	364
Counting footprints and runways	+			+		?	?	?	?	364
Hair tubes and hair catchers	?					*		+		365
Counting seal colonies										366

* Method usually applicable, + method often applicable, ? method sometimes applicable. The page number for each method is given.

Nesting or resting structures

Tree squirrels build drays in trees and beavers build lodges in lakes and streams. Any mammal that builds a visible structure can be censused by counting these structures using standard line-transect or quadrat methods (Chapter 2). Burrowing rodents, such as ground squirrels, leave obvious signs of digging in many habitats, so counting burrows can be used as a census method for these species.

Method

Searches for tree nests, burrows, or other structures built by mammals must be systematic and cover the area of interest, either in its entirety or with a set of random samples. You must know the habitat used by the species you are surveying, and you must determine whether the structures are being used. Animal signs would include digging, fresh droppings and possibly scent. Sites may have to be revisited to determine occupancy. If you wish to obtain an estimate of the absolute density, you will need to count the number of individuals occupying a sample of nesting or resting sites.

Advantages and disadvantages

Few mammal species make obvious nesting or resting sites, but many rodents dig burrows. For species that inhabit these structures, it is not always easy to determine whether a site is active or not without a great deal of observation. Burrow counts are poor estimates of rodent abundance unless one knows that the burrow is active (Van Horne *et al.* 1997). Boonstra *et al.* (1992) developed a tracking method to locate rodents in active burrows. For some species in colder environments infrared imaging can be used to detect active burrows from their heat production (Boonstra *et al.* 1994). For arctic ground squirrels that live in burrows Hubbs *et al.* (2000) developed a powder-tracking tile method that gives results closely correlated with absolute density.

Biases

Occupied burrows or nests may contain more than one individual, or in some cases one individual may inhabit several burrow or nest sites, so some knowledge of natural history may help prevent biases in converting counts into absolute density. The major bias to avoid is that of counting both occupied and unoccupied burrows or nests.

Bat roosts and nurseries

Bats are among the most difficult mammals to census because of their mobility. Both the large fruit bats or flying foxes (Megachiroptera) and the much smaller Microchiropteran bats roost by day and are active at night. Most of the census methods for bats are applied to these roost sites.

Method

Bats may be counted at the roost site or as they emerge from the site (O'Shea *et al.* 2003). A National Bat Monitoring Programme in the UK (http://www.bats.org.uk/nbmp/) has been working for 10 years to standardise monitoring of British bat populations. Three principal methods have been applied: observations at summer maternity roost sites and at winter hibernation sites, together with summer field surveys using bat detectors. Standardised monitoring protocols have been developed to collect UK-wide baseline data for each of the target species. To cross-validate, a double-sampling approach has been applied, whereby each species is monitored by two of the three principal methods:

(1) Maternity colony monitoring. Volunteers stand outside roost sites (generally houses) at sunset and count numbers of bats emerging. Two counts are made in June. Some weather and site details such as habitats surrounding the site are recorded. This survey is appropriate for volunteers with little or no previous experience – including householders.

(2) Field survey monitoring. Volunteers are allocated a randomly selected grid reference along a stretch of waterway or of a 1-km^2 area. They select a route to walk along the waterway or around the square, which includes stopping points spaced out appropriately. The route is walked once during the day to record the habitats present. Two evening surveys walking the route with a bat detector and powerful torch are made in July/August. Bats are recorded while the observer is in transit and while he or she is standing still at the stopping points. This method requires training in the use of a bat detector.

(3) Hibernation-site monitoring. Counts are made of all species encountered at a range of sites selected by surveyors – typically caves, mines and cellars. Two survey visits are made, one in January and one in February each year. Hibernating bats are identified and counted without disturbance. A torch with a deep-red filter will help prevent disturbance in caves or buildings. Photos can be taken and counted later to minimise disruption to colonies.

Warren and Witter (2002) showed that their monitoring programme in Wales could detect a 5% population change over a 5-year period of monitoring. O'Donnell (2002) suggested that 10 years might be a more conservative figure, given the high variability in night-to-night counts for some species roosting in caves.

A variety of ultrasonic bat detectors has been used over the past 30 years to identify free-flying bats (O'Farrell *et al.* 1999). Analyses of recorded echo-location calls with older machines were often slow and typically restricted to few calls, but modern computing power has allowed species identification from calls in real time. Use of the Anabat II detector (http://www.titley.com.au/tanabat.htm) and its associated analysis system allows an immediate examination, via a lap-top computer, of the time–frequency structure of calls as they are detected. These calls can be stored on the computer for later examination. Many bats can be identified as to species by qualitatively using certain structural characteristics of calls, primarily approximate maximum and minimum frequencies. All bat calls are not equally useful for identification. To identify calls precisely, it is important to use a continuous sequence of calls from an individual in normal flight rather than single isolated calls. Counts of free-flying bats with bat detectors are subject to high variability, so with every monitoring programmme it is necessary to reduce this variation by counting sufficient replicates.

Advantages and disadvantages

Counts of bats are essential for determining population trends, especially for species of conservation interest, but there are problems with some monitoring methods both for the bats and for the humans counting them. In warm humid climates, bats in caves may transmit histoplasmosis, a serious fungal disease of the lungs. This fungus grows in soil and material contaminated with bat or bird droppings, and spores become airborne when contaminated soil is disturbed. Breathing the spores causes infection. Caves can be physically dangerous to inexperienced people. In a very few cases rabies appears to have been contracted by visiting bat roosts (Constantine 1988), although most cases of rabies come directly from bites by rabid bats.

Biases

There are no completely censused bat populations from which we can validate the counting methods used to index bat populations. Consequently we have at present no way of determining whether there are systematic biases in the existing counts. Typically, even if one roosting site can be counted precisely, we do not know all the roosting sites of a species. Some species use caves and houses only occasionally, which adds variation to counts made at roosting sites. Larger roosts may be easier to find than smaller roosts, and we do not know whether population trends from small and large roosts are congruent.

Line transects

If a species is relatively large and conspicuous, one of the best methods for estimating abundance is the line transect. This method is described in Chapter 3 and in much more detail in Buckland *et al.* (1993); see http://www.ruwpa.st-and.ac.uk/distance/. Line transects can be walked, driven, swum, or flown, and they have become more popular during the past decade.

Method

The essential feature of line transects is that one walks along a straight path and records the individuals seen and their perpendicular distance from the transect line (Figure 10.2). Some individuals to the side of the path being walked escape detection by the observer, and the critical assumption is that all animals on the path are seen. The data gathered on perpendicular distances are used to construct a detection function for distances off the transect line. It is assumed that the detection function has some smooth shape, but a variety of shapes is possible (Krebs 1999,

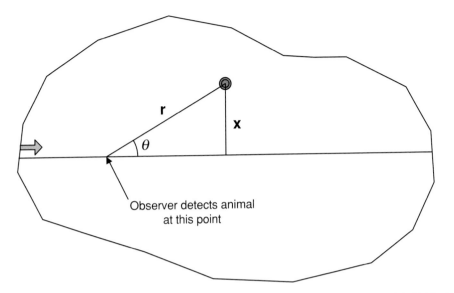

Figure 10.2   An illustration of the basic measurements that can be taken for each individual sighted along a line transect in the direction indicated by the arrow. The key measurement is the perpendicular distance (x_i). If the sighting distance (r_i) is easier to record in the field, the sighting angle (θ) must also be measured. The perpendicular distance $x = r \sin \theta$. Note that not all individuals need to be seen for this method to work.

p. 161). Line transects are best used for visible mammals in open habitats. The sample size should be at least 40 sightings, and better results are obtained from 60–80 sightings.

Advantages and disadvantages

Roads and tracks are often used as the path of the line transect, and there is a general problem of non-random sampling of the study area if convenience sampling is used instead of random sampling. Individuals must be located before they move or the detection function will be biased. If too many animals are observed, there may be problems with recording perpendicular distances accurately. Observers vary greatly in their ability to see animals, particularly if the habitat is less open. Perpendicular distances should be measured accurately rather than just estimated. If the species being studied is rare, it might be difficult to obtain the recommended sample size of 40 sightings. If habitats are interspersed, it might not be possible to get habitat-specific estimates of abundance.

Line transects can be used to estimate the abundance of indices of mammal activity, such as dung piles and burrows. A line transect may be converted to a rectangular sampling quadrat of fixed size if the observer can be certain of sighting all individuals within the fixed strip. This approach is not recommended for most mammals because detectability is not usually 100% and, in constraining sightings to a fixed strip width, data are lost for sightings more distant from the path of the line transect.

Biases

The most important biases arise from non-random sampling from roads or other paths of convenient travel. Perpendicular distances may be underestimated if they are not measured accurately. It is critical that all individuals on the centreline of the path are detected. For sea mammals such as whales or dolphins that surface only momentarily, it may be difficult to determine the sighting distance.

Aerial surveys

Large mammals that are readily seen from the air can be censused most effectively from aerial surveys that utilise light aircraft or helicopters. The most common method involves flying strip transects and counting all the individuals within the strip. An alternative approach is to use the line-transect methodology discussed above to estimate abundance (Burnham & Anderson 1984).

Method

The details of the methods of aerial surveys are covered in Norton-Griffiths (1978) and Krebs (1999). The aircraft used must allow the observers unlimited downward vision, and there must be precise control of the aircraft's altitude to permit a strip of defined width to be counted. Streamers on the wing struts of the aircraft can be used to delineate the width of the strip to be counted.

Advantages and disadvantages

There is no other method for covering very large areas rapidly for mammals that are relatively sparse on the ground or that range over very large spatial areas. Much experience in doing aerial counts has been accumulated, and the statistical analysis is well defined. The major disadvantage is the cost of aircraft usage, together with a very slight danger of aircraft accidents while surveying in remote regions. Aerial surveys can be difficult to use in areas with high relief, although helicopters can be useful in these situations if blocks defined by topography are counted instead of strips.

Large groups of animals can be photographed if they cannot be counted quickly. There is great variation among observers in their ability to count from aircraft, and this must be standardised as much as possible. Observers grow tired after a few hours of counting, and counting becomes less precise the faster one has to count.

Biases

Most observers in aerial censuses tend to undercount, and this negative bias typically ranges from zero when there are few animals to a very high fraction when many animals have to be counted quickly (Marsh & Sinclair 1989). Additional biases with aerial surveys can occur when strict control of the aircraft's altitude is not maintained, so that the actual strip width becomes too wide or too narrow. Observer training can reduce the differences among individuals in the ability to detect animals, but even trained observers differ dramatically in their counts (LeResche & Rausch 1974).

Individual recognition

Many large mammals have individual markings that lend themselves to being visually recognised. Killer whales exhibit individual colour variation and in addition they can be distinguished from one another by the shape and size of their dorsal fins. Wild dogs in Africa have coat-colour variation, so individuals can be identified from photographs. The use of individual recognition marks makes it possible to use all the methods of mark–recapture (Chapter 3) on populations of mammals that are never captured in traps or subjected to a tagging or marking procedure. These types of non-invasive studies are clearly desirable for species of high conservation interest.

Method

A catalogue of photographs of individuals must be maintained for the population of interest, and, if the public is involved, a mechanism for obtaining and copying photos taken by amateur observers is needed. The use of digital cameras with data and time stamps can now make the electronic collection of such photos possible. As in all population studies, good record keeping is critical to the success of these projects. Typically the species chosen are long-lived, and there is relatively little turnover from year to year.

Advantages and disadvantages

One advantage of this method is that it can make use of untrained observers to increase sample sizes. Photos must be of high quality, and a well-defined screening process must be defined in order to eliminate photos that are too poor to allow certain identification. Because this method can be very time-consuming, it can rarely be done solely by paid observers, so amateur input can be essential to successful sampling.

With these methods it is important to define the population being sampled and to cover the area occupied by the population. Most individuals in the population must be identified to provide the most precise population estimates. By providing spatial locations of the photographs, it is also possible to map the movements of individuals when there is good coverage of the areas occupied.

Biases

Care must be taken to ensure that individuals with similar markings are not confused in the counts. For humpback whales, Blackmer *et al.* (2000) showed that colouration patterns on the flukes changed with age, and that dorsal-fin shape was a more stable characteristic for individual recognition. If a total count is to be achieved, a high level of sampling effort must be expended. If mark–recapture methods are to be used on these data, all the assumptions of the particular method must be evaluated (Chapter 3). In particular, some individuals may be more easily located than others, so equal detectability cannot be assumed in a mark–recapture model.

Counting calls

Bats make ultrasonic calls that can be recorded by a bat detector. Lions roar, and seals and sea lions can be heard roaring even in dense fog. Whale and seal calls can be recorded underwater with a hydrophone. The idea is to count calls or bursts of calls for a standard unit of time, and to use these counts as an index of population size.

Method

In these cases it is not possible to recognise individuals from their calls, but species can be distinguished. Stirling *et al.* (1983) showed that it was possible to record and distinguish calls of ringed seals, bearded seals and walruses under the ice in the Canadian arctic. Bat detectors can catalogue species calls automatically.

Advantages and disadvantages

The major disadvantage of all index methods is that they must be calibrated relative to absolute abundance before they can be used reliably. Anderson (2003) has argued strongly against the use of indices like call counts to estimate relative abundance for wildlife species. The problem is that for many species like bats there might not be any other feasible method to use, and you then have to decide whether a poor data set is better than no data set.

Biases

There are many sources of variation in calling rates, ranging from seasonal effects associated with breeding (Cleator *et al.* 1989) to diurnal effects caused by weather factors. If bat detectors are used, there is a problem of standardising different machines, and the key is to develop a standard protocol to control as many of these sources of variation as possible.

Trapping

Small mammals are readily live-trapped in box traps, which is the most common method for assessing the abundances of many mammal species. Many trap designs can be found, but three types of live traps are most commonly used. The Longworth live trap consist of two parts, a tunnel (the trapping mechanism) and a nest box for food and bedding (see http://www.alanaecology.com/). Longworth traps are very expensive but have the advantage of being able to be locked in the open position so that they can be prebaited. They can also be ordered with an optional shrew hole to permit small shrews to escape (most shrews die within a few hours in a live trap). Sherman live traps are lightweight aluminium traps that are all one piece. Four sizes and folding and rigid styles are available. They have the advantage of being light in weight and the folding types of traps are easy to store and carry in the field. They have the disadvantage of being difficult to clean and, because it is difficult to use bedding and food in them, they may cause more inadvertent trap mortality, particularly among young animals. The thin aluminium of the folding traps can be easily chewed through by some rodents. Sherman traps are relatively inexpensive and available from H. B. Sherman Traps, Inc., 3731 Peddie Drive, Tallahassee, Florida 32303 (http://www.shermantraps.com/). Wire-mesh traps are typically used for larger species of small mammals, such as squirrels, and they are manufactured in a variety of sizes and with one door or two doors for entry. The two largest manufacturers are Havahart (http://www.havahart.com/) and the Tomahawk Live Trap Company (http://www.tomahawklivetrap.com/) but other local suppliers produce a variety of wire-trap designs.

In addition to simple box traps, small mammals can also be captured in pitfall traps dug into the ground. These must be deep enough for the species involved and have smooth sides so that animals cannot climb out. They should be covered with a board, to prevent rain from flooding them, and should have drainage holes in the bottom as well.

There are several designs of multiple-catch traps for rodents, but they are mainly used for pest-control purposes rather than population studies. Multiple-catch traps do not work with all small mammals, and may lead to individuals fighting and killing one another in the confined space of the trap.

Method

A variety of trapping designs has been suggested. Typically a trapping grid is surveyed in a checkerboard configuration, and, depending on the species, traps are placed at the checkerboard intersections, which are often 15–30 m apart. The size of the trapping grid must be adjusted to the size of the home range of the species being studied (Bondrup-Nielsen 1983). The typical trapping grid is 10 by 10 but in most studies a larger configuration would be better, particularly for density estimation. Anderson *et al.* (1983) recommend a trapping-web design instead of a checkerboard grid for estimating density more precisely, and the authors of a recent intensive evaluation of trapping grids versus webs prefer the web design for density estimation (Parmenter *et al.* 2003).

To protect small mammals from cold and rain, boards are often used to cover live traps. Food is usually provided to lure animals into the traps, but also to sustain them once they have been caught. Peanut butter mixed with oats is often used as a bait, but the best bait will depend on the species under study, and some experimentation may be needed.

Traps have usually been visited twice a day in early morning and evening, but accumulating evidence suggests that four checks a day would be better, to reduce the stress of capture. For some species overnight trapping should not be used. If night trapping is required, night checks should be carried out if possible, or traps should be set as late as possible and checked as early as possible in the morning. Pitfall traps should be visited as often as possible, and, since they are multiple-catch traps, animals can injure one another by fighting.

Population size is usually estimated by mark–recapture methods (Chapter 2), and a variation of line-transect methods can be applied to trapping-web data. To estimate population density for a trapping grid, one needs to know the effective trapping area of the grid as well as the estimated population size, since some individuals that live at the edge of the trapping area get caught in the edge traps. A boundary strip is usually added to checkerboard grids with a strip width of one-half the average movement of the individuals captured within the grid (Stenseth & Hansson 1979), but this is only an approximate method.

Advantages and disadvantages

The major disadvantage of live-trapping is that it requires a great deal of sustained work in order to achieve good population estimates. A high fraction ($>50\%$) of individuals in a population must be marked in order to achieve estimates with narrow confidence bands (Pollock *et al.* 1990). With small rodents that have life spans of the order of months, frequent live trapping (every 2–3 weeks) is essential to obtain a good description of demographic events. The best estimators of population size, like the programs CAPTURE and MARK, demand multiple trapping sessions (typically five or more) and this demand conflicts with the requirement not to disturb the individuals being studied by confining them in traps too often. The advantage of live-trapping studies is that one obtains data on reproductive status, weight and movements of individuals at the same time as one gets density estimates.

Care must be taken in handling small mammals because of diseases that can be transmitted to humans (Begon 2003). The list is long – from plague to hantavirus, leptospirosis, cowpox, rabies and a host of viral and bacterial diseases not yet described. On the positive side, many ecologists have handled thousands of small mammals and been bitten numerous times, with no apparent damage to their health. Caution is essential, however, particularly with regard to tropical species that have been little studied for transmissible diseases.

Biases

Individuals are not all equally catchable, no matter what type of trap is being used, so methods of population estimation that assume equal catchability must be treated carefully. To avoid bias, authors of many small-mammal studies have attempted to catch a large fraction ($>80\%$) of the individuals in the population, thus reducing the margin for errors of estimation. However, it is important to remember that traps are selective – juvenile animals are rarely captured and breeding males that move over large areas are often captured all the time. The resulting population estimates should be noted as referring to the trappable population, not the entire population. In extreme cases some subordinate individuals may never be caught in traps, and would be invisible in the analysis of trapping data.

Estimation programs such as CAPTURE and MARK have methods for estimating population size in the presence of varying probabilities of capture, and these should be used to alleviate the major problem of individual variation in trappability.

Counting dung

If the species of interest has characteristic dung, it may be easier to find and count dung than to try to capture the animals themselves. Dung can be used simply as an index of the presence or absence of a species, but attempts to use the amount of dung as an index of population abundance have also been made.

Method

Dung can be counted with quadrats (circular, rectangular, square) or by line-transect methods. Some species deposit dung in piles and a clear operational definition of a sampling unit of dung must be made before counts can be undertaken. If quadrats are to be used in sampling, effort can be minimised by determining the optimal quadrat size and shape for the particular species (Krebs 1999, p. 105).

There is a major problem with dung counts in determining the age of dung. In some studies only fresh dung is to be counted, and, since the appearance of dung will change depending on temperature and moisture conditions, it is essential to define carefully what is to be counted. The best method for avoiding this problem is to re-count the same quadrats and clear them of all dung each time they are counted. The counting interval must depend on the rate of decay of dung. In tropical areas with dung beetles, this could be a matter of hours and consequently these methods could not be used. In polar regions dung may last for tens of years.

Attempts to estimate absolute abundance from dung rely on known rates of production of faecal pellets. Since rates will typically vary with diet, values from zoo animals are often unreliable estimates of this parameter. An alternative approach is to calibrate dung counts with live-trapping estimates of population size on the same area, and thus derive an empirical relationship between dung counts and population size (Krebs *et al.* 2001).

Advantages and disadvantages

Life produces dung, and consequently the presence of a species in an area can be ascertained by dung counts. In habitats where dung persists this method can be used very effectively to index population abundance with a minimum of effort. As techniques for identification of individuals from DNA in dung are developed, fresh dung can also potentially become a way of estimating population size by mark–recapture of dung from specific individuals (Eggert *et al.* 2003).

If a long-term study is needed, it is important to have permanent quadrats that are cleared at each sampling so that the issue of dung age does not arise. If no independent estimate of population abundance is available, all of the cautions about using indices need to be considered carefully.

For some species of larger mammals the type of dung produced varies between the growing and non-growing seasons, and this can be used to advantage to index seasonal habitat use.

Biases

The main problem in sampling dung is that it is typically not randomly distributed in the landscape, and that it can decompose at quite different rates in various micro-habitats.

Feeding signs for herbivores

If a herbivorous species leaves characteristic marks on their food plants, these feeding marks or scars can be used both to determine presence and absence and to index the relative abundance of a population.

Method

Counting feeding signs depends on detailed knowledge of the natural history of the species and the food plants it utilises. Since diets of most mammals change seasonally, there may be one season in which the feeding signs are most easily counted.

Voles in grasslands cut tillers of grasses and leave these fresh cuttings in their runways. A census of these clippings was used over 80 years ago to index vole numbers in Britain (Chitty 1996). Ungulate browsing surveys can both index population sizes and measure the pressure these herbivores are exerting on their food plants. In all these cases standard methods of quadrat sampling can be used to estimate feeding marks.

Advantages and disadvantages

Like dung counts, the measurement of feeding signs is easily done during the day and does not depend on capturing or seeing animals. For species that have very broad diets and large seasonal shifts in diet, this method will not usually be the best way to index abundance. The main problem is to validate independently whether the particular feeding sign that you are measuring is a good index of abundance.

Biases

If similar species have similar diets, it might not be possible to separate species in the analysis of relative abundance. If diets vary markedly from year to year depending on the availability of alternative foods, what works well in one year may give a biased representation of abundance in other years.

Counting footprints and runways

Tracks of species in soft ground or snow are an excellent way of determining presence and absence. If standard methods are employed, these counts can be used as an index of relative abundance.

In Finland, Korpimäki *et al.* (2002) used snow tracking to index stoat and weasel abundance 1–2 days after fresh snows. O'Donoghue *et al.* (1997) used snow-track transects to index population changes in coyotes and lynx in northern Canada. Sand tracking stations have been used to index dingo populations in Australia (Allen *et al.* 1996).

Method

Footprints can be assessed actively or passively. Active assessment typically uses scents or baits to bring animals into the tracking plot, whereas passive assessment uses tracks made by animals in their daily travels. These indices measure activity as well as population abundance, and are best used on species for which activity levels are relatively constant at the times of year the counts are made. The method is to set out plots large enough to provide good tracks, to rake them clean and revisit them at fixed intervals. For snow tracking the intervals are usually set by fresh snowfalls. The important point is that old tracks are not confounded with new tracks.

A similar kind of approach can be used for small mammals that make runways in grassland or woodland habitats. The number of active runways intersecting a line transect of fixed length can be used as an index of abundance. Activity of runways can be determined by the presence of fresh scats or fresh plant clippings.

Advantages and disadvantages

The advantage of track counts is that they are a cheap way of determining the relative abundance of wide-ranging carnivores that live at low density. The disadvantage is that all these measures confound activity with abundance; so, if the index increases, it could be because of higher activity, higher abundance, or both. Territoriality in some carnivores may affect the use of trails, and the spatial design of the sampling programme for tracks must take into account the size of the home range of the species under study. Wilson and Delahay (2001) review these methods in more detail.

Biases

All indirect methods of census are best if they can be validated with a population of known size. This has been impossible with most of the cases in which track counts have been used, and consequently it is not possible to estimate bias. The best recommendation is to keep the design of the surveys constant with respect to environmental and seasonal conditions, and to train observers carefully in track identification.

Hair tubes and hair catchers

Method

Hair tubes are long tubes slightly larger in diameter than the species being studied, with sticky tape on the inside so that hairs are left on the tape as the animal passes through the tube (Lindenmayer *et al.* 1999). For larger mammals barbed wire or other sticky devices can be used to sample hair from individuals without needing to capture them (Mowat & Paetkau 2002).

Advantages and disadvantages

The critical assumption is that the hair can be identified as to species or species-groups (Harrison 2002). Keys to mammalian hair are available for many groups (Brunner & Coman 1974; Day 1966; Staines 1958; Wallis 1993). These keys are not necessarily complete, and we recommend that a reference collection be prepared for the species being studied. Keys may represent only hairs from the dorsal surface and colour patterns of species may vary geographically. In addition, there is variation in hair samples from different parts of the body of mammals. Some species cannot be distinguished from hair alone.

DNA can now be extracted from hair samples and thus hair sampling can be used as a mark–recapture method (Mowat & Paetkau 2002). Rigorous methods must be used to identify individuals with minimal typing errors (Paetkau 2003) but, once this has been achieved, these methods of using hair DNA hold great promise for answering population questions that could not be studied previously.

Biases

The major potential biases are in the identification of hair and, if DNA typing is used, in the precise identification of individuals.

Counting seal colonies

Seals and sea lions are a special case of mammals that must haul themselves out outo land to reproduce. Because adults come and go from land colonies, counts of adults present will always be undercounts unless some correction is applied. In some cases it is easier to count the pups since they do not go to sea for several weeks.

Method

Aerial counts can be made from high-quality photographs. Ground counts are more difficult without disturbing the colony, and many individuals are hidden in rock crevices and cannot be seen from a vantage point. If some individuals are marked or carry radio-tags, a correction for the proportion missed can be applied. Methods of mark–recapture can be applied to pups by marking them with fur clips or water-soluble paints.

Advantages and disadvantages

Counts can give information on breeding success as well as abundance.

Biases

Not all adults will come ashore and this part of the population may be missed totally. At any given instant some animals are at sea feeding, so a total count is not possible even with photographs. Some breeding colonies may be missed completely if they are in isolated regions. Recolonisation of islands occurs and one cannot assume that any particular island is a population isolate (Pyle *et al.* 2001; Wilson 2001).

Conclusions

Censusing mammals is an evolving art. The major pitfalls are now well identified, and the statistical methods for dealing with census data are well developed. For mark–recapture estimates, individuals vary in their probability of capture or detection. The higher the fraction of animals captured or marked, the more certain the estimates. For all indices of population abundance, the general warning is not to assume that there is a linear relationship between the index and absolute population size. For many species there is a growing literature of clever census methods that are highly specific and most useful as we try to study and conserve mammals around the world.

References

Allen, L., Engeman, R. & Krupa, H. (1996). Evaluation of three relative abundance indices for assessing dingo populations. *Wildlife Research* **23**, 197–206.

Anderson, D. R. (2003). Index values rarely constitute reliable information. *Wildlife Society Bulletin* **31**, 288–291.

Anderson, D. R., Burnham, K. P., White, G. C. & Otis, D. L. (1983). Density estimation of small-mammal populations using a trapping web and distance sampling methods. *Ecology* **64**, 674–680.

Begon, M. (2003). Disease: health effects on humans, population effects on rodents. In *Rats, Mice and People: Rodent Biology and Management*, ed. G. R. Singleton, L. A. Hinds, C. J. Krebs & D. M. Spratt. Monograph No. 96. Canberra, Australian Centre for International Agricultural Reseach, pp. 13–19.

Blackmer, A. L., Anderson, S. K. & Weinrich, M. T. (2000). Temporal variability in features used to photo-identify humpback whales (*Megaptera novaeangliae*). *Marine Mammal Science* **16**, 338–354.

Bondrup-Nielsen, S. (1983). Density estimation as a function of live-trapping grid and home range size. *Canadian Journal of Zoology* **61**, 2361–2365.

Bookhout, T. A. (ed.) (1994). *Research and Management Techniques for Wildlife and Habitats*, 5th edn. Bethesda, Maryland, The Wildlife Society.

Boonstra, R., Kanter, M. & Krebs, C. J. (1992). A tracking technique to locate small mammals at low densities. *Journal of Mammalogy* **73**, 683–685.

Boonstra, R., Krebs, C. J., Boutin, S. & Eadie, J. M. (1994). Finding mammals using far-infrared thermal imaging. *Journal of Mammalogy* **75**, 1063–1068.

Brunner, H. & Coman, B. J. (1974). *The Identification of Mammalian Hair*. Melbourne, Inkata Press.

Buckland, S. T., Anderson, D. R., Burnham, K. P. & Laake, J. L. (1993). *Distance Sampling. Estimating Abundance of Biological Populations*. London, Chapman & Hall.

Burnham, K. P. & Anderson, D. R. (1984). The need for distance data in transect counts. *Journal of Wildlife Management* **48**, 1248–1254.

Caughley, G. (1977). *Analysis of Vertebrate Populations*. London, Wiley.

Chitty, D. (1996). *Do Lemmings Commit Suicide? Beautiful Hypotheses and Ugly Facts*. New York, Oxford University Press.

Cleator, H., Stirling, I. & Smith, T. G. (1989). Underwater vocalizations of the bearded seal (*Erignathus barbartus*). *Canadian Journal of Zoology* **67**, 1900–1910.

Constantine, D. G. (1988). Health precautions for bat researchers. In *Ecological and Behavioral Methods for the Study of Bats*, ed. T. H. Kunz. Washington, Smithsonian Institution Press, pp. 491–528.

Day, M. G. (1966). Identification of hair and feather remains in the gut and faeces of stoats and weasels. *Journal of Zoology (London)* **148**, 201–217.

Eggert, L. S., Eggert, J. A. & Woodruff, D. S. (2003). Estimating population sizes for elusive animals: the forest elephants of Kakum National Park, Ghana. *Molecular Ecology* **12**, 1389–1402.

Harrison, R. L. (2002). Evaluation of microscopic and macroscopic methods to identify felid hair. *Wildlife Society Bulletin* **30**, 412–419.

Hubbs, A. H., Karels, T. J. & Boonstra, R. (2000). Indices of population size for burrowing mammals. *Journal of Wildlife Management* **64**, 296–301.

Korpimäki, E., Norrdahl, K., Klemola, T., Pettersen, T. & Stenseth, N. C. (2002). Dynamic effects of predators on cyclic voles: field experimentation and model extrapolation. *Proceedings of the Royal Society of London, Series B* **269**, 991–997.

Krebs, C. J. (1999). *Ecological Methodology*, 2nd edn. Menlo Park, California, Addison Wesley Longman Inc.

Krebs, C. J., Boonstra, R., Nams, V. O. *et al.* (2001). Estimating snowshoe hare population density from pellet plots: a further evaluation. *Canadian Journal of Zoology* **79**, 1–4.

LeResche, R. E. & Rausch, R. A. (1974). Accuracy and precision in aerial moose censusing. *Journal of Wildlife Management* **38**, 175–182.

Lindenmayer, D. B., Incoll, R. D., Cunningham, R. B. *et al.* (1999). Comparison of hairtube types for the detection of mammals. *Wildlife Research* **26**, 745–753.

Marsh, H. & Sinclair, D. F. (1989). Correcting for visibility bias in strip transect aerial surveys of aquatic fauna. *Journal of Wildlife Management* **53**, 1017–1024.

Mowat, G. & Paetkau, D. (2002). Estimating marten *Martes americana* population size using hair capture and genetic tagging. *Wildlife Biology* **8**, 201–209.

Norton-Griffiths, M. (1978). *Counting Animals*, 2nd edn. Nairobi, African Wildlife Leadership Foundation.

O'Donnell, C. F. J. (2002). Variability in numbers of long-tailed bats (*Chalinolobus tuberculatus*) roosting in Grand Canyon Cave, New Zealand: implications for monitoring population trends. *New Zealand Journal of Zoology* **29**, 273–284.

O'Donoghue, M., Boutin, S., Krebs, C. J. & Hofer, E. J. (1997). Numerical responses of coyotes and lynx to the snowshoe hare cycle. *Oikos* **80**, 150–162.

O'Farrell, M. J., Miller, B. W. & Gannon, W. L. (1999). Qualitative identification of free-flying bats using the Anabat detector. *Journal of Mammalogy* **80**, 11–23.

O'Shea, T. J., Bogan, M. A. & Ellison, L. E. (2003). Monitoring trends in bat populations of the United States and territories: status of the science and recommendations for the future. *Wildlife Society Bulletin* **31**, 16–29.

Paetkau, D. (2003). An empirical exploration of data quality in DNA-based population inventories. *Molecular Ecology* **12**, 1375–1387.

Parmenter, R. R., Yates, T. L., Anderson, D. R. *et al.* (2003). Small-mammal density estimation: a field comparison of grid-based vs. web-based density estimators. *Ecological Monographs* **73**, 1–26.

Pollock, K. H., Nichols, J. D., Brownie, C. & Hines, J. E. (1990). Statistical inference for capture–recapture experiments. *Wildlife Monographs* **107**, 1–97.

Pyle, P., Long, D. J., Schonewald, J., Jones, R. E. & Roletto, J. (2001). Historical and recent colonization of the South Farallon Islands, California, by northern fur seals (*Callorhinus ursinus*). *Marine Mammal Science* **17**, 397–402.

Rugh, D. J., Ferrero, R. C. & Dahlheim, M. E. (1990). Inter-observer count discrepancies in a shore-based census of gray whales (*Eschrichtius robustus*). *Marine Mammal Science* **6**, 109–120.

Staines, H. J. (1958). Field key to guard hair of middle western furbearers. *Journal of Wildlife Management* **22**, 95–97.

Stenseth, N. C. & Hansson, L. (1979). Correcting for the edge effect in density estimation: explorations around a new method. *Oikos* **32**, 337–348.

Stirling, I., Calvert, W. & Cleator, H. (1983). Underwater vocalizations as a tool for studying the distribution and relative abundance of wintering pinnipeds in the high arctic. *Arctic* **36**, 262–274.

Van Horne, B., Schooley, R. L., Knick, S. T., Olson, G. S. & Burnham, K. P. (1997). Use of burrow entrances to indicate densities of Townsend's ground squirrels. *Journal of Wildlife Management* **61**, 92–101.

Wallis, R. L. (1993). A key for the identification of guard hairs of some Ontario mammals. *Canadian Journal of Zoology* **71**, 587–591.

Warren, R. D. & Witter, M. S. (2002). Monitoring trends in bat populations through roost surveys: methods and data from *Rhinolophus hipposideros*. *Biological Conservation* **105**, 255–261.

Wilson, G. J. & Delahay, R. J. (2001). A review of methods to estimate the abundance of terrestrial carnivores using field signs and observation. *Wildlife Research* **28**, 151–164.

Wilson, S. C. (2001). Population growth, reproductive rate and neo-natal morbidity in a re-establishing harbour seal colony. *Mammalia* **65**, 319–334.

11 Environmental variables

Jacquelyn C. Jones

School of Biological Sciences, University of East Anglia, Norwich NR4 7TJ, UK

John D. Reynolds

Department of Biological Sciences, Simon Fraser University, Burnaby, B.C., Canada V5A 1S6

Dave Raffaelli

Environment Department, University of York, York YO10 5DD, UK

Introduction

Why measure environmental variables?

When describing the distribution and abundance of plants and animals it is important to describe abiotic features of the environment for two reasons. First, presentation of this information is of enormous help to your audience trying to picture the *environmental context* in which the biological census work was carried out. In other words, it provides the backdrop. Secondly, and more importantly, the physical and chemical variables you measure are often *key explanatory variables* for the biological phenomena you observe. In other words, they may well be the drivers of the spatial and temporal patterns you record in a census. Subsequent management of your target species or communities will often require controlling and manipulating these drivers, whether they be phosphate levels in a lake or shade on a woodland floor.

How do I know which variables to measure?

The answer to this question may be easy if other researchers have done similar work already, but in many censuses and surveys it might not be possible to tie down key variables in advance. In that case, you might be tempted to measure everything you can, just in case you overlook something. However, 'Sutherland's Deadly Census Sins' (Chapter 12) apply equally well to environmental variables: it is all too easy to expend inappropriate time and effort in measuring the wrong variable. First, don't just do what everyone else has done – they may be entirely wrong and slavishly following their example will only compound the problem. Second, don't select a variable simply because it can be measured. You may be lucky to have access to a nice, shiny instrument that you can use to record data with great precision, but if that variable is not relevant, you will have at best wasted a lot of time and effort, and at worst misidentified the key drivers. A little time taken to reflect on how your animal or plant might perceive its environment can often point to what you should really be measuring. In this respect, it is important to measure on the appropriate scale.

Scale is a term used widely in ecology, but rarely defined. There are three basic components of scale: *extent*, *grain* and *lag*. *Extent* describes the limits of the survey or census area, e.g. 10 km^2. *Grain* describes the size of the sampling unit with which you record your data, e.g. 1-m^2 quadrat, 500-m line transect, 100-m resolution on a GIS database. *Lag* is the distance or interval between the sampling units. (Equivalent descriptions can be made for temporal scales.) When designing programmes and taking physical and biological samples, it is a good idea to keep these elements in mind. Often, abiotic and biological variables are recorded on quite different scales and this makes interpretation of any causal relationships tricky. For instance, a marine benthic ecologist might wish to investigate how the distribution of sea-bed animals within the sediment is affected by availability of oxygen. Micro-electrodes can now resolve oxygen concentrations in marine sediments on spatial scales of a fraction of a millimetre. In contrast, the fauna within the sediment is sampled at much larger spatial scales, often tens of centimetres. Whilst the electrode measurements are very precise, there is clearly a mismatch of measurement scales in the environmental and biological variables, which makes interpretation of any associations problematic, some would say heroic. Similar issues can arise for many of the standard measurement procedures detailed here, so think hard about the scale on which you measure a variable. For instance, it might not be feasible to use the same grain for environmental and biological variables, but at least one should take many smaller-grain environmental samples throughout the extent of the larger-grain biological samples. Having captured the spatial variation in the environmental variable on the same scale as that in the biological variable, one can at least take some kind of average or, better still, use the between-sample variance in interpreting the effect of the variable of interest.

What measurement technique can I use?

Having selected your variables, there remains the choice of which of the many techniques available are the best for your survey. The choice is rarely straightforward: even the simplest variables, such as temperature, can be measured using an array of methods differing in precision, practicality and expense. Furthermore, many useful protocols tend to be modified and improved as they are passed around informally among researchers, without these improvements finding their way into readily accessible literature. Here, our emphasis is on the simpler and cheaper methods, but more expensive and complex techniques are described where appropriate. Many ingeniously simple methods have been devised, but since they do not come with an 'owner's manual', they deserve a fuller explanation, which we have tried to present here.

Wind and water flow

Wind and water flow are very similar in their behaviour, their measurement and their effects on living organisms. Both have the same basic parameters: direction and speed. Both can vary dramatically with height above the substratum, reducing to very low velocities at the surface of the substratum, so that environmental conditions within this boundary layer where the species of interest may live may be quite different from those where the flow can be measured. Clearly, the physics of air and water flow, and their ecological effects, have much in common, so don't be afraid to delve into the literature on both. In doing so, you might discover better ways to present your data and to interpret interactions between air or water flow and biology.

Wind

On larger scales, wind direction can be easily measured with a compass by noting the direction of cloud movement or bending of vegetation. Remember, though, that that direction and speed can vary greatly with height, so it is important to measure wind at a height appropriate to your study species (e.g. hawks versus snails, versus moss) and to be consistent in repeated sampling. *Wind vanes* are fairly reliable if they are high enough to avoid eddies (turbulence) created by local obstructions. A good rule of thumb is to place the wind vane at a distance from the nearest obstruction equal to ten times the height of the obstruction. The direction of the wind is recorded as the direction that the wind is coming from, so a southern wind comes from the south.

Wind velocity can be estimated using the description provided by the *Beaufort scale of winds*, which ranges from no wind at Beaufort 0 up to Beaufort number 12, which is hurricane force (Table 11.1). Because it records wind speed in categories, the Beaufort scale is not continuous or precise (e.g. there is not a value of 6.47) and the scale is not strictly linear.

Various *hand-held gauges*, on which you can record speed on a continuous scale, are available. They have the advantage in fieldwork of being lightweight, versatile and small (e.g. 20 cm × 5 cm). One device consists of a transparent cylinder that is held vertically with a hole facing into the wind. The wind forces a plastic float to rise inside the cylinder, and its height is read from a scale calibrated with respect to wind speed. These devices may include compasses to indicate wind direction. They are fairly lightweight, inexpensive and typically accurate to within ± 3 km h^{-1}. *Cup anemometers* are more accurate and these consist of three or four metal cups attached to spokes placed horizontally on a spindle. When the wind blows into the cups, the spokes rotate at a speed that can be calibrated to obtain the wind velocity. Anemometers come in a range of sizes. Some are large enough to require a fixed stand, others can be hand-held and connected to a digital meter about the size and weight of a calculator. These are relatively cheap and have the advantage that you can record wind speed quite close to the substratum, if that is where your organisms live.

Water flow

Water flow in streams and rivers can be measured simply by placing a *float* in the water and measuring the time taken for it to travel a predetermined distance. An orange works well because it does not rise high enough to be affected strongly by wind. This method, though simple and cheap, measures only surface flow rates, and gives crude results when there are eddies and variation in flow rates within the stream. In marine waters it does not work well at all.

A much more accurate method is to use a *flow meter*. This converts the speed of rotation of a small propeller, termed an impeller, into current velocity, and gives readings from specific depths or regions within the water body. However, remember that the flow measured relates only to the local area of the impeller. Deciding where to take readings is important because flow will vary with

Table 11.1. *The Beaufort scale for ecologists*

Beaufort number	Name of wind	Observable features	Field-ecologists' impression	Velocity (km/h)
0	Calm	Smoke rises vertically	You're having a good time	<2
1	Light air	Smoke drifts downwind; wind does not move wind vane	You're still having a good time	2–5
2	Light breeze	Wind felt on face; leaves rustle; vane moved by wind	It's a bit tricky to photograph insects on plants	6–12
3	Gentle breeze	Leaves and twigs in constant motion; wind extends light flag	At least there are no biting insects to contend with	13–20
4	Moderate breeze	Raises dust and loose paper; small branches are moved	It's hard to keep your notes from flapping	21–29
5	Fresh breeze	Small trees in leaf begin to sway; crested wavelets form on inland waters	You prefer to work in sheltered places	30–39
6	Strong breeze	Large branches in motion; whistling heard in telegraph wires; umbrellas used with difficulty	Your tripod is blowing over	40–50
7	Moderate gale	Whole trees in motion; inconvenience felt in walking against wind	You're doing this for the good of science	51–61
8	Fresh gale	Twigs break off trees; progress generally impeded	You're not being paid enough	62–74
9	Strong gale	Slight structural damage occurs (chimney pots and slate removed)	You're thinking about where you've parked your vehicle	75–87
10	Whole gale	Seldom experienced inland; trees uprooted; considerable structural damage occurs	You're wondering how you'll get home	88–101
11	Storm	Very rarely experienced; accompanied by widespread destruction	You're wondering what shape your home is in	102–121
12	Hurricane	At sea, visibility is badly affected by foam and spray and the sea surface is completely white	Time to find a new study site	>121

distance from a shore or river bank and with distance above the substratum. By taking measures along a depth gradient at various distances from the shore or river bank, it is possible to build up a reasonable cross-sectional picture of flow. However, when comparing several locations, and perhaps the work of other researchers, a standardised approach can be adopted, such as taking readings at 40% of the water-column depth at set distances out from the bank. For fine spatial resolution of flow some very sophisticated devices that make use of laser Doppler theory are available, but these are expensive items. Finally, when measuring flow in marine coastal systems, remember that, if the area in which you are working experiences tides, flows are likely to be very complex, ranging from zero at slack water to fast at mid-ebb and flood.

Other kinds of water movement

Littoral species in freshwater and marine environments will experience varying degrees of water movement, from a gentle lapping to heavy wave action. Wave action (often termed exposure) is an important variable, but one that is difficult to measure. Since wave action is highly dependent on *fetch*, the distance over which the wind blows to generate the waves, shores of lakes or seas can be ranked using data that can be derived from maps and knowledge of the prevailing wind direction (Thomas 1986). Sites can also be ranked by securing weighed balls of plaster of Paris to the shore and recording their loss of weight over time through dissolution at the various sites. Finally, wave action can be assessed on rocky shores by examining the zonation patterns of the major taxa present (Raffaelli & Hawkins 1986). Whilst this might seem a little circular, it is a valid procedure when the species of interest do not form part of the obvious zonation patterns. Assessing the wave-action climate on sandy beaches relies on quantifying various physical parameters of sediments and waves (McLachlan 1980).

Rainfall

For a quick, rough estimate of rainfall, a flat-bottomed pan with straight sides can be placed out in an open area for a set time period. At the end of the time, the amount of rain it contains can be measured and converted into millimetres fallen per unit time. Beware of evaporation, though, which can be a serious problem, especially during the day.

Rain gauges are more precise, measuring as little as 0.1 or 0.05 mm of rainfall. They usually consist of a cylinder 10–20 cm in diameter with a funnel at its base. Rain entering the cylinder is then directed by the funnel into a narrow tube with a calibrated scale. You can make your own gauge if you relate the volume of water collected to the area covered by the funnel. With any rain-collecting device you still need to be careful of evaporation, splashing, wind and the effects of surrounding vegetation, which can divert water into or away from your instrument.

Snow can also be measured and converted into rainfall. A general rule is that 10 cm of snow is equivalent to 1 cm of rainfall, but this can vary widely depending on how compact the snow is. To make a more accurate conversion, snow collected in a pan or in a rain gauge (the funnel may be removed for snow collection) should be melted and the depth of the water measured. Watch for snowdrifts at the entrance to the collecting device!

Temperature

This can be measured most simply using a mercury- or alcohol-filled *thermometer*. Thermometers with extended probes in wire casings are also available for places that are difficult to reach.

Box 11.1. **Data loggers**

A data logger stores information from various sensors, including thermometers and light meters. Readings may be taken instantaneously or they may be recorded at predetermined intervals over a set time period. They may include maximum and minimum values, averages, or more complex calculations. Data can often be downloaded directly to a computer.

A *min/max thermometer* is useful for recording minimum and maximum temperatures. It is re-set after each reading, and is typically used for 24-h periods. The best results are obtained by keeping thermometers (and hygrometers; see below) in a meteorological box with wooden slats (a *Stevenson screen*). This maintains a uniform temperature that matches the air outside, avoiding surface heating by incident radiation. Specifications can be found in any meteorological textbook or catalogue.

Electronic thermometers comprising a battery-powered meter and metal probe that can sample air, liquid or soil are also available. The accuracy of such thermometers varies with width of measurable range and price.

The main advantage of these thermometers is that they sample a wider range of temperatures than standard mercury thermometers do and, in fieldwork, are less fragile. Electronic sensors can also be connected to a data logger (Box 11.1).

A *thermograph* is useful for a continuous sequence of temperatures recorded on graph paper.

Consult any general apparatus catalogue for choices of thermometers, and look for carrying cases that will withstand fieldwork. Price differences tend to reflect the temperature range to be measured, the accuracy of the instrument and the substance to be measured (liquid, air, soil). Note that many other instruments such as oxygen, pH and conductivity meters also give temperature readings.

Conversion from Fahrenheit to Centigrade and vice versa is done as follows:

$$°F \text{ to } °C = \frac{5}{9}(°F - 32)$$
$$°C \text{ to } °F = \frac{9}{5}°C + 32$$

Humidity

The simplest instrument for measuring relative humidity is a *hygrometer*. Dial-type hygrometers give direct readings from an arm that is made to rotate by the contraction or expansion of an appropriate material (e.g. hair) with changes in humidity. Readings can be recorded continuously on rotating paper (a *hygrograph*). Hygrometers are sold with tables for converting differences between the readings into a humidity value. The best results are obtained in a meteorological box (see 'Stevenson screen', above).

Greater accuracy is afforded by a *wet/dry bulb thermometer*. This consists of two adjacent thermometers, one with a cloth around the bulb. The cloth is attached to a wick in a bottle of water, and capillary action keeps the cloth around the bulb wet. In 100% humidity, there will be no evaporation from the cloth, and the two thermometers will therefore give the same reading. As humidity decreases, evaporation will cause the wet thermometer to give a progressively lower reading than the dry one, because of the latent heat of vaporisation of water.

A *whirling hygrometer* is more versatile and accurate. The two thermometers are swung around on a shaft like a rattle, causing maximal evaporation and giving accurate readings.

Electronic hygrometers are also available either for min/max readings or for continuous readings of both temperature and humidity. Some will give the vapour-pressure deficit, which is the difference between the partial pressure of water vapour in the air and the saturation vapour pressure, which is the maximum vapour pressure possible at a given temperature (Grace 1983). The vapour-pressure deficit is critical for water loss from plants and animals.

pH

This indicates the acidity or alkalinity of a solution, and is a measurement of hydrogen- or hydroxyl-ion activity. It can be measured by two main methods: by using indicator paper and with an electronic probe. These can be used for water or soil.

Indicator paper

Wide-range indicator paper that measures the full pH range (log scale, 1–14) is available. However, measurements are best fine-tuned with narrow-range paper once the general region has been determined. Indicator paper is cheap, quick and convenient to use, but less precise than electronic methods. Water-testing kits for pH are also available.

Electronic determination

This involves a pH meter and electrode. The electrode is immersed in the solution and the meter reads the pH. The meter has to be calibrated at intervals using buffer solutions that resist change in pH. Buffer tablets to make up solutions of pH 4, 7 and 9 often accompany a pH meter. At least two of these (the ones at the upper and lower values) must be used and the standardisation done with buffers at an equivalent temperature to the samples (usually room temperature).

Meters come in two forms: portable ones, for the field, and laboratory ones. Most portable meters are accurate to within about ±0.01 pH unit, whereas laboratory meters are accurate within about ±0.02 pH units. Bear in mind that the electrodes are fragile and must be kept in distilled water when not in use.

Water pH can be measured directly in the field using a portable meter by dipping the electrode directly into the water. Always take pH paper with you to double-check your meter and note that it is advisable to calibrate the meter frequently since some have a tendency to drift.

Alternatively, samples may be collected in clean glass bottles, rinsed in the sample water and brought back to the laboratory for reading. Fill bottles to the top, avoiding the formation of air bubbles because air contains carbon dioxide, which can alter the pH. You should select a method according to accuracy required, time and equipment available and number of sites or samples to be measured.

Soil pH can be determined by mixing one part soil by volume with two parts distilled water (pH 7), waiting for about 10 min, and taking a reading using one of the methods described above. If you cannot measure the pH when you collect the soil, seal the sample in a container to prevent it from drying before the reading is taken. Spear-like probes are also available.

Duration of sunshine

Meteorological suppliers sell simple sunshine recorders consisting of a hollow cylinder that can be set at an inclination appropriate to the latitude. These are mounted at an elevation sufficient for an uninterrupted view of the sky. Sunlight passes through narrow slits in the cylinder and exposes a photo-sensitive chart inside. The resultant trace consists of segments whose length is proportional to the duration of sunlight. An alternative design consists of a spherical lens that burns light tracks onto paper.

Slope angles and height above shore

The slope and height of a census site can have a drastic effect on plants and animals. This is most striking in the intertidal zone of seashores, but it applies to any shoreline study site. Suppliers of survey equipment offer various instruments for making quick, precise readings, including computerised models. We will describe two low-tech methods.

An *Abney level* is relatively inexpensive and commonly used. It is a pocket-sized device that is held against the top of a pole or stick (e.g. at comfortable eye level) at the shoreline and aimed at another pole, of the same length, which is located at the target height. It measures the angle of elevation, α. This can be converted to height by measuring the distance between the two points (shoreline to target). The height of the target above the shoreline is then given by distance $\times \sin \alpha$. Alternatively, the sighting device can be kept strictly horizontal and focused on a calibrated surveyor's pole (or metre rule). By calculating the difference between the vertical height of the sight and the reading on the pole, a detailed profile of the shore can be made. A variant of this approach much loved by shore ecologists is to use a funnel to fill a 10–15-m length of flexible, clear pipe with water from a bucket, and hold your thumbs over each end. One person holds their end of the pipe on the ground at the study site, with the end pointing up about 5 cm above the substrate (Figure 11.1). The other person walks with their end of the pipe down towards the shore, stopping anywhere before the hose is fully stretched out. The shoreward person holds the pipe high enough that, when both people release their thumbs, the water stays in the tube. Gravity ensures that, when this occurs, the top of the water at each end of the pipe must be at the same

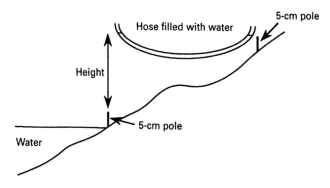

Figure 11.1   A simple method for measuring height above shore.

height. The down-shore person can now measure the vertical distance from the meniscus to the ground using a tape measure. This gives the vertical drop in elevation between the census site and the point towards the shore. It may be necessary to repeat this procedure, with the up-shore person moving to the down-shore site, and the down-shore person moving farther down until the shoreline is reached. The successive vertical measurements are then added together.

Light

The aspect of light which is measured depends on the biological question being addressed. To choose the best method of measurement, ecologists must therefore understand a little physics and a lot about what matters to the organisms being considered. Endler (1990) provides an excellent overview of the measurement of light and colour in an ecological context, and Sheehy (1985) provides a more technical discussion. The following is a distillation of those references as well as discussions with colleagues and our own experience.

Photon irradiance

If you are interested in measuring light available to plants for photosynthesis, or as perceived by animal photoreceptors, you should measure the number of photons striking a unit area. This is the *photon irradiance*, also known as *photon fluence rate* or (almost interchangeably), *photon flux density*. It is measured in mol m^{-2} s^{-1}, where one mole is Avogadro's number of photons (6.02257×10^{23}). As long as you measure photon flux density within the range of wavelengths which the plants or animals use, then all that matters is the number of photons within that total range, not the specific wavelength associated with each photon or the energy of the photons. This is because a photon has the same photochemical effect irrespective of its energy content – the fact that photons at the blue end of the spectrum have more energy than ones at the red end is irrelevant (c.f. 'Energy flux', below).

Measurements of photon flux density are made with *quantum radiometers*. They usually respond to photons over the range of wavelengths 400–700 nm. This is the visible range and also the range of photosynthetically active radiation for plants. Although it is applicable to most vertebrate photoreceptors, you should beware of exceptions, such as birds seeing in the ultraviolet (UV). Arthropods typically span 300 nm (UV) to 650 nm. Inferences from your measurements will be only as good as your knowledge of what is meaningful for your particular study species.

Lightweight, portable radiometers are available for fieldwork. The most useful models have a long, rod-like apparatus (e.g. 1 m long) equipped with an array of sensors to integrate over a large area. This accounts for uneven lighting in vegetation; otherwise, one leaf can have a huge effect on the light level you record. A hand-held digital display unit attached to the sensor rod typically stores readings taken over a period of time, to account for the effects of clouds and other temporal changes, and the data can be downloaded as point readings or averages.

Energy flux

You may be interested in the energy budget of plants, which includes the radiative, convective and conductive exchanges of energy between plants and their environment (photochemical processes such as photosynthesis involve a very minimal part of the total energy budget). In this case, it is the *energy flux* that matters, and this is measured in $W\,m^{-2}$ (irradiance).

Light energy is measured with *radiometers* (for total radiation) or *spectroradiometers* (for specific wavelengths). Hand-held battery-operated versions are available for fieldwork. *Tube solarimeters* can also be used, though these are more often used in crop stands. They measure energy over the full range of wavelengths from solar radiation. They consist of a glass tube containing a flat thermopile painted with an alternating pattern of black and white rectangles along the midline. The thermopile measures the temperature difference between the black and white rectangles. This information can be recorded by a data logger (Box 11.1), or a millivolt integrator, which is much cheaper and gives the accumulated total radiation over a period of time. There are also meteorological standard solarimeters such as the Kipp design, which give an absolute reference because they are cosine corrected (for differing angles of incidence of the light). These are expensive, however, and do not offer the larger area of coverage of tube solarimeters, though they may be useful for standardising your readings for long-term meteorological records.

In the past, equipment for measuring energy was much cheaper than equipment for measuring photons. Thus, energy measurements from devices such as solarimeters were (and still are) used as indirect measurements of light input for photosynthesis. However, the conversion to photon flux density is not straightforward, and corrections for the different range of wavelength sensitivity of solarimeters must be made. Recent technological advances have improved the portability and price of photon-flux equipment, reducing the need to substitute for it equipment that measures energy.

Photometers

Standard photographic light meters, which are often calibrated in *lux*, are useful for studies of human vision, but they have little relevance to anything else. This is because the sensitivity of these meters is based on the spectral responses of 52 pairs of American eyes in 1923! Thus, they are not useful for any animal that does not have the same sensitivity at various wavelengths as humans, and they have no relevance for plants.

Aquatic light

This is commonly measured to determine the depth to which photosynthetic organisms are limited. This occurs where approximately 1% of the surface light still penetrates, and the distance from the surface to this depth is termed the euphotic zone (z_{eu}). This depth can be estimated by examining the visibility of a *secchi disc* lowered into the water. Readings are best taken under consistent lighting conditions. The secchi disc is about 30 cm in diameter with alternating black and white or yellow quarters. The disc is lowered slowly into calm water using a calibrated line (for example, marked every half meter) and, when the disc disappears, the depth of the line is recorded. Then the disc is lowered slightly further and raised slowly until it reappears. This depth is also recorded. The average of these two depths is the final secchi-disc visibility reading (Dowdeswell 1984). This reading itself can be used for comparative purposes for a water body over time, but the depth reading (d) has an approximate relationship to the euphotic zone depth z_{eu}, which can be estimated as $z_{eu} = 1.2$ to 2.7 (where the mean is 1.7) times the secchi-disc depth (Moss 1988). Clearly, this estimate is very rough and the measured depth will depend on the ambient light, water-surface movement and the person using the disc. If more accurate light readings for a variety of depths are required, then an underwater light meter should be used.

Underwater light meters consist of a probe with a light sensor (measuring photon flux) attached to a recorder or data logger (Box 11.1). Light intensity in the range of photosynthetically active radiation (400–700 nm) is measured by lowering the probe into the water to the appropriate depth. By using selective filters, it can also measure the penetration of narrower ranges of wavelengths (red, blue and green). This is important because wavelengths are not all absorbed by the water at the same rate; red light is usually absorbed first, followed by blue, then green. Most light meters will measure intensity as photon flux (μmol m^{-2} s^{-1}), but what is of interest is the percentage of that received at the surface for a given depth, or the percentile absorption:

$$\text{Percentile absorption} = \frac{100(I_o - I_z)}{I_o}$$

where I_o is the irradiance at the surface and I_z is the irradiance at depth z (Wetzel 1975). If this is done for a range of depths, a light-depth profile can be constructed.

Water turbidity

Water turbidity, which is caused by suspended particulate matter, is important in aquatic systems because it reduces the penetration depth of surface light, thereby limiting photosynthesis as well as the visual range of aquatic animals.

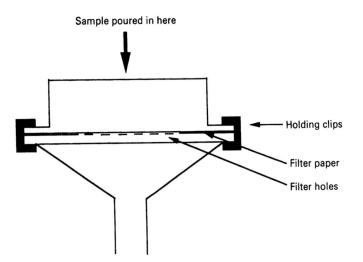

Figure 11.2   A Buchner funnel for filtering water samples.

The cheapest but most time-consuming method of measuring turbidity is to weigh the particulate matter present in a water sample, which indicates the total suspended solids. If a sample from a particular depth is required, this can be obtained using a Niskin or Go Flow Flask (General Oceanics Inc.). This consists of a tube with a lid at either end, each attached to a spring closing mechanism. The tube is lowered into the water in the open position and then triggered to trap water at the chosen depth. In the laboratory, a glass-fibre filter paper is rinsed in distilled or deionised water, dried and then weighed. A standard quantity of the sample is filtered through the dried paper using a Hartley form of a Buchner funnel (Figure 11.2). In this design the base with the holes and the funnel sides are two separate parts held together by clips, so the filter paper covers the entire base, with no gap between the paper and side of the funnel. The used glass-fibre filter is then dried until the weight remains constant, to yield the weight of the particulate matter (original filter weight subtracted from the used filter weight) (Allen 1989).

The water sample may also be analysed using a (portable or laboratory-based) *turbidity meter*, whereby some of the sample is put into a sample cell and the light passing through it is measured.

The quickest and most convenient method of measuring turbidity is to use a portable *suspended-solids monitor*. This consists of a meter with a probe containing a fixed light path. The probe is dropped to the depth required, and the turbidity is then read from a meter. The main advantages of this meter are that water samples do not have to be collected, readings are given directly, various pre-set ranges are available and some devices may take long-term readings if set up to a data logger (Box 11.1).

Conductivity

Conductivity is the measure of the current carried by electrolytes in a solution. In general, seawater has a higher conductivity than freshwater, due to its higher ionic concentration. Many portable and lab-based conductivity meters are available. These may measure either a wide range or a number of narrow ranges that give more accurate readings. Check catalogues for models to suit the accuracy required and your budget.

Salinity

Salinity is the salt content of seawater and is the weight of total salts (g) dissolved in 1 kg of seawater, usually expressed as parts per thousand or ‰. For full-strength seawater, this value is about 35‰, or, more properly, 35 psu (practical salinity units). Originally, salinity was measured by determining the chlorinity (‰) (which is the mass of chlorine in 1 kg of seawater equivalent to the combined mass of chlorine, bromide and iodide) and this is done titrametrically, but is not covered here because newer and faster ways to determine salinity have been developed since.

There is a linear relationship between salinity and chlorinity, which can be expressed as

$$S\text{‰} = 0.030 + 1.8050Cl\text{‰}$$

(Parsons *et al.* 1984).

Salinity can be measured with a range of methods, either electrically or through the use of simple but fairly reliable gadgets. We will not discuss chemical methods here since they are time-consuming and the methods discussed below are accurate enough for most purposes. Readers interested in chemical methods for determination of salinity and chlorinity are referred to Strickland & Parsons (1968).

Conductivity meters

Salinity can be measured by measuring the conductivity of the water (see 'Conductivity', above) and converting the reading to salinity.

Meters that measure over a single wide range tend to be inaccurate, especially at the freshwater end of the scale. A better option is to use meters that can be focused on whichever part of the scale is relevant, to allow more accurate salinity readings within the sub-range.

The conductivity reading (mS) can be converted into salinity (‰) using an equation developed with the aid of Richard Sanders, School of Environmental Sciences, University of East Anglia, Norwich:

$$\text{Salinity} = 0.64 \times \text{Conductivity}$$

This was devised by measuring the conductivity of a dilution series of standard seawater (NaCl) at room temperature. This gives a measurement of salinity that, although not compensated for temperature and pressure (since the error is minimal), is accurate enough for an estimate of salinity when the major ions are Na and Cl. More complicated sets of tables are available for converting conductivity ratio to salinity, with compensation for temperature and depth (NIO and UNESCO, 1966).

Salinometers

A salinometer works on a principle similar to that for the conductivity meter but provides a direct reading of salinity (‰). With some meters the temperature has to be measured separately and then set on the meter when readings are taken. These machines tend to be more bulky than conductivity meters and come with a range of probes and cable lengths. They are suitable for fieldwork, especially for boat use and for giving readings at depth. Other salinometers are laboratory-based.

Specific gravity

Salinity can be measured indirectly by a very simple method whereby the specific gravity of seawater is recorded with a *hydrometer*. This gives a reading ranging from 1.000 (freshwater) to 1.025 (seawater), which can be converted to salinity when temperature is incorporated. A conversion chart is given in Figure 11.3. A simpler cheap variation of this for use at the seawater end of this range is a device called a *Seatest Specific Gravity Meter*. This is a small plastic container with an arm that floats to a reading depending on the water's specific gravity. It is available in many aquarium shops.

Hand-held salinity meters

These work optically, taking advantage of the fact that daylight is refracted to differing degrees depending on the salinity of the water sample. Essentially, a drop of water is placed on a glass window in a device no bigger than a cigar tube, and held up to the sunlight. The salinity is given on a scale of practical salinity units (psu) or parts per thousand. These devices are cheap, fairly accurate, small and require no power supply.

Preamble to water chemistry

Many substances that are important for aquatic ecological studies are present in water, such as nutrients, silicates, metals, gases and plant extracts. It would require an entire book to cover all of these, and we recommend Mackereth *et al.* (1978) and Golterman *et al.* (1978) for freshwater and Strickland & Parsons (1968) and Parsons *et al.* (1984) for seawater. A very comprehensive and accessible manual is provided by Radojevic and Bashkin (1999) for a range of aquatic (and terrestrial) environments. We present methods for quantifying nitrogenous and phosphorus compounds, since these are important to most studies of plant growth, pollution and eutrophication. First, the following notes on safety should be read, and ecologists who have forgotten their basic chemistry may find the subsequent comments helpful.

Safety

Always be aware of the hazards of the chemicals that you intend to use. Although a symbol is usually given on the container (e.g. corrosive, hazardous, toxic), more detailed hazard forms are available on request from the suppliers, and some produce hazard books and CDs providing

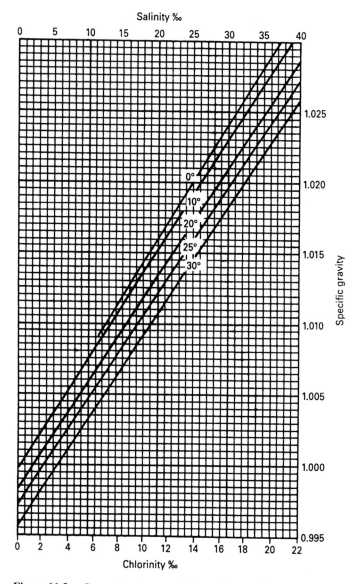

Figure 11.3   Conversion from specific gravity to salinity. (After Harvey (1960).)

safety information. These describe the handling, first-aid and disposal procedures that should be followed. Many institutes require that you sign hazard forms that assess your personal risk on the basis of likely levels of exposure. One general rule of safety is to treat all chemicals as hazardous. This means using disposable gloves, both to protect you and to avoid contaminating your samples (this is especially important for phosphorus). Use goggles and fume cupboards if recommended by hazard forms.

Disposal procedures are normally outlined on the hazard forms. Innocuous chemicals may be flushed down the sink with copious amounts of water. Special disposal procedures for more harmful chemicals and contaminated disposables (gloves, pipette tips, etc.) are available in most institutes. **It is vital that you consult fully with the appropriate Health and Safety representative in your institution before embarking on any of the procedures outlined below.**

Glassware

It is important to use inert glassware that will not contaminate your chemicals (e.g. Pyrex). Glassware should be washed beforehand using a harmless surfactant cleaning agent. Sometimes a dilute acid wash is needed to reduce contamination: soak in 10% HCl for 48 h and then rinse several times in distilled deionised water. This is especially important for determination of nitrogen and phosphorus. After washing, store glassware with stoppers or covered with aluminium foil to reduce atmospheric contamination. Some solutions react to light, so they should be stored in a dark brown bottle or in a clear bottle wrapped in aluminium foil.

Making up reagents

Solutions are best made up using volumetric flasks, which are available in a wide range of volumes (1 ml to 5 l). It is best to put some liquid into the flask before you add a solid chemical (so it dissolves more easily). Then fill the flask to the measured line, rinsing the sides of the flask and the weighing boat (if used).

The quality of water required depends on the analysis to be done. Water may be available as tap water, distilled water, doubly distilled water, distilled deionised water, or Milli-Q water or may be purchased as extra pure. For determination of nitrogen and phosphorus, it is advisable to use the water of highest purity available to you (often distilled deionised water), but always check that the water available to you is free of the substance that you want to measure. For dissolved oxygen, either distilled or distilled deionised water is suitable.

In making up standards, anhydrous chemicals must be used. These are obtained by drying them in a drying oven (usually at 105 °C) overnight and then storing them in a desiccator when not in use.

Chemical abbreviations

w/v or weight per volume; the actual weight of the compound is used instead of the relative molecular mass. For example, for a 100-ml solution of 10% w/v NaCl, 10 g of NaCl is dissolved in 100 ml of water, as described above.

v/v or volume per volume; the actual volume of the liquid is used instead of the relative molecular mass. For example, for a 100-ml solution of 10% v/v HCl, 10 ml of concentrated HCl is added to 90 ml of distilled water.

Dissolved oxygen

Dissolved oxygen (DO) is a critical factor in aquatic ecology. The DO concentration is affected not only by natural factors such as temperature, salinity, plant respiration and amount of organic material present, but also by organic pollution and eutrophication.

Oxygen in water can be measured using two main techniques: chemically using the *Winkler titration* or electronically with an *oxygen electrode*. Although the chemical titration method is very accurate, it is more time-consuming than using an oxygen meter. Your choice of method will be influenced by the accuracy required and the number of samples that you need to analyse. Water-testing kits to measure DO are available. There are also *auto-titration machines*, which add chemicals automatically, and these can be coupled to home-made units that detect when the titration is complete (the end point), although commercial end-point detectors are now available. Oxygen is expressed either as % saturation or in mg l^{-1}, which is the same as ppm (parts per million).

Winkler titration

The original Winkler titration was developed for freshwater but has been modified for many types of water, including seawater (Parsons *et al.* 1984), waters rich in organic matter and waters of high alkalinity (Mackereth *et al.* 1978). The method given in Box 11.2 may be used for both freshwater and seawater samples.

Box 11.2. **Winkler titration for measuring dissolved oxygen**

The oxygen in a water sample is fixed with addition of the manganous chloride and alkaline iodide solutions. At this stage the sample can be left until it is analysed back in the laboratory. The sample is then acidified in the presence of iodine and then titrated with sodium thiosulphate using a starch indicator. The amount of sodium thiosulphate used in this titration is related to the amount of oxygen in the original sample. The method was modified from Parsons *et al.* (1984).

It is advisable to read the safety notes in the 'Preamble to water chemistry' before using this procedure.

Reagents

Manganese chloride
Dissolve 600 g of analytical-reagent (AR)-grade manganous chloride ($MnCl_2.4H_2O$) in distilled water and make up to 1 litre.

Alkaline iodide
Dissolve 320 g of AR-grade sodium hydroxide (NaOH) and 600 g of AR-grade sodium iodide (NaI) in distilled water, leave to cool before making up to 1 l with distilled water.

Sulphuric acid
Slowly add 280 ml of concentrated sulphuric acid (H_2SO_4) to 500–600 ml of distilled water. Once cooled, make up to 1 l with distilled water.

Thiosulphate solution
Dissolve 2.9 g of AR-grade sodium thiosulphate ($Na_2S_2O_3.5H_2O$) and 0.1 g AR-grade sodium carbonate (Na_2CO_3) in distilled water and make up to 1 l with distilled water. Store in a dark bottle.

Potassium iodate
Dissolve 0.3467 g of anhydrous AR-grade potassium iodate (KIO_3) in 200–300 ml of distilled water, warming if necessary. Cool and make up to 1 l with distilled water.

Starch indicator (1 % solution)
Dissolve 1 g of starch in 100 ml of distilled water by warming to 80–90 °C. This solution should be made up fresh if the stock solution is over a week old.

Method

- The water sample is taken using a clean 125-ml reagent bottle and rinsed a few times with the water to be sampled.
- A blank should be prepared in the laboratory using deoxygenated water (purge it with nitrogen) and treated in the same way as the sample.
- Fill the bottle using a displacement sampler (Figure 11.4) or long tube reaching into the bottom of the bottle to avoiding 'glugging' and the formation of air bubbles, which may introduce more oxygen into the sample. The water should be allowed to overflow the bottle before being stoppered.
- Immediately remove the stopper and add 1 ml of the manganous reagent followed by 1 ml of the alkaline iodide solution. In the field this may be done using 1-ml syringes (one for each solution).
- Stopper the bottle firmly and invert to mix the reagents with the water; a yellow precipitate should form.
- If the sample has been taken in the field or is not to be analysed immediately, it can be kept in this state indefinitely by storing the tightly stoppered bottles under water, preferably in dark conditions at ambient temperature.
- Acidify the sample by adding 1 ml of sulphuric acid reagent to the sample.
- Restopper the bottle and mix the sample thoroughly.
- Within 1 h of the acidification transfer 50 ml of the sample to a conical flask.
- Titrate the 50-ml sample immediately with the thiosulphate reagent from a burette, using a magnetic stirrer in the flask (or while swirling the flask), until the solution becomes faintly yellow.
- Add 0.5 ml of the starch solution and the solution will turn blue.
- Continue titrating cautiously with the thiosulphate solution until the blue colour disappears and the liquid just turns clear. It may help to put a sheet of white paper behind the flask so that the colour change can be observed more easily.
- The volume of thiosulphate used is noted and the volume needed for the blank's titration subtracted to give V_{sample}.

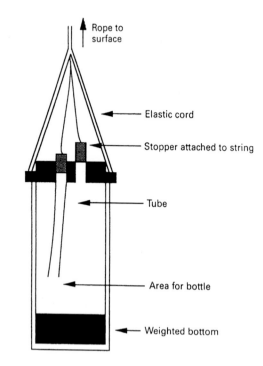

Rope to surface

Elastic cord

Stopper attached to string

Tube

Area for bottle

Weighted bottom

Figure 11.4   A displacement sampler for water oxygen samples. A sampling bottle is placed inside, with the tube inserted into it. The flask is dropped to the required depth and then the rope is jerked. This causes the elastic cord to stretch, pulling out the stoppers and permitting water to flow through the tube and into the bottle.

Before the oxygen content can be calculated, a calibration must be done each time a set of samples is analysed, in order to account for variation in the thiosulphate solution.

Calibration

- Fill a 125-ml bottle with distilled water and to it add 1 ml of concentrated sulphuric acid and 1 ml of the alkaline iodide reagent and then mix thoroughly.
- Add 1 ml of the manganous chloride reagent and mix again.
- Transfer 50 ml of this solution into a conical flask.
- Add 5 ml of the potassium iodate solution and mix gently for 2 min.
- Titrate with the thiosulphate solution as before, noting the volume used (V_{cal}).
- Carry out this procedure three times and calculate f for each replicate:

$$f = \frac{5.00}{V_{cal}}$$

Take the mean of the three values and calculate the actual oxygen content of your samples (mg l^{-1} or ppm):

$$O_2 \text{ content (mg l}^{-1}) = 0.1016 \times f \times V_{sample} \times 16$$

The oxygen content of water depends on temperature and salinity (as the temperature and salinity increase, the oxygen content decreases), so if the percentage saturation of DO is required, a theoretical DO content in mg l^{-1} must be calculated. This is the oxygen content in mg l^{-1} of a fully saturated sample, adjusted for temperature and salinity (see Table 11.2). The percentage DO content is then calculated from

$$\%DO = \frac{\text{Titrated oxygen concentration(as above)}}{\text{Theoretical oxygen concentration}} \times 100$$

Oxygen electrodes

An *oxygen electrode* provides a more convenient and time-saving method of measuring dissolved oxygen. Probes can be portable or laboratory-based and usually consist of a meter (sometimes multipurpose) and a sensor. Although a probe provides direct monitoring, it is less accurate than Winkler titration, which will need to be done occasionally to check that the oxygen probe is calibrated correctly and giving accurate readings. A detailed overview of available oxygen meters has been given by Richardson (1981). When taking readings with an oxygen meter, a water flow must be present or the probe must be kept moving since such probes abstract oxygen from the water. Electrodes must be kept clean and moist.

Dissolved-oxygen meters need to be calibrated frequently. Most modern portable meters can be calibrated easily using zero-oxygen solutions and saturated air or 100%-oxygen solutions (this can be achieved by filling a bottle with water and shaking it for a minute or two). Some expensive meters calibrate themselves automatically relative to air.

Scientific catalogues offer a wide range of meters. Your choice will depend on price, accuracy needed (some are accurate to 0.1%, others only to 1%) and your specific requirements. Some meters can be used with more than one type of probe. For example, oxygen, pH and conductivity can be measured using one meter with three separate probes. The percentage DO concentration depends on temperature and salinity and, although most meters will automatically measure and compensate for temperature, only some have a salinity- (and altitude-) compensation function. Readings from meters lacking temperature- or salinity-compensation functions can be adjusted using Table 11.2.

Nitrogenous compounds

Nitrogen is often determined in water because it is important for plant growth and may be a key limiting nutrient in seawater. If excessive quantities of nitrogen (and phosphorus) are present, due to agricultural run-off, silage, sewage or industrial discharges, eutrophication may result.

In water, total nitrogen exists in two forms, particulate (organic, for example in plants and algae) and soluble (inorganic). Soluble nitrogen exists in many forms and constantly fluctuates among them as it is oxidised or reduced. Ammonium (NH_4^+) and nitrate (NH_3^-) ions are most stable in the environment, and these are important because they are the main forms available for uptake by plants.

Table 11.2. *Solubility of oxygen in water*

Temperature (°C)	Solubility of oxygen in water (mg l^{-1})	Correction to be subtracted for each degree of salinity (‰)
0	14.62	0.0875
1	14.22	0.0843
2	13.83	0.0818
3	13.46	0.0789
4	13.11	0.0760
5	12.77	0.0739
6	12.45	0.0714
7	12.14	0.0693
8	11.84	0.0671
9	11.56	0.0650
10	11.29	0.0632
11	11.03	0.0614
12	10.78	0.0593
13	10.54	0.0582
14	10.31	0.0561
15	10.08	0.0546
16	9.87	0.0532
17	9.66	0.0514
18	9.47	0.0500
19	9.28	0.0489
20	9.09	0.0475
21	8.91	0.0464
22	8.74	0.0453
23	8.58	0.0443
24	8.42	0.0432
25	8.26	0.0421
26	8.11	0.0407
27	7.97	0.0400
28	7.83	0.0389
29	7.69	0.0382
30	7.56	0.0371

Example. At 15 °C, the oxygen content of fully saturated water (100%) contains 10.08 mg l^{-1} of oxygen, with 0.0546 subtracted for each degree of salinity of the sample. HMSO (1988).

There are four main methods of determining the concentrations of nitrogenous compounds in water. The easiest way is to use a sophisticated machine that does the chemical analyses automatically, the *continuous-flow autoanalyser* (Box 11.3). A much cheaper and faster approach, which is far less accurate and constrained to a narrower detection range, is the use of a *water-testing kit* (see p. 399). A middle option, which we present here in detail, involves individual chemical analyses. These are much cheaper than using the autoanalyser and fairly accurate, but

---

Box 11.3. **Continuous-flow autoanalysers**

This type of machine analyses your samples spectrophometrically for dissolved nitrogen and dissolved phosphate using methods similar to those employed manually. A continuous steady flow of reagents is present, into which a small quantity of your sample is introduced. It moves down through the reagents and then the absorbance is read and stored in a computer. These automated analysers yield results that are highly reproducible, although, since reagents are made manually or samples may need to be diluted, some human error still exists.

---

time-consuming. Finally ammonium and nitrate probes that are used in conjunction with a pH meter (mV reading) are available. A set of standards is made up and readings from them plotted to determine the N concentration (in a similar way to the chemical method). Although this is faster than the chemical method, the detection limit of this type of electrode, especially the nitrate electrode, may be too high to pick up the low concentrations that are common in many waters.

Recipes for determining the concentrations of nitrate and ammonium are presented below. We recommend that readers interested in determining particulate matter or total nitrogen refer to Mackereth *et al.* (1978) or Parsons *et al.* (1984). Although the methods presented here may be used both for fresh and for saline waters, since saline water contains lower concentrations of nitrogenous compounds, the concentrations of the standards may need to be reduced to suit their range. Also note that, on mixing salt water with reagents (containing distilled water), some opaqueness will result. The solution should be allowed to settle out before reading the absorbance of a sample. If high accuracy is needed for saline water, the method in Parsons *et al.* (1984) is recommended.

Nitrate

In surface freshwater, NO_3^- is found in concentrations from 0 to almost 10 mg l^{-1}, usually in the range 0–3 mg l^{-1}. Groundwater concentrations can reach 30 mg l^{-1} (Allen 1989; Wetzel 1975). In seawater, the nitrate level is usually less than 0.5 mg l^{-1}. However, be aware that, although levels increase with the ingress of pollutants such as sewage effluent, levels are also very much affected by time of year (there are increases in late autumn and winter due to agricultural run-off) and the geology of the area. If an autoanalyser is not available (Box 11.3), nitrate concentration can be determined using the method in Box 11.4.

Ammonium ions

In surface freshwater, ammonium (NH_4^+) is generally found in the range of 0–3 mg l^{-1}, though for clean waters the range is usually 0–100 µg l^{-1} (Allen 1989). Again, as for nitrate, the level depends on pollution sources, time of year and the geology of the area. A manual method for determining the ammonium concentration is given in Box 11.5, though an autoanalyser may also be used if available (see Box 11.3).

Box 11.4. **Nitrate (NO$_3^-$)**

Nitrate is determined by reducing all of the nitrate to nitrite and then determining this nitrite concentration spectrophotometrically. Any nitrite previously present will also be included, but concentrations of nitrite are minimal and may be disregarded. If high accuracy is required, this nitrite content can be determined using the method found in Mackereth *et al.* (1978). There may be interference from particulate matter, high quantities of dissolved organic matter or sulphide, in which case the reader is referred to Mackereth *et al.* (1978). The method used here for nitrate determination was kindly made available by Professor Brian Moss of the University of Liverpool, and is modified from Mackereth *et al.* (1978).

It is advisable to read the safety notes in the 'Preamble to water chemistry' before using this procedure.

Reagents

Spongy cadmium
Place zinc rods in a solution of 20% w/v AR-grade cadmium sulphate (3CdSO$_4$.8H$_2$O) overnight to build up a cadmium deposit on the rods. Then use a spatula to scrape off the deposits and divide them into small particles. Wash the filings for 15 min with a 2% solution (v/v) of hydrochloric acid (HCl) and then rinse them several times with distilled deionised water to remove all the acid. Filings should be stored under distilled deionised water. Just before use, repeat the HCl wash (10 min) and rinse again with distilled deionised water. Cadmium is poisonous, so wear disposable gloves for handling and dispose of all filings and solutions in a safe manner, as prescribed by the institution at which you are based.

Ammonium chloride
Dissolve 2.6 g of AR-grade ammonium chloride (NH$_4$Cl) in 100 ml of distilled deionised water.

Borax
Dissolve 2.1 g of AR-grade borax (disodium tetraborate (Na$_2$B$_4$O$_7$.10H$_2$O)) in 100 ml of distilled deionised water.

Sulphanilamide
Dissolve 1 g of AR-grade sulphanilamide (NH$_2$C$_6$H$_4$SO$_2$NH$_2$) in 100 ml of 10% v/v hydrochloric acid (HCl).

N-1-naphthylethylene diamine dihydrochloride
Dissolve 0.1 g of AR-grade *N*-1-naphthylethylene diamine dihydrochloride (C$_{10}$H$_7$NHCH$_2$ CH$_2$NH$_2$.2HCl) in 100 ml of distilled deionised water. This solution should be stored in a dark bottle and renewed each month.

Hydrochloric acid (2%)
Add 2 ml of concentrated hydrochloric acid (HCl) to about 50 ml of distilled deionised water and then make up to 100 ml.

Stock nitrate standard

Dissolve 0.722 g of anhydrous AR-grade potassium nitrate (KNO_3) in 1 l of distilled deionised water. This solution contains 0.1 mg ml^{-1} or 100 mg l^{-1} of nitrate N.

Nitrate standards

Make new standards for each use to obtain a calibration curve. Dilute your stock nitrate standard (0.1 mg ml^{-1} of nitrate N) with distilled deionised water as specified below to prepare standards with concentrations of

> 1 mg l^{-1} (10 ml of stock made up to 1 l)
> 0.8 mg l^{-1} (8 ml of stock made up to 1 l)
> 0.6 mg l^{-1} (6 ml of stock made up to 1 l)
> 0.4 mg l^{-1} (4 ml of stock made up to 1 l)
> 0.2 mg l^{-1} (2 ml of stock made up to 1 l)
> 0 mg l^{-1} (use distilled deionised water)

Method

- Filter your water sample as soon as possible and keep it cool. Filtering may be done using a Whatman GF/C glass-fibre filter, handling with forceps to avoid contamination, but bear in mind that the paper may retain or release nitrogen. This can be minimised by first rinsing the paper in distilled deionised water, or using cellulose acetate filters.
- Your sample should be analysed as soon after collection as possible. If it must be stored (which is not recommended since your N content may change), use a filtered sample in a full bottle (excluding air) at 1 °C.
- Prepare your standards and treat them in the same way as your sample.
- Place 10 ml of your sample water into a 30-ml polystyrene bottle with cap. Add 3 ml of your ammonium chloride solution and then 1 ml of your borax solution, followed by 0.5–0.6 g of the spongy cadmium. Screw on the cap and shake in a mechanical shaker for 20 min precisely.
- Transfer 7 ml into a 50-ml volumetric flask and add 1 ml of your sulphanilamide reagent. Mix by swirling.
- After 4–6 min, add 1 ml of your *N*-1-naphthylethylene diamine dihydrochloride reagent and mix again. Make the solution up to 50 ml with distilled deionised water.
- After 10 min (and before 120 min), measure the absorbance of the samples at 535 nm, using a 1-cm spectrophotometer cell. Use your 0-mg l^{-1} standard as your blank; that is, use it to set your zero on the spectrophotometer so that the sample reading automatically has the blank absorbance subtracted from it. This saves you from having to subtract the blank reading from each sample reading.
- Using the absorbance values for your standards, plot a calibration curve (of the concentration of N as mg l^{-1} of nitrate N against the absorbance reading). Your standard graph should be linear. If it levels off, then the top solutions that cause this levelling off should not be included. Determine the best fitting line.

- To determine the concentration of N in your sample, use your fitted-line equation to calculate from the sample's absorbance its related N content (mg l^{-1} of nitrate N). If your sample's absorbance does not fall on your calibration-curve line, then dilute the sample (for example two-fold or ten-fold) until it does and then multiply your calculated N content by the dilution factor.

Box 11.5. **Ammonium (NH_4^+)**

This is measured by reacting the ammonium with phenol and hypochlorite in an alkaline solution to form indophenol blue. The reaction is catalysed by nitroprusside and the absorbance of the end product is read spectrophotometrically. The method used here for ammonium determination was kindly made available by Professor Brian Moss of the University of Liverpool, and is modified from Golterman *et al.* (1978). There may be interference from very calcareous water, in which case consult Golterman *et al.* (1978). This method measures both ammonium and ammonia (total oxidised nitrogen), but ammonia is present in small concentrations so this contamination is minimal. Also note that phenol is highly toxic, so an alternative method for freshwater is that given by HMSO (1981) and one for seawater is given by Parsons *et al.* (1984).

It is advisable to read the safety notes in the 'Preamble to water chemistry' before using this procedure.

Reagents

Sodium nitroprusside solution
Dissolve 1 g of AR-grade sodium nitroprusside ($Na_2[Fe(CN)_5NO].2H_2O$) in 200 ml of distilled deionised water. Store in a dark bottle. This solution is stable for one month. Take care since sodium nitroprusside is toxic.

Phenol solution
Note that phenol is highly toxic, so handle with extreme care (see 'Preamble to chemical analysis').
Dissolve 20 g of AR-grade phenol (C_6H_5OH) in 200 ml of 95% ethanol.

Alkaline solution
Dissolve 100 g of AR-grade trisodium citrate ($Na_3C_6H_5O_7.2H_2O$) and 5 g of sodium hydroxide (NaOH) in 500 ml of distilled deionised water.

Hypochlorite solution
Using 300 ml of 14% w/v available-chlorine hypochlorite solution, make this up to 500 ml using distilled deionised water. Make up this solution fresh for each use.

Oxidising solution

Mix by volume four parts of the alkaline solution to one part of the hypochlorite solution; 5 ml is required for each sample and should be made up fresh before each use.

Stock ammonium standard

Dissolve 0.9433 g of anhydrous AR-grade ammonium sulphate $((NH_4)_2SO_4)$ in 1 l of distilled deionised water. This solution contains 200 µg ml^{-1} or 200 mg l^{-1} of ammonium N.

Working ammonium standard

Take 1 ml of the stock ammonium standard and make up fresh to 500 ml with distilled deionised water for each use. This solution contains 0.4 µg ml^{-1} or 400 µg l^{-1} of ammonium N.

Ammonium standards

Using the working ammonium standard, make up the following standards fresh for each use using distilled deionised water:

> 400 µg l^{-1} (50 ml of the working standard).
> 300 µg l^{-1} (37.5 ml of the working standard made up to 50 ml)
> 200 µg l^{-1} (50 ml of the working standard made up to 100 ml)
> 100 µg l^{-1} (50 ml of the 200-µg-l^{-1} standard made up to 100 ml)
> 50 µg l^{-1} (25 ml of the 100-µg-l^{-1} standard made up to 50 ml)
> 0 µg l^{-1} (use distilled deionised water)

Method

- Filter your water sample as soon as possible. This may be done with Whatman GF/C glass-fibre filter, handling with forceps to avoid contamination, but bear in mind that the paper may retain or release nitrogen. This can be minimised by first rinsing the paper in distilled deionised water, or using cellulose acetate filters.
- Your sample should be analysed as soon after collection as possible and kept cool. If it must be stored (which is not recommended because the N content may change), do so using filtered sample in a full bottle (excluding air) at 1 °C.
- Prepare your standards and treat them in the same way as your filtered sample.
- Add 50 ml of your filtered sample to 2 ml of the phenol solution and mix.
- Add 2 ml of the sodium nitroprusside solution and mix.
- Finally, add 5 ml of the freshly prepared oxidising solution, mix and store the sample in the dark until the absorbance is read.
- After 1.5 h (and before 12 h has elapsed), measure the absorbance of the samples at 640 nm, using a 1-cm glass cell. Use your 0-µg-l^{-1} standard as your blank; that is, use it to set your zero on the spectrophotometer so the sample reading automatically has the blank absorbance subtracted from it. This saves you from having to subtract the blank reading from each sample reading.

- Using the absorbance values for your standards, plot a calibration curve (concentration of N as $\mu g\ l^{-1}$ of ammonium N against absorbance reading). Your standard graph should be linear. If it levels off, then the top solutions that cause this levelling off should not be included. Determine the best fitting line.
- To determine the concentration of N in your sample, use your fitted line equation to work out from the sample's absorbance its related N content ($\mu g\ l^{-1}$ of ammonium N). If your sample's absorbance does not fall on your calibration-curve line, then dilute the sample (for example two-fold or ten-fold) until it does and then multiply your calculated N content by the dilution factor.

Phosphorus compounds

Phosphorus is commonly determined in water because it, like nitrogen, is important for plant growth, and may be limiting in freshwater. Excessive quantities of phosphorus from agricultural run-off, sewage and industrial and household detergents can cause eutrophication.

Phosphorus is present in two forms in water, particulate phosphorus (organic, for example in plants and algae) and dissolved phosphorus (inorganic: soluble reactive phosphorus and soluble unreactive phosphorus). We are usually interested in the soluble form because this is the form that is available to plants and algae. Most of this is made up of soluble reactive phosphorus, which is orthophosphate (PO_4^{3-}) and small reactive compounds that form PO_4^{3-} complexes rapidly. Soluble unreactive phosphorus is the rest of the phosphorus that does not form these complexes rapidly. We present only a method for determining soluble reactive phosphorus since this constitutes the bulk of the dissolved phosphorus and is the most common form of phosphorus pollutant. To measure other forms of phosphorus, for example in studies of nutrient cycling, we recommend Mackereth *et al.* (1978) for freshwater samples and Parsons *et al.* (1984) for seawater samples.

As with nitrogenous compounds, various options are available for measuring phosphorus, including using an autoanalyser (see Box 11.3), use of a water-testing kit and chemical methods (see Box 11.6).

Box 11.6. **Soluble reactive phosphorus (PO_4^{3-})**

Under acidic conditions, phosphate reacts with molybdate to form molybdo-phosphoric acid. This is reduced to a molybdenum-blue complex that is intensely coloured and this is determined spectrophotometrically. Interference from arsenate can occur, in which case refer to Mackereth *et al.* (1978) The method used here for determination of soluble reactive phosphorus was kindly made available by Professor Brian Moss of the University of Liverpool, modified from Mackereth *et al.* (1978).

It is advisable to read the safety notes in the 'Preamble to water chemistry' before using this procedure.

Reagents

Sulphuric acid
It is advisable to wear disposable gloves and eye protection when handling the concentrated sulphuric acid, since it is highly corrosive.

Carefully add 140 ml of concentrated sulphuric acid (H_2SO_4) to 900 ml of distilled deionised water.

Ammonium molybdate solution
Dissolve 15 g of AR-grade ammonium molybdate ((NH_4)$_6Mo_7O_{24}$.$4H_2O$) in 500 ml of distilled deionised water. Do not use the solution if it is more than 6 weeks old.

Ascorbic acid solution
Dissolve 5.4 g of AR-grade ascorbic acid ($C_6H_8O_6$) in 100 ml of distilled deionised water. Make this solution up fresh for each use.

Potassium antimonyl tartrate solution
Dissolve 0.34 g of AR-grade potassium antimonyl tartrate solution ($KSbO.C_4H_4O_6$) in 250 ml of distilled deionised water. Do not use the solution if it is more than 6 weeks old.

Mixed reagent
Make this up fresh each time you analyse samples; 5 ml is required for each sample. Mix the above reagents in these parts and in this order: five parts sulphuric acid; two parts ammonium molybdate solution; two parts ascorbic acid solution and one part potassium antimonyl tartrate solution.

Stock phosphate standard
Make up a stock solution by dissolving 4.39 g of anhydrous AR-grade potassium dihydrogen orthophosphate (KH_2PO_4) in 1 l of distilled deionised water. This standard solution contains 1 g l^{-1} or 1 mg ml^{-1} of phosphate P.

Working standards
Add 2 ml of the stock phosphate standard and make up fresh to 500 ml with distilled deionised water. This solution contains 4000 µg l^{-1} or 4 µg ml^{-1} of phosphate P.

Standards
Using the working standard, make up fresh the following standards using distilled deionised water:

> 200 µg l^{-1} (15 ml of working standard, made up to 300 ml)
> 150 µg l^{-1} (75 ml of the 200-µg-l^{-1} standard made up to 100 ml)
> 100 µg l^{-1} (100 ml of the 200-µg-l^{-1} standard made up to 200 ml)
> 50 µg l^{-1} (50 ml of the 100-µg-l^{-1} standard made up to 100 ml)
> 0 µg l^{-1} (use distilled deionised water)

Method

- Filter your water sample as soon as possible and keep it cool. Filtering may be done using a Whatman GF/C glass-fibre filter, handling with forceps to avoid contamination, but bear in mind that the paper may retain or release phosphorus. This can be minimised by first rinsing the paper in distilled deionised water, or using cellulose acetate filters.
- Your sample should be analysed as soon after collection as possible and kept cool. If you must store it (which is not recommended because the P content may change), do so using a filtered sample in a full bottle (excluding air) at 1 °C.
- Soak all glassware in a weak solution of hydrochloric acid, before rinsing with distilled deionised water. This is done because many detergents contain phosphorus and could contaminate your sample.
- Filter the water sample and standards. Place 50 ml of this filtered sample or standard into a 100-ml conical flask.
- To each flask, add 5 ml of the mixed reagent.
- After 30 min (and preferably well before 12 h has elapsed), measure the absorbance of the samples at 885 nm, using a 4-cm glass cell. Use your 0-μg-l^{-1} standard as your blank; that is, use it to set your zero on the spectrophotometer so the sample reading automatically has the blank absorbance subtracted from it. This saves you from having to subtract the blank reading from each sample reading.
- Using the absorbance values for your standards, plot a calibration curve (concentration of P as μg l^{-1} of phosphate P against absorbance reading). Your standard graph should be linear. If it levels off, then the top solutions that cause this levelling off should not be included. Determine the best fitting line.
- To determine the concentration of P in your sample, use your actual fitted-line equation to work out from the sample's absorbance its related P content (μg l^{-1}). If your sample's absorbance does not fall on your calibration-curve line, then dilute the sample (for example two-fold or ten-fold) until it does and then multiply your calculated N content by the dilution factor.

Soluble reactive phosphorus

In freshwater, PO_4^{3-} can be found in the range 0–500 mg l^{-1}, but in most unpolluted waters it is commonly found in concentrations of less than 100 μg l^{-1} (Allen 1989; Wetzel 1975). Seawater concentrations are smaller, usually less than 30 μg l^{-1}. Again, as with nitrate and ammonium, the concentration will vary depending on pollutants, time of year and the geology of the surrounding area. A manual method for determining soluble reactive phosphorus is given in Box 11.6, though an autoanalyser may also be used if available (see Box 11.3).

Water-testing kits

Kits to test the above water conditions (dissolved oxygen, ammonium, nitrate, phosphate) as well as a wide range of other characteristics such as hardness, pH, calcium, iron and silicate are available. Examples include water-testing tablets (Fisons), the Water Quality Test Kit (Fisons) and Water checkits (BDH). These kits are handy for quick, rough estimates of many water-chemistry parameters and may be used to double-check measurements made using other methods.

Most of these testing kits have been developed to measure drinking water, so beware of the range they measure. Ammonium should be <100 μg l^{-1} and nitrate should be <5 μg l^{-1}. The detection limits may be too high for the nitrate kit to read the low concentrations contained in most freshwater.

Many quick individual tests are available for some of the other field measurements discussed in this chapter, such as pH and dissolved oxygen, but check their accuracy against your requirements. These methods have the benefit of being simple, quick and cheaper than electric equipment but are not as accurate. Check scientific catalogues to see what is available and for further details.

Soil and sediment characteristics

Terrestrial soils and freshwater and marine sediments have much in common, but also many important differences. Many of the techniques that are distinctive to terrestrial and aquatic systems are a simple reflection of differences in ease of access to the soil material – it is more difficult to dig a deep hole in a stream than on land! Other differences are more fundamental – flow and wave action has a huge effect on the physical characteristics of freshwater and marine sediments, which thus experience much greater mobility and transport than do terrestrial soils. In addition, salinity interactions in estuaries are largely responsible for the distribution of fine sediments, with no obvious parallel in terrestrial soils. Given such differences in the origin and behaviour of soils and sediments, it is perhaps not surprising that different techniques have been developed by terrestrial, freshwater and marine ecologists to measure and describe these variables.

Soil profiles

These can be examined by digging a hole with clean sides, a *soil pit*, and using a measuring tape to record the depth of each layer. Taking samples with an *auger* is less destructive and time-consuming, so more samples can be taken to obtain a better representation. The simplest way is to twist a screw-shaped auger a short distance into the ground, noting the maximum penetration. The auger is then pulled out carefully (watch your back as well as the sample!) and laid on a pan so that the soil can be removed in its original vertical profile. Then successive, deeper samples are taken.

Using a *gouge auger* or *cylindrical core sampler* is less time-consuming and more accurate because it can remove an entire core at once. A large metal cylinder (e.g. 22 cm long, 4 cm in diameter) can be sharpened at one end and twisted into the ground by rotating a cross-bar inserted through the top (Dowdeswell 1984). Various photographic guides are available to allow classification of the soil profile into a soil type (podzol, gleysol, etc.).

Soil surface hardness, or 'penetrability'

Penetrability can be important for seed germination and for animals that dig or scrape the ground for food, including the many birds which probe the substrate for invertebrates. The easiest way to measure penetrability, especially for mudflats and other soft soils, is the '*BJPS*', or 'Bob James Pointy Stick', method (Figure 11.5). Hold a sharpened stick or rod inside a cylinder, drop it straight down, and measure how far it penetrates! This is easiest if the stick is marked in centimetres. Be consistent with the dropping height, and expect to use the average of five or more measurements, depending on variation among readings. Remember to account for temporal variation due to soil moisture content.

More technologically minded researchers can use a *penetrometer*. The following specifications are for a home-made version that we have used (Figure 11.5), but the exact dimensions and materials are not critical. The version illustrated consists of a length of PVC pipe 1 m long and 3.5 cm in diameter. The top of the pipe meets a T-bar of similar diameter and 30 cm long. This T-bar forms the handle, for pushing the instrument downwards. A second pipe fits inside the first and can move freely, protruding 1 cm above the T-bar and 13 cm below the bottom end. The bottom of the inner pipe has a sharpened rod attached to it, and this is the part which will be pushed into the soil. The top of the inner rod has the top of a spring scale attached to it. The bottom of the spring scale is mounted onto the outer pipe. Thus, the only connection between the outer and inner pipes is the spring scale, so, when the outer pipe is pushed downwards, the weight required for the inner pipe to penetrate the soil is recorded on the scale. It is important that the same person takes several samples, pushing evenly and consistently. The values are then averaged to give soil penetrability in weight (e.g. kg). This will be specific to the diameter of the sharpened rod.

There are two important refinements. First, you cannot read the spring scale at the same time as you are pushing the rod into the ground. Therefore, you should wrap a plastic slide around the scale, so that it is pushed down by the plunger inside the spring scale when you push the instrument downwards, and then holds its position after you relax (Figure 11.5). Second, it is useful to have both the spring scale and the penetrating rod readily detachable. Then you can use a scale with whatever weight range and accuracy is appropriate for the substrate you are measuring, as well as an appropriate diameter of penetrating rod (e.g. matching many birds' bills).

Sediment shear strength

Whilst penetrometers measure the vertical penetrability of soil and sediment, a major interest of marine benthic ecologists is how difficult it is for animals, such as worms and clams, to burrow through the sediment horizontally. This can be measured using a shear vane, a device that inserts a flat blade into the sediment for the whole profile or for a predetermined section

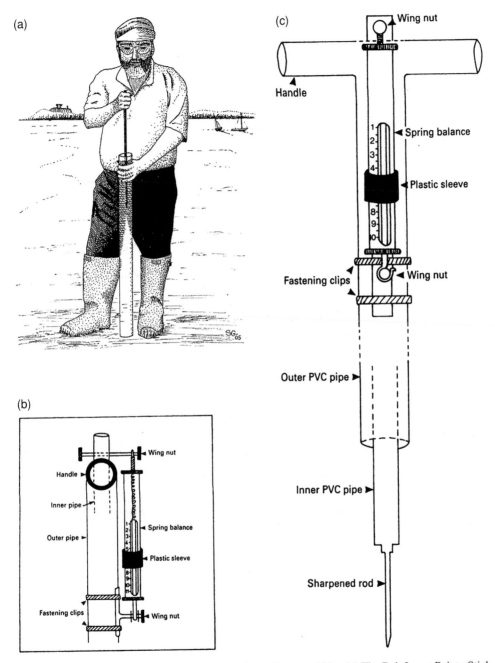

Figure 11.5 Methods for measuring soil penetrability. (a) The Bob James Pointy Stick method. Drawing by Simon Gillings. (b) Front view of a penetrometer. (c) Side view of a penetrometer.

Table 11.3. *Fractions of soil particles based on diameter, using the USDA classification*

Gravel	> 2000 μm
Very coarse sand	2000–1000 μm
Coarse sand	1000–500 μm
Medium sand	500–250 μm
Fine sand	250–125 μm
Very fine sand	125–50 μm
Silt	50–2 μm
Clay	< 2 μm

and measures the force required to rotate the blade. These devices are also used for terrestrial soils.

Soil texture

This is determined according to the percentages of various particle sizes. By convention, any particles greater than 2 mm in diameter are excluded beforehand, by using a sieve. Various texture scales are available, but the United States Department of Agriculture system (Table 11.3) is most widely used, and has been recommended by the United Nations Food and Agriculture Organization. For most biological purposes, coarse divisions of sand, silt and clay are used. For very sandy soil, the subdivisions in Table 11.3 can be helpful.

Particle size can be measured after passing dried samples through successive sieves, though this is less accurate for fine particles (silt and clay) than for sands. It is much better to separate soil particles by sedimentation, whereby a sample of soil is mixed in a clear cylinder filled with water. Loose organic material usually floats to the top, and the various particle types settle according to size. The percentages of each fraction can be measured using the *Bouyoucos method* (Box 11.7). The resulting percentages are expressed on a dry-weight basis, and can be used to classify soil texture (Figure 11.6). This classification refers to mineral fractions of soil only (see below for organic matter).

Particle-size measurements in aquatic sediments

The procedures are very similar to those described above. Full details of these procedures can be found in Holme and McIntyre (1984) and Raffaelli and Hawkins (1986); only a summary is provided here. First, the fine particles are removed by wet-washing a weighed sample of dried sediment through a 63-μm mesh, drying the sediment retained on the sieve and calculating the silt content as the difference in weight. (This fine fraction is very important for benthic organisms and is occasionally the only fraction that needs to be assessed.) The remaining (>63-μm) fraction is

Box 11.7. **The Bouyoucos method for determining fractions of silt and clay in soil**

The following procedure is from notes kindly provided by Dr David Dent, University of East Anglia.

Place 50 g of soil into a beaker, and add approximately 100 ml of distilled water. Add a dispersing agent such as 10 ml of 10% w/v of 'Calgon'. Alternatively, a solution of sodium hydroxide (NaOH) can be used: dissolve 100 g of sodium hydroxide pellets in 200 ml of water, dilute to 2.5 l, and add 10 ml of this to the soil sample. Stir and leave for 15 min, make up the volume to 600 ml, and record the temperature. Shake the suspension thoroughly for 1 min, set the cylinder on the bench, and record the time.

Reading 1: After 30 s, insert a *soil hydrometer* into the solution. This is similar to a hydrometer used to measure salinity, but is calibrated in g/l. Be careful not to stir up the mixture. After precisely 40 s, take a reading to obtain the weight of silt and clay in suspension.

Reading 2: After 6.5 h take the final reading.

Calculations

To correct for temperature, for each degree above or below 20 °C, add or subtract, respectively, 0.4 g per litre. The percentage moisture (% moisture) used in the calculations below is the percentage by weight, obtained as outlined in the section 'Soil moisture'.

$$\text{Percentage sand } (> 50\,\mu\text{m}) = \frac{\text{corrected Reading 1} \times 100}{50 - (\% \text{ moisture}/2)}$$

$$\text{Percentage clay } (< 2\,\mu\text{m}) = \frac{\text{corrected Reading 2} \times 100}{50 - (\% \text{ moisture}/2)}$$

$$\text{Percentage silt } (2 - 50\,\mu\text{m}) = 100 - (\% \text{ } sand + \% \text{ } clay)$$

The results are expressed on an oven-dry weight basis, not including organic content (see below).

then dried and further fractionated through a tower of sieves of decreasing mesh size, as described above. Whilst the techniques are basically the same for marine and terrestrial 'soils', there are differences in the way the data are presented and in the properties of interest (Raffaelli & Hawkins 1986).

Soil moisture

This is broken down for ecological purposes into categories that differ in their availability to plants: (1) free-draining (or gravitational) water drains away soon after a rain, but may be available to plants for several days; (2) plant-available (capillary) water occupies the spaces between soil

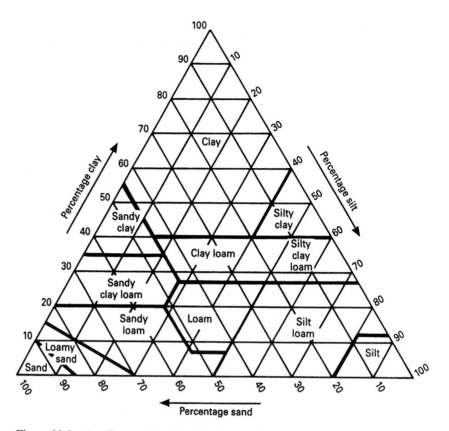

Figure 11.6   A soil-texture classification chart. (After FAO (1977).)

particles and is most important to plants; and (3) plant-unavailable (hygroscopic) water forms a thin film on the surface of particles.

Plant-available water can be approximated by the loss in weight of soil at field capacity (soil that had been saturated and allowed to drain for a couple of days) when air-dried (e.g. spread out on a newspaper for a week). This method extracts slightly more than the plant-available water, but the difference is not great, except in clay and peat soils.

Soil organic content

This is important as a source of nutrients to plants and invertebrates. It is measured by drying a sample of soil and examining its percentage weight loss after it has been ignited: (1) oven-dry a sample of 5–10 g overnight, or air-dry until the weight is stable; (2) grind the soil with a pestle and mortar; (3) ignite the sample at 375 °C in a furnace for 16 h; (4) allow to cool in a desiccator; and (5) re-weigh the sample. Higher temperatures can be used (e.g. 450 °C), but this will result in the loss of carbonates, and structural water will also be lost from clays. If pH tests indicate that the soil is low in carbonates, and clay constitutes less than approximately 10% of the sample, reasonable results can be obtained if the sample is ignited for 2 h at 850°C, thereby speeding up the processing.

The procedure is essentially the same for freshwater and marine systems, except that in marine systems there is often a high proportion of shell material present, i.e. calcium carbonate. This dissociates at the high temperatures reached in a furnace and the CO_2 lost can be a significant fraction of the total weight loss on ignition. Pre-treatment with dilute HCl is enough to remove the shell material in order to obtain a true organic-carbon weight. Finally, some beaches contain coal particles, and even oil, which will inflate the estimates of organic carbon. A quick scan of the sediment under a low-power microscope (coal and oil particles are jet black) will tell you whether there is likely to be a problem.

Redox potential

The redox or oxidation–reduction potential is a measure of how readily a medium will donate electrons to (reduce) or accept electrons from (oxidise) any reducible or oxidisable substance. Solutions with high redox potentials are highly oxidising. This is important in waterlogged soils, for example, because at low redox potential nitrate (an important plant nutrient) is reduced to nitrogen and sulphates are reduced to toxic H_2S.

To measure the redox potential, a platinum electrode is connected to a millivolt meter. Many pH meters also have a millivolt scale. It is most convenient to use a combined platinum–KCl electrode. Tables to relate E_h to the ion forms of interest are available, but redox alone is a useful index of the extent of oxidation or reduction in the system.

It is also important to measure redox potential in marine sediments because it can be a major driver of the distribution and abundance of the invertebrate fauna. Marine sediments are permanently waterlogged at some depth, close to the surface for fine-particle mudflats and a long way down for coarse sandy beaches. They are also rich in sulphates, which are easily reduced by bacteria to sulphides, which are toxic to the invertebrate fauna. The high-sulphide environment is revealed as a black layer and smells of rotten eggs due to H_2S. It is therefore important to record the location of the transition (which is very rapid) from reduced (black) to oxic (brown) conditions. This can be done crudely by breaking the sediment vertically and measuring the depth from the surface at which the sediment changes its colour, the so-called *chemocline*. A more accurate method is to use a platinum combination electrode, which can be slowly drilled down through the sediment a few millimetres at a time to create a redox profile (Pearson and Stanley 1979).

Oxygen in soils and sediments

Oxygen content in waterlogged soils or soft sediments may be measured simply, using the following method. If an unglazed, earthenware tube is filled with water, corked and then placed in the ground, an equilibrium of oxygen between the water and the sediment will be reached. For this to occur the tube should be left for at least a week. Then the tube is recovered and the oxygen level of the water (now equivalent to that in the sand or mud) can be measured by any of the aforementioned methods (Dowdeswell 1984). An alternative used routinely by marine ecologists is to take a core of sediment and carefully insert oxygen electrodes through ports in the side of the corer at various depths, or to measure the concentration at a standard depth (e.g 2 or 4 cm) with an electrode, if many sites are to be compared.

Acknowledgements

John Reynolds and Bill Sutherland dedicate this chapter to Dr Bob James, mentor to countless field ecologists, on the occasion of his retirement after 39 years at the University of East Anglia.

We appreciate valuable comments on various parts of this chapter by R. Boar, A. J. Davy, D. L. Dent, T. D. Jickells, R. Sanders and D. C. Wildon. We had additional helpful discussions with M. Ausden, C. A. Davenport, J. Farquar and B. Moss. B. Moss kindly permitted us to use his modified protocols for measuring nitrogen and phosphorus compounds, and D. L Dent provided his protocol for determining soil particle fractions.

References

Allen, S. E. (1989). *Chemical Analysis of Ecological Materials*, 2nd edn. Oxford, Blackwell Scientific Publications.

Dowdeswell, W. H. (1984). *Ecology: Principles and Practice*. Norwich, Fletcher and Sons Ltd.

Endler, J. A. (1990). On the measurement and classification of colour in studies of animal colour patterns. *Biological Journal of the Linnean Society* **41**, 315–352.

FAO (1977). *Guidelines for Soil Profile Description*. Rome, Food and Agriculture Organization of the United Nations.

Golterman, H. L., Clymo, R. S. & Ohnstad, M. A. M. (1978). *Methods for Physical and Chemical Analysis of Fresh Waters*, 2nd edn. Oxford, Blackwell Scientific Publications.

Grace, J. (1983). *Plant–Atmosphere Relationships*. London, Chapman and Hall.

Harvey, H. W. (1960). *The Chemistry and Fertility of Sea Waters*. Cambridge, Cambridge University Press.

HMSO (1981). *Ammonia in Waters*. London, Her Majesty's Stationery Office.

(1988). *Dissolved Oxygen in Waters Amendment*. London, Her Majesty's Stationery Office.

Holme, N. A. & McIntyre, A. D. (1984). *Methods for Studying the Marine Benthos*. Oxford, Blackwell Scientific Publications.

Mackereth, F. J. H., Heron, J. & Talling, J. F. (1978). *Water Analysis: Some Revised Methods for Limnologists*. Ambleside, FBA.

McLachlan, A. (1980). The definition of sandy beaches in relation to exposure, a simple rating system. *South African Journal of Science* **76**, 137–138.

Moss, B. (1988). *Ecology of Fresh Waters*. Oxford, Blackwell Scientific Publications.

NIO and UNESCO (1966). *International Oceanographic Tables*. London, National Institute of Oceanography of Great Britain and UNESCO.

Parsons, T. R., Maita, Y. & Lalli, C. M. (1984). *A Manual of Chemical and Biological Methods for Seawater Analysis*. Oxford, Pergamon Press.

Pearson, T. H., & Stanley, S. O. (1979). Comparative measurements of the redox potential of marine sediments as a rapid means of assessing the effect of marine pollution. *Marine Biology* **53**, 371–379.

Radojevic, M. & Bashkin, V. N. (1999). *Practical Environmental Analysis*. Cambridge, Royal Society of Chemistry.

Raffaelli, D. & Hawkins, S. J. (1986). *Intertidal Ecology*. Dordrecht, Kluwer.

Richardson, J. (1981). Oxygen meters: some practical considerations. *Journal of Biological Education* **15**, 107–116.

Sheehy, J. E. (1985). Radiation. In *Instrumentation for Environmental Physiology*, ed. B. Marshall & F. I. Woodward. Cambridge, Cambridge University Press, pp. 5–28.

Strickland, J. D. H. & Parsons, T. R. (1968). *A Practical Handbook of Seawater Analysis*. Ottawa, Fisheries Board of Canada.

Thomas, M. L. H. (1986). A physically derived exposure index for marine shorelines. *Ophelia*, **25**, 1–13.

Wetzel, R. G. (1975). *Limnology*. Philadelphia, Pennsylvania, W. B. Sanders Company.

12 The twenty commonest surveying sins

William J. Sutherland

Centre for Ecology, Evolution and Conservation, School of Biological Sciences, University
of East Anglia, Norwich NR4 7TJ, UK

1. NOT KNOWING YOUR SPECIES
Understanding the study species is essential for considering biases and interpreting data.

2. NOT KNOWING EXACTLY WHY YOU ARE SURVEYING
Think exactly what the question is and what data are required to answer it. How will the data be
presented and analysed?

3. COUNTING IN ONE OR A FEW LARGE AREAS RATHER THAN A LARGE NUMBER OF SMALL ONES
A single count gives no measure of the natural variation and it is then hard to see how significant
any changes are. This also applies to quadrats.

4. NOT GIVING PRECISE INFORMATION AS TO WHERE SAMPLING OCCURRED
Give the precise date and location. 'Site A, behind the large tree' or 'near to the road' may be
sufficient now but might mean nothing later. Use of the GPS is the easy solution to this.

5. ONLY SAMPLING SITES WHERE THE SPECIES IS ABUNDANT
It seems obvious to concentrate upon sites where the species is known to occur. However, without
knowing the density where it is scarce, it is impossible to determine the total population size.

6. CHANGING THE METHODS IN MONITORING
Unless there is a careful comparison of the different methods, changing the methods prevents
comparisons between the years.

7. PRETENDING THAT THE SAMPLES TAKEN WITHIN A SITE ARE REPLICATES
For example, if the project involves comparing logged and unlogged forest, but you have just
collected a number of samples in one area of each, then there is only one replicate of each
treatment and it is impossible to obtain statistics. It is unacceptable to compare the samples within
the logged forest with the samples from the unlogged forest by pretending that each sample is a

Ecological Census Techniques: A Handbook, ed. William J. Sutherland.
Published by Cambridge University Press. © Cambridge University Press 1996, 2006.

replicate (this is known as pseudoreplication). Either make sure that there are several samples from logged and several from unlogged forest or, if this is impossible, do something else to provide data that can be analysed.

8. NOT HAVING CONTROLS IN MANAGEMENT EXPERIMENTS
This is the greatest problem in interpreting the consequences of management.

9. NOT BEING HONEST ABOUT THE METHODS USED
If you survey moths only on warm still nights or place pitfall traps in the locations that are most likely to be successful then this is fine, but say so. Someone else surveying on all nights or randomly locating traps may otherwise conclude that the species has declined.

10. BELIEVING THAT THE DENSITY OF TRAPPED INDIVIDUALS IS THE SAME AS THE ABSOLUTE DENSITY

11. ASSUMING THAT THE SAMPLING EFFICIENCY IS SIMILAR IN DIFFERENT HABITATS
Differences in physical structure or vegetation structure will influence almost every surveying technique and thus confound comparisons.

12. DEVIATING FROM TRANSECT ROUTES
On one reserve the numbers of green hairstreaks *Callophyrs rubi* seen on the butterfly-monitoring transect increased markedly one year. It turned out that this was because the temporary warden that year climbed through the hedge to visit the colony on the far side.

13. NOT KNOWING THE ASSUMPTIONS OF THE SURVEY TECHNIQUES
Each technique has assumptions and it is important to consider these. For example, many mark–release–recapture methods assume the population to be closed (i.e. no gains or losses) yet are often applied in situations where this is clearly not the case.

14. THINKING THAT SOMEONE ELSE WILL IDENTIFY ALL YOUR SAMPLES FOR YOU
Most taxonomists have a huge backlog of samples and, contrary to widespread belief, are not just waiting around for more to arrive.

15. ASSUMING THAT OTHERS WILL COLLECT DATA IN EXACTLY THE SAME MANNER AND WITH THE SAME ENTHUSIASM
Everyone collects data in a slightly different way, which will affect the results, including setting traps, erecting mist nets or counting plants within quadrats. It is essential to standardise and test.

16. BEING TOO AMBITIOUS
A common problem is to start an extensive project that could never be completed. The partly completed project is usually far less worthwhile than a smaller, completed project would be.

Collecting far more samples than can possibly be analysed is a common problem. Careful planning of the time necessary (Chapter 1) can reduce the risk of this.

17. NOT KNOWING THE DIFFERENCE BETWEEN ACCURACY AND PRECISION
Ideally one would like the result to be accurate and precise, but this is not always possible. A precise but biased (inaccurate) measure may be sufficient if one is looking for changes over time or in comparing sites. A precise but inaccurate measure of a population size for assessing threat is usually not of great use.

18. BELIEVING THE RESULTS
Practically every survey has biases and inaccuracies. The secret is to evaluate how much these matter.

19. NOT STORING INFORMATION WHERE IT CAN BE RETRIEVED IN THE FUTURE
The new warden of a national reserve in England could find out from old work programmes the days on which his predecessor had counted a rare orchid but could find no record of the actual numbers.

20. NOT TELLING THE WORLD WHAT YOU HAVE FOUND
There is no point in doing work unless the results are presented to the appropriate people by publishing the results or feeding results back to any key audiences, ranging from local villagers to national governments.

Acknowledgements

Many thanks are due to Aldina Franco, Richard Gregory and Matthew Linkie for useful suggestions.

Index

Note: page numbers in *italics* refer to figures, tables and boxes.